AF352783

Advances in Modeling and Management of Urban Water Networks

Advances in Modeling and Management of Urban Water Networks

Editors

Alberto Campisano
Enrico Creaco

MDPI • Basel • Beijing • Wuhan • Barcelona • Belgrade • Manchester • Tokyo • Cluj • Tianjin

Editors
Alberto Campisano
University of Catania
Italy

Enrico Creaco
University of Pavia
Italy

Editorial Office
MDPI
St. Alban-Anlage 66
4052 Basel, Switzerland

This is a reprint of articles from the Special Issue published online in the open access journal *Water* (ISSN 2073-4441) (available at: https://www.mdpi.com/journal/water/special_issues/water_networks).

For citation purposes, cite each article independently as indicated on the article page online and as indicated below:

LastName, A.A.; LastName, B.B.; LastName, C.C. Article Title. *Journal Name* **Year**, *Volume Number*, Page Range.

ISBN 978-3-03943-789-4 (Hbk)
ISBN 978-3-03943-790-0 (PDF)

Contents

About the Editors vii

Preface to "Advances in Modeling and Management of Urban Water Networks" ix

Alberto Campisano and Enrico Creaco
Advances in Modeling and Management of Urban Water Networks
Reprinted from: *Water* **2020**, *12*, 2956, doi:10.3390/w12112956 . . . . **1**

Amir Nafi and Jonathan Brans
Cost–Benefit Prediction of Asset Management Actions on Water Distribution Networks
Reprinted from: *Water* **2019**, *11*, 1542, doi:10.3390/w11081542 . . . . **9**

Peyman Yousefi, Gregory Courtice, Gholamreza Naser and Hadi Mohammadi
Nonlinear Dynamic Modeling of Urban Water Consumption Using Chaotic Approach (Case
Study: City of Kelowna)
Reprinted from: *Water* **2020**, *12*, 753, doi:10.3390/w12030753 . . . . **29**

**Alexandros Mentes, Panagiota Galiatsatou, Dimitrios Spyrou, Achilleas Samaras and
Panagiota Stournara**
Hydraulic Simulation and Analysis of an Urban Center's Aqueducts Using Emergency
Scenarios for Network Operation: The Case of Thessaloniki City in Greece
Reprinted from: *Water* **2020**, *12*, 1627, doi:10.3390/w12061627 . . . . **51**

**Enrico Creaco, Giacomo Galuppini, Alberto Campisano, Carlo Ciaponi
and Giuseppe Pezzinga**
A Bi-Objective Approach for Optimizing the Installation of PATs in Systems of
Transmission Mains
Reprinted from: *Water* **2020**, *12*, 330, doi:10.3390/w12020330 . . . . **81**

Parima Mirshafiei, Abolghasem Sadeghi-Niaraki, Maryam Shakeri and Soo-Mi Choi
Geospatial Information System-Based Modeling Approach for Leakage Management in Urban
Water Distribution Networks
Reprinted from: *Water* **2019**, *11*, 1736, doi:10.3390/w11081736 . . . . **97**

**Daniel Manzi, Bruno Brentan, Gustavo Meirelles, Joaquín Izquierdo and
Edevar Luvizotto Jr.**
Pattern Recognition and Clustering of Transient Pressure Signals for Burst Location
Reprinted from: *Water* **2019**, *11*, 2279, doi:10.3390/w11112279 . . . . **109**

Camillo Bosco, Alberto Campisano, Carlo Modica and Giuseppe Pezzinga
Application of Rehabilitation and Active Pressure Control Strategies for Leakage Reduction in
a Case-Study Network
Reprinted from: *Water* **2020**, *12*, 2215, doi:10.3390/w12082215 . . . . **123**

Yu Shao, Huaqi Yao, Tuqiao Zhang, Shipeng Chu and Xiaowei Liu
An Improved Genetic Algorithm for Optimal Layout of Flow Meters and Valves in Water
Network Partitioning
Reprinted from: *Water* **2019**, *11*, 1087, doi:10.3390/w11051087 . . . . **135**

**Ulrich A. Ngamalieu-Nengoue, Pedro L. Iglesias-Rey, F. Javier Martínez-Solano,
Daniel Mora-Meliá and Juan G. Saldarriaga Valderrama**
Urban Drainage Network Rehabilitation Considering Storm Tank Installation and
Pipe Substitution
Reprinted from: *Water* **2019**, *11*, 515, doi:10.3390/w11030515 . . . . **151**

**Ulrich A. Ngamalieu-Nengoue, F. Javier Martínez-Solano, Pedro L. Iglesias-Rey
and Daniel Mora-Meliá**
Multi-Objective Optimization for Urban Drainage or Sewer Networks Rehabilitation through
Pipes Substitution and Storage Tanks Installation
Reprinted from: *Water* **2019**, *11*, 935, doi:10.3390/w11050935 . . . . **173**

Zhonghao Mao, Guanghua Guan and Zhonghua Yang
Suppress Numerical Oscillations in Transient Mixed Flow Simulations with a Modified
HLL Solver
Reprinted from: *Water* **2020**, *12*, 1245, doi:10.3390/w12051245 . . . . **187**

**Olivia Bailey, Ljiljana Zlatanovic, Jan Peter van der Hoek, Zoran Kapelan, Mirjam Blokker,
Tom Arnot and Jan Hofman**
A Stochastic Model to Predict Flow, Nutrient and Temperature Changes in a Sewer under Water
Conservation Scenarios
Reprinted from: *Water* **2020**, *12*, 1187, doi:10.3390/w12041187 . . . . **207**

Martin Rinas, Jens Tränckner and Thilo Koegst
Sediment Transport in Sewage Pressure Pipes, Part I: Continuous Determination of Settling and
Erosion Characteristics by In-Situ TSS Monitoring Inside a Pressure Pipe in Northern Germany
Reprinted from: *Water* **2019**, *11*, 2125, doi:10.3390/w11102125 . . . . **227**

Martin Rinas, Alexander Fricke, Jens Tränckner, Kurt Frischmuth and Thilo Koegst
Sediment Transport in Sewage Pressure Pipes, Part II: 1 D Numerical Simulation
Reprinted from: *Water* **2020**, *12*, 282, doi:10.3390/w12010282 . . . . **243**

Qianxun Wang, Wenqi Peng, Fei Dong, Xiaobo Liu and Nan Ou
Simulating Flow of An Urban River Course with Complex Cross Sections Based on the MIKE21
FM Model
Reprinted from: *Water* **2020**, *12*, 761, doi:10.3390/w12030761 . . . . **259**

About the Editors

Alberto Campisano, Ph.D., Associate Professor, Associate Professor at the University of Catania since 2014. He has been an Assistant Professor at the same university since 2001. He is a qualified Civil engineer with a Ph.D. degree in Hydraulic Engineering from the University of Naples, Italy. He has 20 years of teaching and research experience in the field of water engineering. His teaching experience includes courses in hydraulic constructions, hydrology, and river engineering. His research activities include design, modeling and control of urban water systems, hydrology of natural and urban catchments, modeling of sediment transport in river and drainage systems, reservoir sedimentation, design of sediment control structures in rivers, sustainable urban drainage, rainwater harvesting systems. He is an author of more than 150 papers published in scientific journals and books, national and international conferences, author of books and manual chapters. He has been the supervisor of more than 70 diploma and master theses, and 8 Ph.D. theses in the field of hydraulic engineering. Research consultant for municipalities and other public institutions for problems related to water engineering. He has been chairman of the IWA/IAHR Urban Stormwater Harvesting Working Group.

Enrico Creaco, Ph.D., Associate Professor, Enrico Creaco obtained his Ph.D. in Hydraulic Engineering in 2006 and has researched topics pertinent to water and environmental systems for over ten years. His career began at the Universities of Catania and then Ferrara, Italy. From May 2014 to June 2015, he took up a Research Fellow post at the University of Exeter and became an Assistant Professor at the University of Pavia in September 2015. He has been an Associate Professor at the University of Pavia since September 2018. He has been Honorary Senior Reseach Fellow at the University of Exeter and Adjunct Senior Lecturer at the University of Adelaide. In September 2018, he obtained the Full Professor Qualification in the Hydraulics, Hydrology and Hydraulic Infrastructure Sector. He has been a lecturer on hydraulic infrastructures at both undergraduate and postgraduate levels and has published more than 100 papers in a variety of Scopus and ISI indexed international journals. He is the Associate Editor of the *Journal of Water Resources Planning and Management*—ASCE and participated in/coordinated various national and international research projects. Research interests: design and management of water distribution; sewer and irrigation systems; numerical modeling of shallow waters and sediment transport; demand analysis; protection of water distribution systems from contamination.

Preface to "Advances in Modeling and Management of Urban Water Networks"

Technical progress in sensors, control systems, and computational resources has breathed new life into the research on urban water networks (UWNs), which has broadened from the well-established topics of simulation and design to aspects of optimization, real-time monitoring and management, and water quality. Following an introductory paper, this book presents a collection of fifteen works that well represents the current research lines in the field of UWNs.

Alberto Campisano, Enrico Creaco

Editors

Editorial

Advances in Modeling and Management of Urban Water Networks

Alberto Campisano [1] and Enrico Creaco [2,*]

[1] Dipartimento di Ingegneria Civile e Architettura, University of Catania, Via Santa Sofia 64, 95123 Catania, Italy; alberto.campisano@unict.it

[2] Dipartimento di Ingegneria Civile e Architettura, University of Pavia, Via Ferrata 3, 27100 Pavia, Italy

* Correspondence: creaco@unipv.it; Tel.: +39-0382-985317

Received: 16 October 2020; Accepted: 20 October 2020; Published: 22 October 2020

Abstract: This Editorial presents a representative collection of 15 papers, presented in the Special Issue on Advances in Modeling and Management of Urban Water Networks (UWNs), and frames them in the current research trends. The most analyzed systems in the Special Issue are the Water Distribution Systems (WDSs), with the following four topics explored: asset management, modelling of demand and hydraulics, energy recovery, and pipe burst identification and leakage reduction. In the first topic, the multi-objective optimization of interventions on the network is presented to find trade-off solutions between costs and efficiency. In the second topic, methodologies are presented to simulate and predict demand and to simulate network behavior in emergency scenarios. In the third topic, a methodology is presented for the multi-objective optimization of pump-as-turbine (PAT) installation sites in transmission mains. In the fourth topic, methodologies for pipe burst identification and leakage reduction are presented. As for the Urban Drainage Systems (UDSs), the two explored topics are asset management, with a system upgrade to reduce flooding, and modelling of flow and water quality, with analyses on the transition from surface to pressurized flow, impact of water use reduction on the operation of UDSs and sediment transport in pressurized pipes. The Special Issue also includes one paper dealing with the hydraulic modelling of an urban river with a complex cross-section.

Keywords: water distribution system modeling; urban drainage system modeling; asset management; emergency scenarios; leakage; demand; energy; water quality; sediment transport

---

1. Introduction

The technical progress in sensors, control systems and computational resources have recently breathed new life into the research on urban water networks (UWNs), which has broadened from the well-established topics of simulation and design to aspects of optimization, real-time monitoring and management, and water quality.

As regards water distribution systems (WDSs), the aging of the system causes the deterioration of elements, thus resulting in such problems as the increase in pipe bursts and the decrease in hydraulic capacity. Therefore, the asset management for WDS renewal has been the subject of much research, e.g., [1–4]. In this context, there is still curiosity about the application of single- and multi-objective optimization techniques to new case studies. Sticking to WDSs, the increasing availability of data measured at high frequency at flow meters and smart meters installed in WDSs has supplied researchers with numerous and large databases for the set-up and calibration of demand modelling/forecast methodologies, e.g., [5–8]. The abundance of methodologies in the scientific literature calls for comparative studies and for sensitivity analysis of such parameters as the forecast lead time. Another well-established research topic in the scientific literature of the most recent decade is the modelling of WDS hydraulics, which has seen the development of numerous algorithms able to model the behavior of WDSs with increasing computational efficiency, e.g., [9–12]. However, the matching of these algorithms to the real behavior of WDSs, with special focus on emergency scenarios, requires further investigation. The potential of energy recovery through the installation of pumps operating as turbines (PATs), able to convert pressure surpluses into electric energy, has also attracted the attention of numerous researchers, e.g., [13–16]. In this context, the application of multi-objective techniques for the identification of optimal sites for PAT installation has been explored in few works to date. Finally, the need to reduce water waste and to keep a high efficiency of service has spurred researchers to develop methodologies for fast pipe-burst identification, e.g., [17–19], and for effective leakage reduction, e.g., [20–22]. Though numerous works in the scientific literature have dealt with these topics, there is still need for more effective and efficient algorithms for application in real case studies.

As for UDSs, asset management is focused not only on the replacement of old and deteriorated elements, but also on updating the system to tackle the increase in peak water discharges under wet weather conditions, caused by urbanization processes and climate change in progress worldwide. Therefore, to prevent the occurrence of flooding in urban areas, methodologies have recently been developed to plan interventions of pipe replacement and storage tank installation, e.g., [23–25]. However, there is curiosity about the results obtainable in this field through the application of modern optimization techniques, such as those based on the multi-objective approach. Other research was carried out to analyze flow conditions in sewer channels, with specific focus on localized phenomena such as the transition from free surface to pressurized flow, e.g., [26–28]. Nevertheless, some problems of numerical instability have not yet been fully solved, calling for more research efforts. Finally, numerous experimental and numerical works have focused on water quality, e.g., [29–31], and sediment transport, e.g., [32–34], in sewers. However, some aspects, such as the sediment transport in pressurized pipes and the impact of water use reduction on the operation of UDSs, need to be analyzed in more depth.

Other research has been carried out on the topic of urban rivers, with special focus on the restoration of natural water courses, e.g., [35–37]. However, few works have dealt with the flow characteristics of the urban channelized rivers with complex cross-sections.

The papers in this Special Issue attempt to find an answer to some open questions in the research topics presented above. In the following section, the main novelties of these papers are presented. The paper continues with a discussion to comment the results achieved, followed by the conclusions.

2. Overview of the Special Issue

The Special Issue was established to point out the recent trends in UWNs, with emphasis on the opportunities introduced by technical progress for modelling, management, forecast, and performance improvement. The collected papers are grouped in the following subsections, related to the research topics inherent to WDSs, UDSs and urban rivers.

2.1. Water Distribution Systems

2.1.1. Asset Management

Nafi and Brans [38] predict the potential costs and benefits of a combination of asset management actions of maintenance and renewal in the WDS. The actions are treated as decisional variables in a bi-objective optimization aimed at finding solutions in the trade-off between cost and efficiency, to be minimized and maximized, respectively. Both objective functions are calculated through an artificial neural network (ANN) trained on real data. The selection of the ultimate solution in the Pareto front is made based on budget constraints or efficiency targets. The applications to a French WDS show that the methodology can effectively help decision-makers in selecting the most suitable interventions in the WDS.

2.1.2. Modelling of Demand and Hydraulics

Yousefi et al. [39] propose the simulation of urban water consumption using chaos theory, with the aim to help optimize system management. They test consumption forecast models optimized through chaos theory techniques against an urban consumption dataset obtained from the City of Kelowna (British Columbia, Canada). The analysis of the results proves that the non-linear local approximation model performs best, and the phase space reconstruction improves the accuracy of the models.

Mentes et al. [40] develop an extended period simulation model for the WDSs of the Thessaloniki city in Greece to model the current operating state of the networks, as well as its response to emergency conditions resulting from failure in one of them. The model is calibrated under both normal and critical operating conditions. Failure in the WDSs is analyzed through the model in five emergency scenarios of network operation, two of which consider possible interconnections of the studied WDSs. The results of the work prove the model is able to give reliable insights into the management of limited water reserves in some areas of the network when considering the interconnections.

2.1.3. Energy Recovery

Creaco et al. [41] investigate the benefits associated with the installation of PATs in systems of transmission mains, which convey water from source(s) to WDS tanks at an almost constant rate. They apply their bi-objective methodology to an Italian case study made up of nine systems of transmission mains, to find solutions in the trade-off between installation costs and generated hydropower, to be minimized and maximized, respectively. Due to the large geodesic elevation variations available in the case study, the post processing of solutions in terms of long-life profit show that, in all systems, the optimal solution with the highest values of installation cost and generated electric power, which can be as high as 83 KW, is the most profitable under usual economic scenarios. Furthermore, payback periods lower than 3 years are always obtained for these solutions.

2.1.4. Pipe Burst Identification and Leakage Reduction

Mirshafiei et al. [42] present a Geospatial Information System (GIS)-based methodology for partially isolating a leaking pipeline from the remainder of a WDS. Specifically, a web GIS application based on the traceability concept is developed to find the optimal valves close around the pipeline. The algorithm is applied and tested against one of the districts of the Teheran WDS, proving to be accurate and handy.

Manzi et al. [43] present a methodology aimed at processing transient signals for the detection and location of bursts in WDSs. Disaggregation, ANN, and clustering techniques are the methodological elements. The application to two real WDSs proves the methodology to be accurate and effective at locating pipe bursts and at characterizing them in terms of leaked flow.

By making use of a calibrated numerical model, Bosco et al. [44] investigate the potential of rehabilitation measures and active pressure control strategies for leakage reduction in a water distribution system (WDS) in southern Italy. Three different scenarios, namely pipe rehabilitation (S1), implementation of pressure local control (S2), and introduction of remote real-time pressure control (RTC) (S3), are analyzed and compared with the current operational scenario (S0). The results point out that 16.7%, 35.0%, and 37.5% leakage reductions (as compared to S0) can be obtained under scenarios S1, S2, and S3, respectively.

Shao et al. [45] present a novel methodology to be used in the context of WDS partitioning, which is a management practice beneficial for consumption monitoring and leakage mitigation. Their methodology operates by identifying the optimal locations of flow meters and valves to separate the DMAs by means of an improved genetic algorithm, applied for minimizing the number of installed flow meters and the behavioral variations of the WDS in terms of hydraulics and water quality, in comparison with the unpartitioned WDS. The applications show the effectiveness of the methodology in a real WDS.

2.2. Urban Drainage Systems

2.2.1. Asset Management

Ngamalieu-Nengoue et al. [46] present a methodology to rehabilitate UDSs to cope with increased peak flows due to the development of urbanization and the climate change currently in progress. UDS rehabilitation is carried out by combining actions of pipe replacement and storm tank installation. The sites of intervention are the decisional variables of a pseudo-genetic heuristic algorithm aimed at minimizing the overall installation costs, while the UDS is modelled using the Storm Water Management Model (SWMM) model. The effectiveness of the methodology is applied to the UDS of Bogotà, Colombia.

Ngamalieu-Nengoue et al. [47] enhanced the methodology presented above by implementing multi-objective optimization, to obtain optimal solutions in the trade-off between installation costs and flooding damage, which are simultaneously minimized.

2.2.2. Modelling of Flow and Quality

Mao et al. [48] analyze the transition between free-surface and pressurized flow, which is a crucial phenomenon in UDSs. In the simulation of this phenomenon, severe numerical oscillations may appear behind filling-bores, causing unphysical pressure variations and computation failure. After reviewing various oscillation-suppressing methods, among which only one provides stable results under a realistic acoustic wave speed, the authors present a new oscillation-suppressing method with first-order accuracy based on two easily evaluable parameters. Besides being able to suppress numerical oscillations under an acoustic wave speed of 1000 ms^{-1}, it yields numerical results in good agreement with experimental data.

Bailey et al. [49] present the analysis of the impact of water use reduction on the operation of UDSs. The methodology used for the analysis is made up of two elements, namely SIMDEUM WW®and InfoWorks®ICM. The former is used for the generation of stochastic appliance-specific discharge profiles for wastewater flow and concentration, which are fed into the latter to quantify the impacts within the sewer network. After being calibrated by using measured field data from a sewer system in Amsterdam serving 418 households, the model is used to analyze the effects of three water conservation strategies (greywater reuse, rainwater harvesting and water-saving appliances) on flow, nutrient concentrations, and temperature in sewer networks. Results show both the reduction in sewer flow up to 62% and the increase in COD, TKN and TPH concentrations by up to 111%, 84% and 75%, respectively, offering more favorable conditions for nutrient recovery.

Rinas et al. [50] present the results of an experimental campaign on solids transport in a pressurized sewer pipe. Data with a very fine temporal resolution are obtained from the one year in-situ turbidity/total suspended solids (TSS) monitoring inside a pipe (600 mm diameter) in an urban region in northern Germany. This enables the determination of solid sedimentation (within pump pauses) and erosion behavior (within pump sequences). The measurements point out a change in the sedimentation and erosion behaviors as a function of the inflow rate, with faster settling solids as the inflow increases.

Rinas et al. [51] calibrate a numerical transport model to simulate the sedimentation and erosion behavior of solids in a pressurized sewer pipe, using the data collected in the work of Rinas et al. [50]. The model is applied to investigate sediments transport under low flow velocities (due to energy saving intentions). The simulation of the 30-day-long pumping operation shows that sediments can be transported even under variable inflow conditions with low flow velocities. As a result, low-energy pump operation can be applied without increasing the risk of deposition formation.

2.3. Urban Rivers

Wang et al. [52] present the application of the software MIKE21 FM for analyzing the flow in a planned urban river course with a complex cross-section. The first part of the work concerns the verification of the rationality and feasibility of the planning scheme through a two-dimensional numerical model. The second part of the work is dedicated to the analysis of the river course, to give several suggestions and improvement measures for the follow-up of the river planning. The third and last part of the work derives some general rules to be followed for the modelling of rivers with a complex cross-section.

3. Discussion

All the papers of the Special Issue are focused on topics that are at the forefront of the research in Urban Water Networks. They are valuable contributions from the viewpoint of methodological development, objectives achieved and review of the scientific literature.

In the field of WDSs, all works are numerical analyses, though most of them, i.e., [38–40,43,44], present models calibrated on the basis of experimental data. The methodologies adopted are multi-faceted, ranging from physically based modelling, [40,41,44,45], to GIS algorithms, [42], data-driven techniques, [38,39,43], graph-theory, [43], and optimization algorithms, [38,41,45].

In the field of UDSs, almost all the works, i.e., [46–49,51], are numerical. In this context, the methodologies adopted include physically based modelling, [46–49,51], stochastic, [49], and empirical, [51], modelling, optimization algorithms, [46,47]. The only experimental work [50] reports on and analyses the data of a measurement campaign on hydraulics and sediment transport in a pressurized pipe.

The paper on the topic of urban rivers, i.e., [52], makes exclusive use of the physically based numerical modelling.

The kinds of paper published in this Special Issue are representative of current research trends in the scientific literature of urban water systems, which features an evident prevalence of numerical over experimental works. In fact, although most papers use literature data for the validation of numerical models, only one of fifteen papers reports novel experimental results. Indeed, this represents a drawback of the current trends in research. In fact, data obtained through the modern and reliable measurement devices available nowadays could significantly enrich the literature, as they could enable the more exhaustive validation of models and the development of more refined numerical models. As a result, new experimental campaigns both in laboratories and in the field are needed in the future.

4. Conclusions

The Editorial presents an analysis of the papers published in the Special Issue on Advances in Advances in Modeling and Management of Urban Water Networks.

Starting with WDSs, the first analyzed topic is asset management, which was dealt with by Nafi and Brans [38] by applying the bi-objective optimization to find optimal solutions in the trade-off between costs and efficiency in the WDS to a French WDS. ANNs trained on real data are used for the assessment of both objective function within the optimization process. The second analyzed topic is the modelling of demand and hydraulics. As for demand, Yousefi et al. [39] propose simulation of urban water consumption in a Canadian city using chaos theory, with the aim to help optimize system management. As for the modelling, Mentes et al. [40] show the extent to which a well-calibrated extended period simulation model can reproduce the hydraulics of real WDSs in Greece, while giving reliable insights into WDS management in emergency scenarios. The third explored topic lies in the exploitation of WDSs for energy recovery. In this context, Creaco et al. [41] show that transmission mains in a real Sicilian case study featuring large geodesic elevation variations can be used to generate electric powers up to 83 KW. The fourth explored topic concerns operational management in terms of pipe burst identification and leakage reduction. In this context, Mirshafiei et al. [42] present a Geospatial Information System (GIS)-based methodology for partially isolating a leaking pipeline from the remainder of a WDS and its application to an Iranian WDS. Manzi et al. [43] present a methodology based on disaggregation, ANN, and clustering techniques for the identification and location of bursts in real WDSs. Thanks to the well-calibrated model of a WDS in Southern Italy, Bosco et al. [44] show the extent to which WDS rehabilitation and active pressure control can help in reducing leakage.

Finally, Shao et al. [45] improves a genetic algorithm present in the scientific literature to optimize the positions of flow meters and valves in the context of WDS partitioning, which is a management practice beneficial for consumption monitoring and leakage mitigation.

In the field of UDSs, the first explored topic is asset management, in which Ngamalieu-Nengoue et al. [46,47] show how optimization can help decision-makers in selecting the most suitable sites for pipe replacement and storage tank installation, with the objective to improve UDS performance during intense rain events. In the context of the modelling of flow and quality, second topic for the UDSs, Mao et al. [48] propose an effective oscillation-suppressing method to be used for simulating the transition between free-surface and pressurized flow in sewer channel. Through the stochastic simulation of users' water and pollutant discharges and through the hydraulic modelling of sewer channels, Bailey et al. [49] present the analysis of the impact of water use reduction on the operation of UDSs. The results in a real Dutch case study point out the reduction in water discharges and the increase in pollutant concentrations, which are favorable conditions for nutrient recovery. Finally, Rinas et al. [50,51] present experiments and numerical simulations on solids transport in a pressurized sewer pipe, providing valuable results for a topic that is probably underrepresented in the scientific literature.

The Special Issue also includes a paper, i.e., [52], on the analysis of the flow in a planned urban river course with a complex cross-section, offering some general modelling rules to be followed in this context.

Each of these papers is a valuable contribution to the research on urban water networks and paves the way for further developments in the future.

Author Contributions: Writing—Original Draft Preparation, E.C.; Review & Editing, A.C. All authors have read and agreed to the published version of the manuscript.

Funding: This research received no external funding.

Acknowledgments: This research was conducted using the funds supplied by the University of Catania and by the University of Pavia.

Conflicts of Interest: The author declares no conflict of interest.

References

1. Dandy, G.C.; Engelhardt, M. Optimal scheduling of water pipe replacement using genetic algorithms. *J. Water Resour. Plan. Manag.* **2001**, *127*, 214–223. [CrossRef]

2. Kleiner, Y.; Adams, B.J.; Rogers, S. Water Distribution Network Renewal Planning. *J. Comput. Civ. Eng.* **2001**, *15*, 15–26. [CrossRef]

3. Giustolisi, O.; Laucelli, D.; Savic, D. Development of rehabilitation plans for water mains replacement considering risk and cost-benefit assessment. *Civ. Eng. Environ. Syst.* **2006**, *23*, 175–190. [CrossRef]

4. Alvisi, S.; Franchini, M. Multi-objective optimization of repair and leakage detection scheduling in water distribution systems. *J. Water Resour. Plan. Manag.* **2009**, *135*, 426–439. [CrossRef]

5. Cutore, P.; Campisano, A.; Kapelan, Z.; Modica, C.; Savic, D. Probabilistic prediction of urban water consumption using the SCEM-UA algorithm. *Urban Water J.* **2008**, *5*, 125–132. [CrossRef]

6. Creaco, E.; Blokker, E.J.M.; Buchberger, S.G. Models for Generating Household Water Demand Pulses: Literature Review and Comparison. *J. Water Resour. Plan. Manag.* **2017**, *143*, 04017013. [CrossRef]

7. Pacchin, E.; Gagliardi, F.; Alvisi, S.; Franchini, M. A Comparison of Short-Term Water Demand Forecasting Models. *Water Resour. Manag.* **2019**, *33*, 1481–1497. [CrossRef]

8. Zubaidi, S.; Abdulkareem, I.; Hashim, K.; Al-Bugharbee, H.; Ridha, H.; Gharghan, S.; Al-Qaim, F.; Muradov, M.; Kot, P.; Al-Khaddar, R.G. Hybridised Artificial Neural Network Model with Slime Mould Algorithm: A Novel Methodology for Prediction of Urban Stochastic Water Demand. *Water* **2020**, *12*, 2692. [CrossRef]

9. Creaco, E.; Franchini, M. Comparison of Newton Raphson Global and Loop Algorithms for Water Distribution Network Resolution. *J. Hydraul. Eng.* **2014**, *140*, 313–321. [CrossRef]

10. Farina, G.; Creaco, E.; Franchini, M. Using EPANET for modelling water distribution systems with users along the pipes. *Civ. Eng. Environ. Syst.* **2014**, *31*, 36–50. [CrossRef]

11. Gorev, N.B.; Gorev, V.N.; Kodzhespirova, I.F.; Shedlovsky, I.A.; Sivakumar, P. Simulating control valves in water distribution systems as pipes of variable resistance. *J. Water Resour. Plan. Manag.* **2018**, *144*, 06018008. [CrossRef]

12. Piller, O.; Elhay, S.; Deuerlein, J.W.; Simpson, A.R. A Content-Based Active-Set Method for Pressure-Dependent Models of Water Distribution Systems with Flow Controls. *J. Water Resour. Plan. Manag.* **2020**, *146*, 04020009. [CrossRef]

13. Carravetta, A.; Del Giudice, G.; Fecarotta, O.; Ramos, H. Energy production in water distribution networks: A PAT design strategy. *Water Resour. Manag.* **2012**, *26*, 3947–3959. [CrossRef]

14. Fecarotta, O.; McNabola, A. Optimal location of pump as turbines (PATs) in water distribution networks to recover energy and reduce leakage. *Water Resour. Manag.* **2017**, *31*, 5043–5059. [CrossRef]

15. Fontana, N.; Giugni, M.; Glielmo, L.; Marini, G.; Zollo, R. Hydraulic and Electric Regulation of a Prototype for Real-Time Control of Pressure and Hydropower Generation in a Water Distribution Network. *J. Water Resour. Plann. Manag.* **2018**, *144*, 04018072. [CrossRef]

16. Fontana, N.; Giugni, M.; Glielmo, L.; Marini, G.; Zollo, R. Operation of a Prototype for Real Time Control of Pressure and Hydropower Generation in Water Distribution Networks. *Water Resour. Manag.* **2019**, *33*, 697–712. [CrossRef]

17. Romano, M.; Kapelan, Z.; Savić, D. Automated Detection of Pipe Bursts and Other Events in Water Distribution Systems. *J. Water Resour. Plann. Manag.* **2014**, *140*, 457–467. [CrossRef]

18. Capponi, C.; Ferrante, M.; Zecchin, A.C.; Gong, J. Leak Detection in a Branched System by Inverse Transient Analysis with the Admittance Matrix Method. *Water Resour. Manag.* **2017**, *31*, 4075–4089. [CrossRef]

19. Cheng, W.; Xu, G.; Fang, H.; Zhao, D. Study on Pipe Burst Detection Frame Based on Water Distribution Model and Monitoring System. *Water* **2019**, *11*, 1363. [CrossRef]

20. Campisano, A.; Modica, C.; Reitano, S.; Ugarelli, R.; Bagherian, S. Field-oriented methodology for real-time pressure control to reduce leakage in water distribution networks. *J. Water Resour. Plan. Manag.* **2016**, *142*, 04016057. [CrossRef]

21. Creaco, E.; Campisano, A.; Fontana, N.; Marini, G.; Page, P.R.; Walski, T. Real time control of water distribution networks: A state-of-the-art review. *Water Res.* **2019**, *161*, 517–530. [CrossRef] [PubMed]

22. Galuppini, G.; Magni, L.; Creaco, E. Stability and Robustness of Real-Time Pressure Control in Water Distribution Systems. *J. Hydraul. Eng.* **2020**, *146*, 04020023. [CrossRef]

23. Duan, H.-F.; Li, F.; Yan, H. Multi-objective optimal design of detention tanks in the urban stormwater drainage system: LID implementation and analysis. *Water Resour. Manag.* **2016**, *30*, 4635–4648. [CrossRef]

24. Cunha, M.C.; Zeferino, J.A.; Simões, N.E.; Saldarriaga, J.G. Optimal location and sizing of storage units in a drainage system. *Env. Model. Softw.* **2016**, *83*, 155–166. [CrossRef]

25. Starzec, M. A Critical Evaluation of the Methods for the Determination of Required Volumes for Detention Tank. Available online: https://www.e3s-conferences.org/articles/e3sconf/abs/2018/20/e3sconf_infraeko2018_00088/e3sconf_infraeko2018_00088.html (accessed on 21 October 2020).

26. Vasconcelos, J.G.; Wright, S.J.; Roe, P.L. Numerical oscillations in pipe-filling bore predictions by shock-capturing models. *J. Hydraul. Eng.* **2009**, *135*, 296–305. [CrossRef]

27. Malekpour, A.; Karney, B.W. Spurious numerical oscillations in the Preissmann slot method: Origin and suppression. *J. Hydraul. Eng.* **2016**, *142*, 04015060. [CrossRef]

28. An, H.; Lee, S.; Noh, S.J.; Kim, Y.; Noh, J. Hybrid numerical scheme of Preissmann slot model for transient mixed flows. *Water* **2018**, *10*, 899. [CrossRef]

29. Métadier, M.; Bertrand-Krajewski, J.-L. The use of long-term on-line turbidity measurements for the calculation of urban stormwater pollutant concentrations, loads, pollutographs and intra-event fluxes. *Water Res.* **2012**, *46*, 6836–6856. [CrossRef]

30. Creaco, E.; Berardi, L.; Sun, S.; Giustolisi, O.; Savic, D. Selection of relevant input variables in storm water quality modeling by multiobjective evolutionary polynomial regression paradigm. *Water Resour. Res.* **2016**, *52*, 2403–2419. [CrossRef]

31. Kozak, C.; Fernandes, C.V.S.; Braga, S.M.; Do Prado, L.L.; Froehner, S.; Hilgert, S.G. Water quality dynamic during rainfall episodes: Integrated approach to assess diffuse pollution using automatic sampling. *Environ. Monit. Assess.* **2019**, *191*, 402. [CrossRef]

32. Seco, I.; Valentín, M.G.; Schellart, A.; Tait, S. Erosion resistance and behaviour of highly organic in-sewer sediment. *Water Sci. Technol.* **2014**, *69*, 672–679. [CrossRef] [PubMed]

33. Regueiro-Picallo, M.; Anta, J.; Suárez, J.; Puertas, J.; Jácome, A.; Naves, J. Characterisation of sediments during transport of solids in circular sewer pipes. *Water Sci. Technol.* **2018**, *2017*, 8–15. [CrossRef]

34. Shahsavari, G.; Arnaud-Fassetta, G.; Campisano, A. A field experiment to evaluate the cleaning performance of sewer flushing on non-uniform sediment deposits. *Water Res.* **2017**, *118*, 59–69. [CrossRef] [PubMed]

35. Follstad Shah, J. Standards for ecologically successful river restoration. *J. Appl. Ecol.* **2005**, *42*, 208–217.

36. Beechie, T.; Imaki, H. Predicting natural channel patterns based on landscape and geomorphic controls in the Columbia River basin, USA. *Water Resour. Res.* **2014**, *50*, 39–57. [CrossRef]

37. Chen, X.; Wang, D.; Tian, F.; Sivapalan, M. From channelization to restoration: Sociohydrologic modeling with changing community preferences in the Kissimmee River Basin, Florida. *Water Resour. Res.* **2016**, *52*, 1227–1244. [CrossRef]

38. Nafi, A.; Brans, J. Cost–Benefit Prediction of Asset Management Actions on Water Distribution Networks. *Water* **2019**, *11*, 1542. [CrossRef]

39. Yousefi, P.; Courtice, G.; Naser, G.; Mohammadi, H. Nonlinear Dynamic Modeling of Urban Water Consumption Using Chaotic Approach (Case Study: City of Kelowna). *Water* **2020**, *12*, 753. [CrossRef]

40. Mentes, A.; Galiatsatou, P.; Spyrou, D.; Samaras, A.; Stournara, P. Hydraulic Simulation and Analysis of an Urban Center's Aqueducts Using Emergency Scenarios for Network Operation: The Case of Thessaloniki City in Greece. *Water* **2020**, *12*, 1627. [CrossRef]

41. Creaco, E.; Galuppini, G.; Campisano, A.; Ciaponi, C.; Pezzinga, G. A Bi-Objective Approach for Optimizing the Installation of PATs in Systems of Transmission Mains. *Water* **2020**, *12*, 330. [CrossRef]

42. Mirshafiei, P.; Sadeghi-Niaraki, A.; Shakeri, M.; Choi, S. Geospatial Information System-Based Modeling Approach for Leakage Management in Urban Water Distribution Networks. *Water* **2019**, *11*, 1736. [CrossRef]

43. Manzi, D.; Brentan, B.; Meirelles, G.; Izquierdo, J.; Luvizotto Jr., E. Pattern Recognition and Clustering of Transient Pressure Signals for Burst Location. *Water* **2019**, *11*, 2279. [CrossRef]

44. Bosco, C.; Campisano, A.; Modica, C.; Pezzinga, G. Application of Rehabilitation and Active Pressure Control Strategies for Leakage Reduction in a Case-Study Network. *Water* **2020**, *12*, 2215. [CrossRef]

45. Shao, Y.; Yao, H.; Zhang, T.; Chu, S.; Liu, X. An Improved Genetic Algorithm for Optimal Layout of Flow Meters and Valves in Water Network Partitioning. *Water* **2019**, *11*, 1087. [CrossRef]

46. Ngamalieu-Nengoue, U.A.; Iglesias-Rey, P.L.; Martínez-Solano, J.; Mora-Meliá, D.; Saldarriaga Valderrama, J.G. Urban Drainage Network Rehabilitation Considering Storm Tank Installation and Pipe Substitution. *Water* **2019**, *11*, 515. [CrossRef]

47. Ngamalieu-Nengoue, U.A.; Martínez-Solano, J.; Iglesias-Rey, P.L.; Mora-Meliá, D. Multi-Objective Optimization for Urban Drainage or Sewer Networks Rehabilitation through Pipes Substitution and Storage Tanks Installation. *Water* **2019**, *11*, 935. [CrossRef]

48. Mao, Z.; Guan, G.; Yang, Z. Suppress Numerical Oscillations in Transient Mixed Flow Simulations with a Modified HLL Solver. *Water* **2020**, *12*, 1245. [CrossRef]

49. Bailey, O.; Zlatanovic, L.; van der Hoek, J.P.; Kapelan, Z.; Blokker, M.; Arnot, T.; Hofman, J. A Stochastic Model to Predict Flow, Nutrient and Temperature Changes in a Sewer under Water Conservation Scenarios. *Water* **2020**, *12*, 1187. [CrossRef]

50. Rinas, M.; Tränckner, J.; Koegst, T. Sediment Transport in Sewage Pressure Pipes, Part I: Continuous Determination of Settling and Erosion Characteristics by In-Situ TSS Monitoring Inside a Pressure Pipe in Northern Germany. *Water* **2019**, *11*, 2125. [CrossRef]

51. Rinas, M.; Fricke, A.; Tränckner, J.; Frischmuth, K.; Koegst, T. Sediment Transport in Sewage Pressure Pipes, Part II: 1 D Numerical Simulation. *Water* **2020**, *12*, 282. [CrossRef]

52. Wang, Q.; Peng, W.; Dong, F.; Liu, X.; Ou, N. Simulating Flow of An Urban River Course with Complex Cross Sections Based on the MIKE21 FM Model. *Water* **2020**, *12*, 761. [CrossRef]

Publisher's Note: MDPI stays neutral with regard to jurisdictional claims in published maps and institutional affiliations.

Article

Cost–Benefit Prediction of Asset Management Actions on Water Distribution Networks

Amir Nafi [1,2,*] and Jonathan Brans [1]

[1] Unité Mixte de Recherche Gestion Territoriale de l'Eau et de l'Environnement (GESTE) IRSTEA-ENGEES, 1 quai Koch, 67070 Strasbourg, France
[2] CSIP-ICube, Université de Strasbourg, 3-5, rue de l'Université, 67084 Strasbourg, France
* Correspondence: amir.nafi@engees.unistra.fr; Tel.: +33-388-248-293

Received: 19 June 2019; Accepted: 15 July 2019; Published: 25 July 2019

Abstract: The potential costs and benefits of a combination of asset management actions on the water distribution network are predicted. Two types of actions are considered: maintenance actions and renewal actions. Leak detection and reparation of failures on connections and pipes define the set of potential maintenance actions to be carried out. Renewal actions concern connections, pipes, and meters. All these actions represent the model's decision variables in order to determine a trade-off between two objectives: (i) the maximization of the water efficiency rate and (ii) the minimization of the total cost of actions to be carried out on the water system. The assessment of objective functions is ensured by an artificial neural network (ANN) trained on a French mandatory database «SISPEA». A non-dominated sorting genetic algorithm (NSGA-II) is coupled to the ANN to reach the set of compromised solutions representing potential actions to achieve. Applied to a real water distribution system in the southeast of France, the proposed decision model indicates that the improvement of water efficiency rate (*WER*) in the short term requires increasing operation expenditures (*OPEX*), which represent 99% of the total cost. Results show the existence of a threshold effect that implies to use the budget in a certain way to improve performance. A potential solution can be chosen by the decision maker among the generated Pareto front with regard to the constraint on the budget and the targeted *WER*.

Keywords: actions; asset management; ANN; prediction; performance; water utility; water system; NSGA-II

---

1. Introduction

Water utility performance monitoring is widely addressed in the literature. IWA initiative carried out by Ref. [1] to build key performance indicators (KPIs) led to the emergence of national mandatory databases in several countries in order to improve the management of water utility and ensure transparency against stakeholders and users. However, KPIs are generally measured on an ex-post basis in order to assess the ability of conducted policy to achieve planned goals; otherwise, corrective actions can be planned in the case of a mismatch. This way of management could be expensive in terms of time and money.

One possible improvement to avoid this mismatch is the use of a decision-aiding model to predict KPIs based on potential decisions and a set of explanatory data. A possible shortfall concerns the absence of data collection at the scale of the water utility, which renders it difficult to train and fit a prediction model. The existence of an information system (IS) seems to be a prerequisite for the assessment and the prediction of KPIs. This shortfall tends to be solved. In fact, in the last 2 decades, we observe the development of sensors technologies and information and communications technology (ICT) that encourage water utility to install smart devices in order to monitor water systems in real time and collect information about their operation. The relevance of adopting smart water systems and the potential benefits in terms of leak management, water quality monitoring, and energy savings are discussed in Ref. [2]. Smart systems generate an important quantity of data which are not always exploited in the decision-making process. Data gathering improves the water utility information system (IS) and constitutes a prerequisite for prospective analysis. The current research addresses the assessment of KPIs in an ex-ante way based on the exploitation of data due to the emergence of mandatory databases and the deployment of smart devices in the water systems. The current paper aims at answering the following question: How can the existing data collections or IS be exploited for prospecting asset management actions and assessing their costs and benefits in an ex-ante way?

For any planning of asset management actions, the assessment of expected costs and benefits is recommended because it allows decisions mitigation. The importance of cost–benefit quantification in the determination of optimal maintenance time is underlined in Ref. [3]. Models for asset management of water pipes seem to be driven by the estimation of the optimal date of renewal based on the deterioration of the asset, the assessment of whole life costing [4], the achievement of a critical threshold for the number of breaks [5] or the rendered service (pressure, flow, quality) under economic or technical constraints [6,7]. Pipes renewal planning considering multiple objectives can be achieved by genetic algorithms [8]. The problem of water pipe renewal planning based on a cost–benefit approach is addressed in Ref. [9]. Authors define five items of benefit. Items calculate the benefit of reduction of the repair cost, the benefit from avoiding potential damages of water suspension for domestics and non-domestics, and the benefit from avoiding the social cost in case of roads unavailability. The optimal time for pipe renewal is reached when expected benefits are greater than costs.

The use of genetic programming for pipe breaks prediction is discussed in Ref. [10]. Authors develop an economic-based model for pipes replacement. They assume that there exist two categories of models for pipe breaks prediction: The physically-based models that aim at identifying physical causes of breaks and statistical models that analyze historical data to identify explanatory variables.

The use of machine learning seems relevant to tackle prediction problems. Between 2006 and 2016, the use of Artificial Neural Networks (ANNs) has increased in the drinking water sector, particularly for modeling the infrastructure and water quality [11]. ANNs address water quality problems by modeling chlorine concentration [12]. To improve leakage management, hydraulic and water quality data collected from sensors are used to fit ANNs for detecting and locating leakage in Yorkshire Water's Keighley distribution system [13]. A principal component analysis (PCA) and ANN was carried out to predict the leakage ratio in the drinking water system using six effective parameters: pipe deterioration ratio, the volume of water supplied, pipe length, mean pipe diameter, the number of leaks, and an energy ratio [14]. Authors show the advantage of coupling ANN with PCA. To estimate the magnitude and the location of leaks, ANNs were trained on different sets of input data (pressure and flow rate) collected from sensors installed in the piping network [15].

It appears from the literature review that despite the output variable to predict, the training of ANNs in the drinking water sector is done at the local scale by using a series of monitoring data collected by sensors disseminated in the network. What can be done in case of the absence of monitoring data? A partial answer is given by Ref. [16], who investigated the training of ANNs not on monitoring data but on aggregated data or KPIs, representing high-level data gathered in mandatory databases. Authors establish cause–effect relationships between KPIs. They compare the use of ANN or multiple regression analysis (MRA) for calibrating a decision model that is able to predict the water efficiency ratio from a set of nine mandatory indicators considered as input variables.

In the context of absence or paucity of low-level monitoring data, the current work improves the model developed in Ref. [16] by prospecting asset management actions based on high-level data represented by ex-post KPIs measured at the scale of the water utility.

We assume that the proposed model can be adapted in the context of smart water systems where monitoring data are available at a low-level scale. The main added value of the proposed model is its ability to prospect asset management actions by measuring KPIs in an ex-ante way using an adaptation of ANNs and a multi-objective genetic algorithm. The prediction model can be fitted with a multiset of data from several water utilities or a national database of mandatory KPIs as SISPEA (French context) and the IS of the water utility. This can be very helpful in case of absence of enough monitoring data at the scale of the water utility.

The paper is organized into five sections. The current section proposes a literature review of asset management of water pipes and the use of ANN for KPI's prediction and genetic algorithm for problem optimization. Section 2 defines the objective functions and the mathematical formulation of the considered problem. The characteristics of the ANN and NSGA II are also detailed. Section 3 illustrates the use of the developed model on a real case study and shows how it is carried out. Section 4 discusses the results and the main added value of the model. Finally, the last section concludes the paper.

2. Materials and Methods

This paper focuses on the prediction of two KPIs considered as objective functions: (1) the water efficiency rate considered as a benefit and (2) the total cost obtained by the sum up of *OPEX* and capital expenditures (*CAPEX*). Considered costs are the result of the implementation of asset management actions: renewal of pipes, connections and meters on one hand; and leak detection, connections and pipes reparation on the other hand. The prediction of KPIs is ensured by an adaptation of ANNs coupled with a multi-objective genetic algorithm NSGA II [17].

2.1. The Water Efficiency Rate (WER)

In the French context, the *WER* is a mandatory KPI calculated for each water utility according to the decree of May 2007 [18]. It measures the ratio between the billed and distributed water. The prediction model uses the theoretical model developed in Ref. [16] to establish relationships between *WER* (output) and nine other mandatory KPIs (Input) considered as explanatory variables. Table 1 lists the explanatory variables with their corresponding code (taken from SISPEA) and their link with asset management actions.

Table 1. Explanatory variables for efficiency rate.

Asset Management Actions	SISPEA Code	Explanatory Variables—Indicators
Metering and metering error	VP.056	Number of users
	VP.228	Linear density of users
	VP.063	Billed metered domestic consumption
	VP.221	Volume of unmetered consumption
	VP.232	Billed metered consumption
	VP.234	Volume produced + Volume imported
Leakage and water losses	VP.225	Average network efficiency rate over last 3 years
	P106.3	Linear leakage index on distribution mains (LLI)
Pipes renewal	P107.2	Average renewal rate of water mains over the last 5 years

The assessment of *WER* requires the analysis of the yearly hydraulic balance of the whole network. Table 2 lists the required variables.

Table 2. List of variables required for hydraulic balance.

Symbol	Definition of the Variable	Unit
W_b	Annual volume of water-billed metered consumption	m^3
W_m	Annual volume of water loss due to metering error	m^3
W_p	Annual volume of water loss due to leaks on main pipes	m^3
W_c	Annual volume of water loss due to leaks on connections	m^3
W_i	Annual volume of water loss due to invisible leaks	m^3
ε_m	Metering error in percentage	%
$\overline{Age}_m$	Average age of meters	#
$MTTR_{vl}$	Mean time to repair visible leak	s
$MTTR_{inv}$	Mean time to repair hidden leak	s
d_p	Average flow rate for a leak on pipe	L/s
d_c	Average flow rate for a leak on connection	L/s
d	Average flow rate for a hidden leak	L/s
n_{inv}	Number of invisible breaks/leaks	#
n_p	Number of breaks/leaks on pipes per year	#
n_{lc}	Number of breaks/leaks on connections per year	#
n_c	Number of connections	#
r_b	Pipe breakage rate	#/km
r_{cb}	Connection breakage rate	#/km
r_d	leak detection efficiency rate	%
α	Invisible leakage rate on main pipes/connections	%
L_{net}	Network length	km

To be able to calculate *WER*, the listed explanatory variables in Table 1 should be calculated or estimated. *WER* can be indirectly estimated from the linear leakage, which encompasses four types of losses: losses due to metering errors W_m, losses due to leaks on main pipes W_p, losses due to leaks on connections W_c, and losses due to invisible leaks W_i. We assume that losses due to metering errors $W_m(t)$ can be calculated by Equation (1):

$$W_m(t) = W_b(t)(t) \times \varepsilon_m (t) \tag{1}$$

with:

$$\varepsilon_m (t) = \varepsilon_m (t-1) \times \frac{\overline{Age}_m(t)}{\overline{Age}_m(t-1)} \tag{2}$$

By considering the meter renewal rate, $\varepsilon_m (t)$ is calculated by Equation (3):

$$\varepsilon_m (t) = \varepsilon_m (t-1) \times (1 - r_m) \tag{3}$$

where r_m is the rate of annual meter renewal in percentage per year as listed in Table 3.

Table 3. List of required variables for cost calculations.

Symbols	Definition of the Variable
C_{rep}	Cost of a leak reparation in € per unit
C_{det}	Cost of leak detection in € per km
C_{meter}	Cost of a meter in € per unit
C_{con}	Cost of a connection renewal in € per unit
C_p	Cost of pipe renewal in € per km
n_{lc}	Number of leaks on connections per year
n_p	Number of leaks on pipes per year
n_d	Number of leaks detected by leak detection
n_m	Number of installed meters
l_{net}	Length of the network
l_{det}	Length of the network investigated by leak detection
r_c	Rate of annual connections renewal in percentage per year
r_p	Rate of annual pipe renewal in percentage of the length renewed per year
r_m	Rate of annual meter renewal in percentage per year

Losses due to leaks on pipes are computed by taking into account the estimated number of leaks on pipes from which the effect of the pipe renewal is subtracted:

$$W_p(t) = MTTR_{vl}(t) \times d_p(t) \times \left[n_p(t) - r_p(t) \times L_{net}(t) \times r_b(t) \right] \tag{4}$$

Analogously, losses due to leaks on connections at a given year $W_c(t)$ are computed by taking into account the estimated number of leaks on connection minus the effect of connections renewal:

$$W_c(t) = MTTR_{vl}(t) \times d_c(t) \times [n_{lc}(t) - r_c(t) \times n_{ct}(t) \times r_{cb}(t)] \tag{5}$$

The model also involves water losses $W_i(t)$ caused by invisible leaks. Equation (6) indicates how they are calculated:

$$W_i(t) = MTTR_{inv}(t) \times d(t) \times \left[n_{inv}(t) \times \left(1 - \alpha \times r_p(t) - (1 - \alpha) \times r_c(t)\right) - r_d(t) \times L_{net}(t) \right] \tag{6}$$

Asset management actions in terms of renewal (pipe, connections) and leak detection have an impact on leaks. Actions decrease the number of invisible leaks and the mean time to repair; this assumption is introduced by Equation (6). The total water loss for year t, $W_l(t)$, is obtained by the sum up of all types of water losses as shown in Equation (7):

$$W_l(t) = W_m(t) + W_p(t) + W_c(t) + W_i(t) \tag{7}$$

Based on previous equations, it is possible to compute the linear leakage index according to Equation (7).

$$LLI(t) = \frac{W_l(t)}{L_{net}(t)} \tag{8}$$

The average renewal rate of water mains over the 5 last years $\overline{r_p}(t)$ (code: P107.2) measures the mean value of the annual renewal rate of water pipes (without connections) over the last 5 years. This includes renewed, reinforced and rehabilitated pipes but does not take into account maintenance actions as pipes reparation. The average renewal rate of water mains over the last 5 years is calculated by Equation (9):

$$\overline{r_p}(t) = \sum_{i=0}^{3} \frac{r_p(t-i-1) + r_p(t)}{5} \tag{9}$$

with $r_p(t-i-1)$ for $i \in [0,3]$ being the annual renewal rate of pipes from the previous 4 years (known); and $r_p(t)$ is the annual renewal rate envisaged.

The remaining explanatory variables: number of users (VP.056), linear density of users (VP.228), billed metered domestic consumption (VP.063), volume of unmetered consumption (VP.221), billed metered consumption (VP.232), volume produced + volume imported (VP.234) are estimated based on water utility manager opinion, historical data and Monte Carlo analysis using a uniform distribution function as explained in Ref. [16].

In the context of a lack of low level data, we advise to use Equation (7) to estimate the mean and standard deviation of the following parameters: leakage flow rate, the number of hidden leaks and repair time for both pipes and connections over an observation period of at least 5 years. Obtained values represent a set of feasible solutions that satisfy the yearly hydraulic balance on the observation period.

The number of visible breaks and leaks on pipes and connections are supposed to be available as local data from the water utility. To involve the uncertainty of estimation, a Monte Carlo analysis is implemented using Equation (7), where a set of parameters and variables of the equation are randomly generated as shown in Figure 1. In the absence of data concerning the characteristics of leaks, normal distribution functions are used to randomly generate the flow rate, the number of leaks and time to repair. The achievement of this analysis provides a potential range of values for parameters of Equation (7) that make the estimation of water losses possible for prediction purposes. Figure 1 illustrates the required steps to estimate annual water losses.

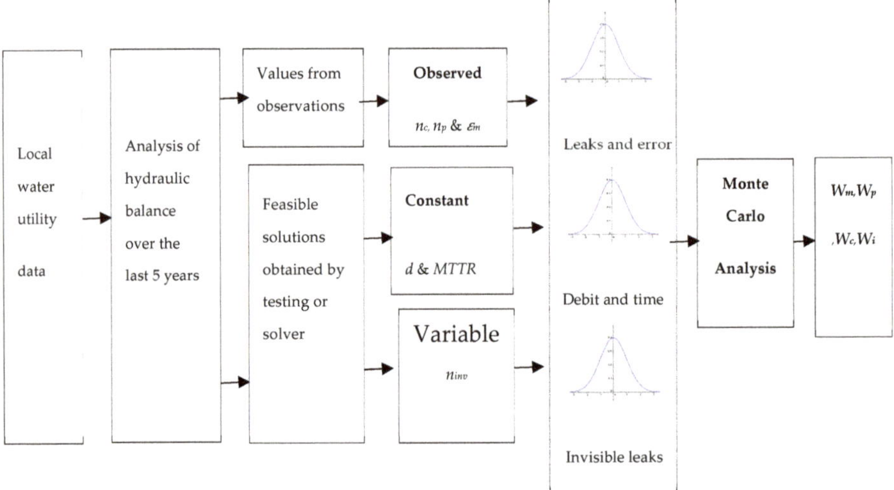

Figure 1. Steps for annual water losses estimation, adapted from Ref. [16].

2.2. The Total Annual Cost

The total annual cost (C_{Tot}) of decisions or a policy defined by asset management actions is calculated by Equation (10). Required variables for cost calculations are resumed in Table 3.

$$C_{Tot} = CAPEX + OPEX \tag{10}$$

OPEX are derived from curative maintenance actions of repairing pipes and connections, on the one hand, and preventive maintenance actions of leak detection, on the other hand; Equation (11) summarizes the annual maintenance costs as follows:

$$OPEX = C_{pipe\_reparation} + C_{connection\_reparation} + C_{leak\_detection} \tag{11}$$

Each component of the maintenance cost is displayed in Equation (12) as follows:

$$OPEX = C_{rep} \times (n_p + n_d) + C_{rep} \times n_{lc} + C_{det} \times l_{det} \tag{12}$$

CAPEX measure the cost of asset management actions in terms of pipes, connections and meters renewal as indicated in Equation (13):

$$CAPEX = C_{pipe\_renewal} + C_{connection\_renewal} + C_{meter\_renewal} \tag{13}$$

Equation (13) becomes as follows when each component of investment cost is displayed:

$$CAPEX = C_p \times r_p \times l_{net} + C_{con} \times r_c \times n_{lc} + C_{meter} \times r_m \times n_m \tag{14}$$

2.3. The Artificial Neural Network (ANN)

A neural network is composed of multiple perceptron and is called a deep neural network when the number of hidden layers is greater than or equal to 2 [19]. We use a multiple layers neural network in order to predict the *WER* based on nine KPIs considered as input [16]. Figure 2 illustrates a perceptron representing a layer in an ANN.

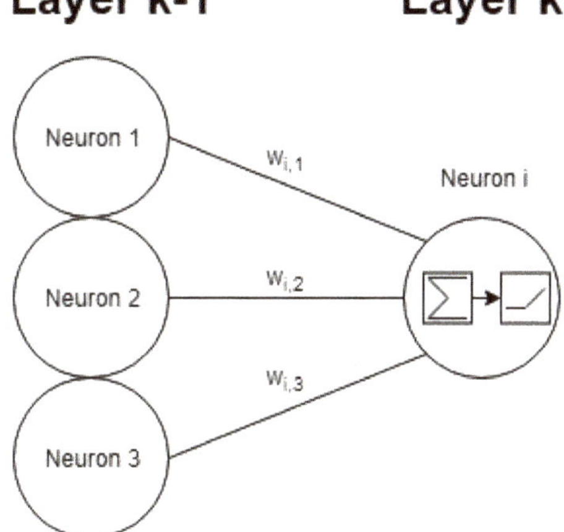

Figure 2. Example of a single perceptron.

The value assigned to neuron *i* in Figure 2 can be calculated by Equation (15) as follows:

$$neuron_i^k = relu\left(w_{i,1} \times neuron_1^{k-1} + w_{i,2} \times neuron_2^{k-1} + w_{i,3} \times neuron_3^{k-1} + b_i^{k-1}\right) \tag{15}$$

The rectified linear unit function *relu* is given by Equation (16):

$$relu(x) = \{0 \ for \ x < 0 \ ; x \ for \ x \geq 0\} \tag{16}$$

The vector *neuron^k* that groups all the values assigned to the neurons in layer k is calculated as follows:

$$neuron^k = relu\left(\begin{bmatrix} w_{0,0} & \cdots & w_{0,n} \\ \cdots & \ddots & \vdots \\ w_{i,0} & \cdots & w_{i,n} \end{bmatrix}\begin{bmatrix} neuron_0^{k-1} \\ \vdots \\ neuron_n^{k-1} \end{bmatrix} + \begin{bmatrix} b_0^{k-1} \\ \vdots \\ b_n^{k-1} \end{bmatrix}\right) \tag{17}$$

Equation (17) becomes:

$$neuron^k = relu\left(M_w^{k-1} \times neuron^{k-1} + b^{k-1}\right) \tag{18}$$

where:

- M_w is the matrix of weights;
- *neuron* is the vector of neuron values;
- k is the index of the layer;
- i is the number of neurons in the kth layer;
- n is the number of neurons in the $(k-1)$th layer;
- b is the bias vector.

The output value of the ANN can be computed by Equation (18). In our case, it is a single neuron which produced the water efficiency rate *WER*. The value of this neuron depends on the values of the previous neuron layers and the associated weights and biases.

Values of the previous layers also depend on weights and biases as well as input variables. The input variables are known, the objective is to determine the optimal values of weights and biases to give a good prediction.

To do this, during the learning phase, the prediction is compared to the real value. Weights and bias are adjusted until a satisfactory error is obtained. Error is commonly calculated with a Loss function noted L. For regression problems, the function L corresponds to the mean square error which computes the square difference between the observed and predicted value:

$$L(y_i, \hat{y}_i) = \frac{1}{n}\sum_{i=1}^{n}(y_i - \hat{y}_i)^2 \tag{19}$$

where n is the number of input values, y_i is the value of input i, and $\hat{y}_i$ is the corresponding predicted value.

To minimize the loss function L, we use an optimization function *Adagrad* which modifies weights and bias in order to minimize the error. *Adagrad* was introduced by Ref. [20] and it is called so for adaptive gradient algorithm. During the learning process, the weights are updated considering Equation (20):

$$\Delta w_i(t) = -\frac{\eta}{\sqrt{G_i(t)} + \varepsilon} \times \frac{\partial L}{\partial w_i}(t) \tag{20}$$

with:

$$\begin{cases} G_i(t) = G_i(t-1) + \left(\frac{\partial L}{\partial w_i}(t)\right)^2 \\ G_i(0) = 0 \end{cases} \tag{21}$$

The term $\frac{\eta}{\sqrt{G_i(t)}+\varepsilon}$ is the effective learning rate, with η being the initial learning rate. The term $\frac{\partial L}{\partial w_i}(t)$ is the gradient (partial derivative of loss function with respect to weights). By this definition, G_i is a monotone increasing function. So, the effective learning rate is monotonously decreasing. Note that G_i and the effective learning rate are different for each weight.

Figure 3 illustrates the ANN built for our prediction model. It is designed for the nine input explanatory variables representing KPIs (with French mandatory codes), two hidden layers with the same number of neurons as the input layer. The output layer considered as the output of the model is only composed of the neuron corresponding to the water efficiency rate, *WER* (code: P104.3).

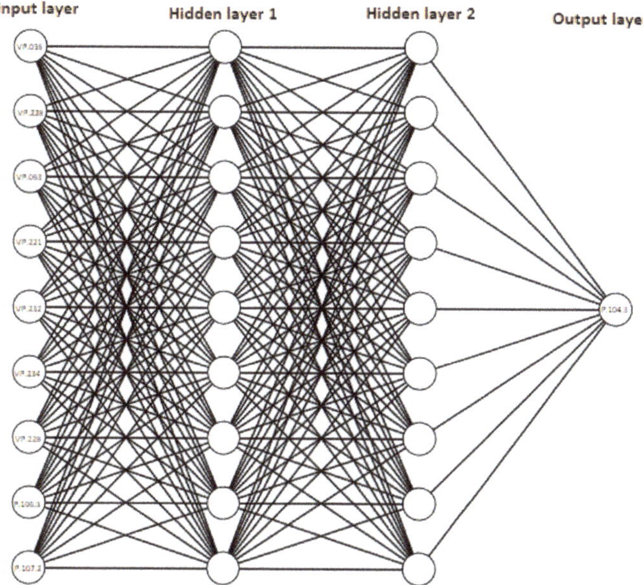

Figure 3. Neural network configuration with two hidden layers.

Since the number of input samples is more than 10,000, the model is trained with a batch size of 200. The batch size corresponds to the number of samples that will be propagated through the neural network. After propagation, weights and biases are updated in order to decrease the error. Once all training samples are passed once through the network, this counts as 1 epoch. The network training is done by performing multiple epochs.

2.4. NSGA II and the Problem Formulation

The problem to solve concerns the optimization of asset management actions in an ex-ante way in order to maximize the *WER* and minimize the annual total costs. The decision variables measure the level of actions in terms of pipes, connections, and meters maintenance and renewal. We consider that the following variables l_{det}, r_c, r_p, and r_m are the most relevant for the decision maker in terms of asset management. The problem can be formulated as the following:

$$\text{Maximize } f_1(l_{det}, r_c, r_p, r_m) = WER(t) \tag{22}$$

$$\text{Maximize } f_2(l_{det}, r_c, r_p, r_m) = \frac{1}{C_{Tot}(t)} \tag{23}$$

constrained by:

$$l_{det\_min} \leq l_{det} \leq l_{det\_max} \tag{24}$$

$$r_{p\_min} \leq r_p \leq r_{p\_max} \tag{25}$$

$$r_{c\_min} \leq r_c \leq r_{c\_max} \tag{26}$$

$$r_{m\_min} \leq r_m \leq r_{m\_max} \tag{27}$$

The value of upper and lower limits of decision variables are defined according to the water utility manager expectations. By considering the two fitness functions f_1 and f_2, NSGA II will attempt to find the best 4-tuple (l_{det}, r_c, r_p, and r_m) from a population of potential solutions. The population size is set in advance and the values of the 4-tuple elements are generated randomly between the upper and lower boundaries to initialize the population as shown in Figure 4.

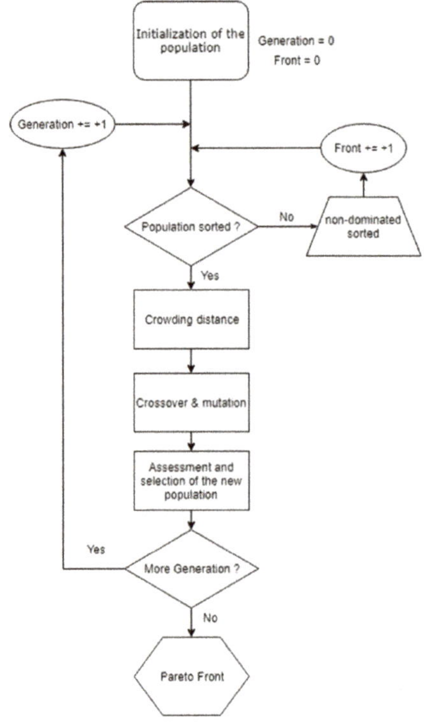

Figure 4. Flowchart of the fast, elitist, non-dominated sorting genetic algorithm (NSGA-II).

2.4.1. The Concept of Non-Dominance

The NSGA II implements the concept of dominance to reach potential solutions. The concept of dominance is well defined in Ref. [21]. Two definitions can be considered. The first one considers two solutions, that solution X_1 dominates solution X_2 if both conditions are true: (i) solution X_1 is not worse than X_2 for all the objectives, and (ii) solution X_1 is strictly better than X_2 for at least one objective. Conditions are resumed in Equations (28) and (29).

$$\forall\, i \in \{1,2\}: f_i\,(X_1) \geq f_i\,(X_2) \tag{28}$$

$$\exists\, j \in \{1,2\}: f_j\,(X_1) > f_j\,(X_2) \tag{29}$$

The second definition considers as non-dominated solutions those that are not dominated by any member of the considered population.

2.4.2. The Crowding Distance

To sort solutions, NSGA II uses a crowding distance [17,22,23]. It is used to estimate the density of solutions surrounding an individual in the population by considering the difference of the objective values of the nearest neighbor as shown in Figure 5. It is an estimate of the size of the largest cuboid enclosing point k, without including any other point in the population. In the following sections, the term individual designates a potential solution.

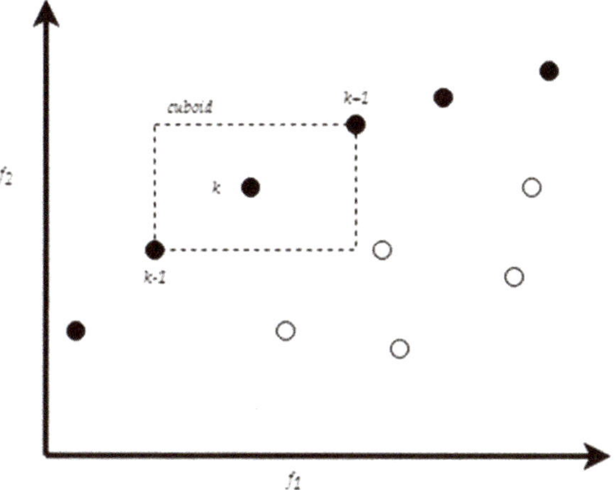

Figure 5. The crowding distance of individual in the front. Amended from Ref. [17].

Let's consider F the size of the front, for individuals, the crowding distance is calculated by the difference between the objective values of the two nearest neighbors:

$$d_i = \sum_{m=1}^{M} \frac{f_{i+1}^m - f_{i-1}^m}{f_{max}^m - f_{min}^m} \tag{30}$$

The edge, the first individual and the last individual in the rank, are assigned with a large distance to ensure that boundary points will always be selected as shown by Equation (31).

$$d_0 = d_{F-1} = \infty \tag{31}$$

where M is the number of objectives, f_i^m is ith fitness values in the mth objective, and f_{max}^m and f_{min}^m are the maximum and minimum objective values of the mth objective (in the non-dominated set).

This formulation maintains diversity in the population by eliminating redundant individuals but suffers from a loss of both vertical and horizontal diversity as explained by Ref. [22]. To improve the diversity in the final front, an improvement of the crowding distance has been proposed by Ref. [23] by defining a dynamic crowding distance:

$$D_{d_i} = \frac{d_i}{\log\left(\frac{1}{V_i}\right)} \tag{32}$$

with:

$$V_i = \sum_{m=1}^{M} \left(\left|f_{i+1}^m - f_{i-1}^m\right| - d_i\right)^2 \tag{33}$$

The dynamic crowding distance is computed for each individual in the non-dominated set. The individual which has the lowest dynamic crowding distance is removed. The dynamic crowding distance is updated after each removal. These operations are repeated until the size of the non-dominated set is equal to the population size.

2.4.3. The Selection Method

Once individuals have been assessed and sorted, k Elements of the population are taken as candidates for the mating pool, where k designates the tournament size [17]. Random selection is a particular case of tournament selection when $k = 1$. For $k > 1$, the selection method is called tournament selection. The k individuals are compared to each other based on their rank and crowding distance. The best individual is added to the mating pool. The operation is repeated a second time to obtain two individuals in the mating pool as shown in Figure 6. Selected solutions are subject to crossover and mutations to create offspring. Tournament selection is repeated until the number of created offsprings is sufficient.

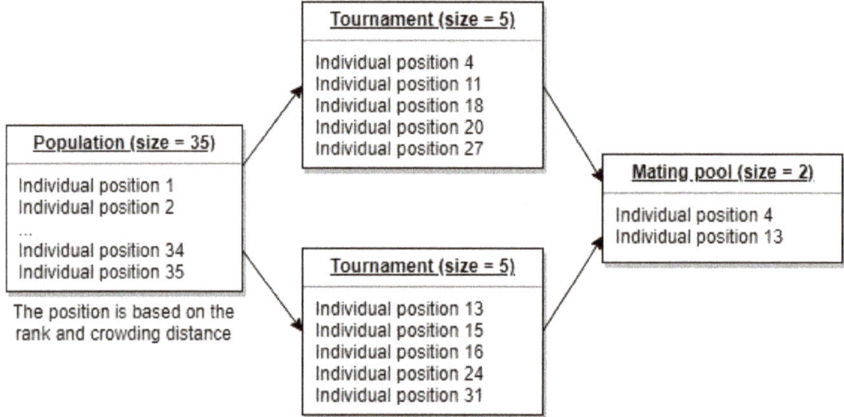

Figure 6. The tournament selection.

There exists other selection operators where individuals are chosen based on their proportional fitness value, as the roulette wheel selection (RWS). The individual is selected according to a probability of selection calculated by the ratio between its fitness value and the sum up of fitness values of individuals in the mating pool [24].

2.4.4. The Crossover

Realizing a crossover is a way of using the information of two parents in the population to obtain one child [25]. There are different possible recombinations and several authors have compared them to each other in different problems [26,27]. There is no consensus in the literature concerning the effectiveness of single point crossover or multi-point crossover. This depends on the particularities of the problem. The danger of algorithms comparison on a small sample according to their performance is underlined in Ref. [28]. The authors advise to integrate problem-specific knowledge into the functioning of the algorithm; this integration can also concern crossover operators. In our case, there are two objective functions and the only constraints in this problem are upper and lower bounds of the 4-tuple variable. Hence, we choose to use the flat crossover which is a widely used crossover method [29]. Considering two parents in the current population:

$$Parent_1 = \left(l_{det1}, r_{p1}, r_{c1}, \ r_{m1}\right) \tag{34}$$

$$Parent_2 = \left(l_{det2}, r_{p2}, r_{c2}, r_{m2}\right) \tag{35}$$

and a random vector:

$$r = (r_1, r_2, r_3, r_4) \tag{36}$$

with random values $r_i \in [0,1]$. The ith child $\left(l_{det}{}^i,\ r_p{}^i,\ r_c{}^i,\ r_m{}^i\right)$ is a linear combination of the two parents:

$$l_{det}{}^i = r_1 \times l_{det1} + (1 - r_1) \times l_{det2} \tag{37}$$

$$r_p{}^i = r_3 \times r_{p1} + (1 - r_3) \times r_{p2} \tag{38}$$

$$r_c{}^i = r_2 \times r_{c1} + (1 - r_2) \times r_{c2} \tag{39}$$

$$r_m{}^i = r_4 \times r_{m1} + (1 - r_4) \times r_{m2} \tag{40}$$

Table 4 shows an example of the offspring that two parents can give by applying this crossover.

Table 4. Crossover and offspring generation.

Individual	l_{det}	r_p	r_c	r_m
Parent$_1$	74.18	0.0042	0.0083	0.028
Parent$_2$	82.07	0.0031	0.0068	0.035
Random vector, r	0.61	0.77	0.75	0.55
Offspring$_1$	77.27	0.0040	0.0079	0.031
Offspring$_2$	78.98	0.0033	0.0072	0.032

2.4.5. The Mutation

The mutation is an operator that modifies an individual to explore the entire search space [25] and to escape from local optima thanks to small changes in the values of the 4-tuple variables. It is used to maintain diversity in the population of potential solutions. We use the polynomial mutation introduced by Ref. [30]. For each variable, there is a mutation probability. The mutation probability is set at 1/4 since each solution is represented by a 4-tuple. There is one mutation per offspring on average.

2.4.6. The Selection of Offspring

Once the crossovers and mutations have been achieved, we end up with a population of P individuals and a population of P offspring. The total size of the selection is $2P$ and this must be reduced to P individuals. This selection is made by keeping the best individuals as requested by NSGA II [17]. In this way, the next generation will be better than the previous generation or equivalent if no individual from the descendants is better than the current population. This is called elitism selection. To select the best individual, we defined an operator $\geq_n$ basis on individual domination rank $rank_p$ and dynamic crowding distance D_{dp}. The partial order $\geq_n$ is defined as:

$$p \geq_n q \text{ if } \left(rank_p < rank_q\right) \text{ or } ((rank_p = rank_q) \text{ and } D_{d_p} > D_{d_q}) \tag{41}$$

The individual with the lower rank, according to the non-dominated sorting algorithm, is preferred. If two individuals have the same rank, the one which is located in the lower density of solutions is preferred.

The selection of $2P$ individual is first sorted in the ascending order with respect to their rank obtained by the non-dominated sorting algorithm. Then, individuals are sorted with respect to the dynamic crowding distance in descending order. The next generation is thus generated until there are P individuals in the new population.

2.4.7. Performance Metrics

The effectiveness of the model depends on its ability to ensure diversity, a good distribution and spread of solutions. To evaluate the distribution, we use the Spacing index (SP) introduced by Ref. [31]. To be able to assess the spread in a population of P individuals (potential solutions), we need to calculate d_i which is the minimum of the sum of the absolute difference in objective fitness values between the ith solution and any other solution as shown in Equation (42), and $\bar{d}$, the mean value of d_i calculated by Equation (43):

$$d_i = \min_{i,\, i \neq k} \left\{ \sum_{m=1}^{M} \left| f_i^m - f_k^m \right| \right\} \tag{42}$$

$$\bar{d} = \sum_{i=1}^{|P|} \frac{d_i}{|P|} \tag{43}$$

Therefore, SP is obtained by Equation (44):

$$SP = \sqrt{\frac{1}{|P|-1} \sum_{i=1}^{|P|} \left(d_i - \bar{d} \right)^2} \tag{44}$$

SP is used to evaluate the spacing between the different solutions. If the distance between each solution is the same, then the SP value will be zero. Thus, a value of zero or near zero indicates a good distribution of solutions on the Pareto front. The spread index was proposed by Ref. [17]:

$$\Delta = \frac{d_f + d_l + \sum_{i=1}^{|P|} \left(d_i - \bar{d} \right)}{d_f + d_l + (|P|-1) \times \bar{d}} \tag{45}$$

d_f and d_l are the Euclidean distances between the extreme solutions and the obtained Pareto solutions. A Δ value close to 0 means that the solutions are well dispersed along the Pareto front.

3. Case Study

The model is implemented on a real water distribution network in the south of France. According to the data of the year 2016, the water system delivers 700,000 m^3 of drinking water for about 6300 users with a network length of 82 km. We consider the actual asset management actions implemented by the water utility as a baseline solution. It can be resumed by a leak detection of the entire network once (l_{det} = 82 km) that allows detecting 14 leaks on average. Annual renewal rates are: r_p = 0.71%, r_c = 2.5% and r_m = 10%. Thanks to these actions, WER = 76.90% with a total cost equals to 551,493 €, shared between 70% in $CAPEX$ and 30% in $OPEX$. We aim at improving WER by conducting alternative asset management actions at lower costs than commonly used strategies. Before searching compromise solutions, ANN is fitted thanks to the SISPEA database. SISPEA is a mandatory French database that gathers 26 KPIs from more than 12,000 water utilities between 2006 and 2016. Data were split into two samples, 70% of the data is split to fit the Ann model, and the remaining 30% is used for validation.

3.1. Artificial Neural Network Fitting

The calibration of the ANN requires the definition of a set of parameters that improve its accuracy. As discussed in Ref. [16], many simulations are carried out in order to determine the most appropriate values for the number of hidden layers, the number of neurons per layer, and the type of activation number. The selected ANN is built by three hidden layers with 144, 36, and 9 neurons at each layer, respectively. The chosen activation function is the function *relu* for all neurons. The estimation of required variables for water losses estimation for the year (N + 1) at the local scale (see Equation (7)) is generated based on expert opinion and Monte Carlo analysis. Table 5 compares the observed and predicted values of WER for the period between 2010 and 2016.

Table 5. Comparison between observed and predicted *WER* between 2010 and 2016.

Year	Observed	Predicted-ANN (%)	Estimation Error (%)
2010	65.9	65.7	−0.30%
2011	63.2	62.4	−1.27%
2012	61.4	60.7	−1.14%
2013	75	75.1	0.13%
2014	75.1	75.1	0.00%
2015	76.9	76.8	−0.13%
2016	76.6	76.6	0.00%

According to Table 5, ANN seems to predict the *WER* with a high accuracy; the estimation error oscillates between −1.27% and +0.13%.

3.2. Parameters of the NSGA II

The implementation of NSGA II requires the definition of the type of tournament to consider and to set the population size of potential solutions. According to performance indicators resumed in Table 5, we compare between the tournament selection method with tournament size $k = 2$, random selection, and roulette wheel selection. The mean values and standard deviation are calculated on 10 tests for each selection method; results are resumed in Table 6. The population size is fixed at 200, and the number of generations is set at 25.

Table 6. Performance comparison of three selection methods in terms of distribution and spread.

Selection Method	Performance Metrics	Space Index	Spread Index
Random	Mean	0.127	0.313
	Std	0.025	0.091
Roulette Wheel	Mean	0.113	0.259
	Std	0.023	0.074
Tournament	Mean	0.115	0.286
	Std	0.035	0.130

Tournament selection and roulette wheel selection seem more efficient than random selection in terms of both distribution and spread.

The method with the best results is roulette wheel selection. Both the average and the standard deviation of the two performance metrics are the lowest. Note that the standard deviation for tournament selection is greater than for random selection.

3.3. Population Evolution

The objective of the model is to get closer to the true and unknown Pareto front. Over the generations, the population should move closer to the Pareto front. The starting population is generated randomly between the limits set for the 4-tuple values of the decision variables as presented in Table 7.

Table 7. Definition of decision variable constraints.

Decision Variables	Lower Bound	Upper Bound
l_{det}	0	$12 \times L_{net}$
r_p	0	5%
r_c	0	5%
r_m	0	10%

Figure 7 shows the evolution of potential solutions composing the population (size $P = 200$) using the roulette wheel selection after 25 generations. The optimal front is quickly reached; the population improves significantly in the first generations, and then very slowly over the last five generations, the front stabilizes for the last generations. The obtained front confirms the relevance of using the roulette wheel selection.

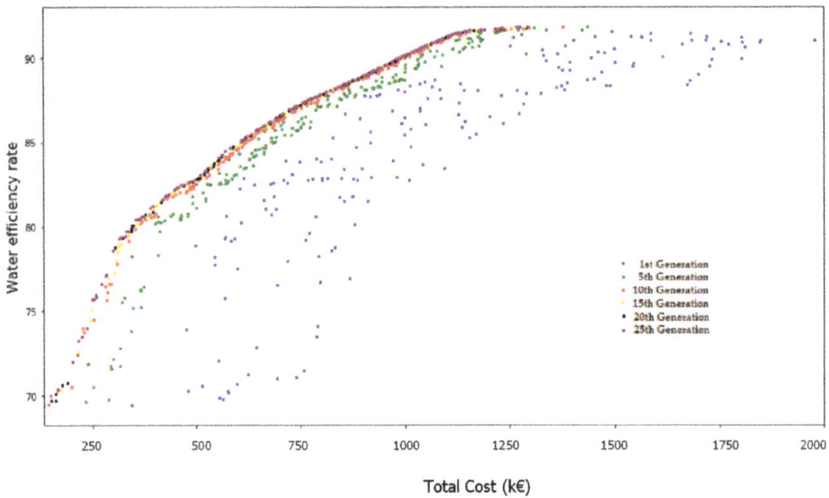

Figure 7. Evolution of Pareto front after 25 generations.

3.4. Problem Resolution

Performed tests guide the choice of the type of selection, crossover and mutation operators. To solve the considered problem (see Equations (22) and (23)), the NSGA II is implemented with an initial population generated randomly with 1000 individuals. The crossover probability is set to 90% to generate the offspring using the flat crossover. The polynomial mutation is used with an index polynomial mutation of 1 and a mutation probability of 25%. The new population is selected from roulette wheel selection. The number of performed generations is 25.

Figure 8 illustrates the Pareto front obtained for the considered objectives. The blue dots forming the front in the middle correspond to the average value of the water efficiency rate predicted by the ANN model. The red dots forming the upper and lower fronts define the limits of the 95% confidence interval. Each point of the front represents the 4-tuple of the constrained decision variables of the problem: the rate of pipes renewal, the rate of connections renewal, meters renewal, and length proven by leak detection. Table 8 details some of the solutions composing the Pareto front.

Table 8. Trade-off solutions with regard to considered constraints and objective.

r_p	r_c	r_m	L_{net} (km)	Total Cost (€)	CAPEX (€)	OPEX (€)	WER (%)
4×10^{-5}	8.1×10^{-5}	4.1×10^{-3}	86.5	223,302	2837	220,465	74.5
4.6×10^{-5}	8.2×10^{-4}	7.3×10^{-3}	162.6	309,838	3935	305,903	79.0
5.4×10^{-5}	9.5×10^{-4}	8.4×10^{-3}	260.2	413,868	4536	409,332	81.6

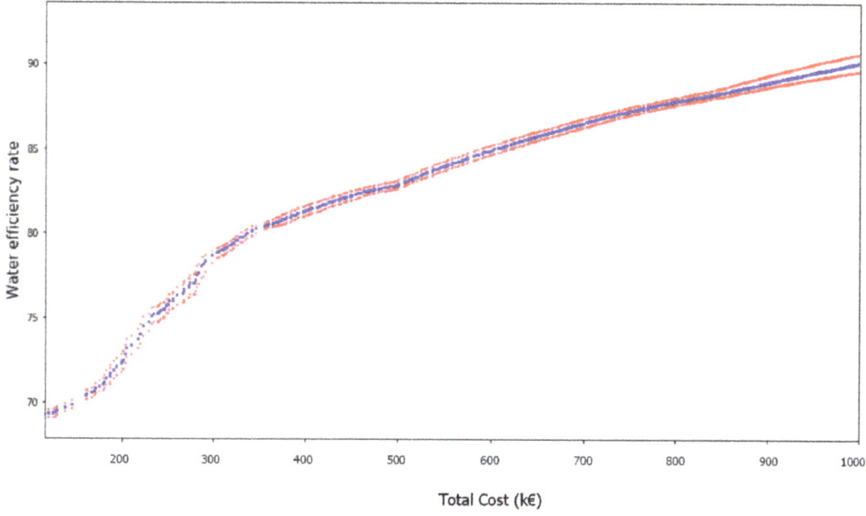

Figure 8. The Pareto front obtained from 1000 individuals and 25 generations.

The comparison of the baseline solution (actual practice) to the proposed solutions indicates that actual practice does not offer a compromise between cost and performance. Its costs more than all solutions listed in Table 8 with a value of $WER = 76.9\%$. Another interesting analysis concerns the repartition of expenditures between $CAPEX$ and $OPEX$. Water utility privileges investment by increasing $CAPEX$ (70 %) where our model advises to increase $OPEX$ to 99% of total expenditures (according to Table 8).

Results show a significant influence of the leak detection and reparation actions on the WER. This is an intuitive result but the main advantage of the proposed model is its capacity to predict the effects of actions on the WER. The length of the water system under consideration is about 86 km. The values of length proven by leak detection per year contained in Table 8 correspond to approximately one, two and three times the total length of the network. The total cost is shared into two parts, $CAPEX$ and $OPEX$. $OPEX$ are largely due to the leak detection and asset reparations while $CAPEX$ value is low due to low investments. In the short term period, the model shows that the main way to improve WER when the value of the efficiency is already high is the investigation for leaks. Indeed, leak detection allows improving more efficiently the water efficiency rate for an acceptable cost, compared to renewal actions. Renewal actions start to have an impact when performance values and asset condition are low.

4. Discussion

Predictions of WER (outputs) obtained for management actions (inputs) seem to be coherent with practice. In fact, Table 8 shows a positive correlation between WER and total cost, which confirms that it is required to spend more money to enhance performance. There seems to exist a threshold effect between expenditures and performance, even if we double the budget (from 2.23 k€ to 4.13 k€) performance increases only by 7%. This result is important because it indicates that even if the budget is available, it has to be spent in a certain manner and shared adequately between investment and maintenance actions. Another point of interest concerns the share of $OPEX$ in the total expenditures. $OPEX$ represents 99% of the total costs, which implies that if we aim at improving WER in the short term, it is recommended to spend more money for maintenance actions than investment. The advantage of the proposed approach is to drive the decision by indicating the type of maintenance actions to implement. For the studied case, Table 8 indicates that the leak detection and reparation of leaks seem to be efficient. The values of upper limits for decision variables were defined as the following: The limit of leak detection rate corresponds to a total inspection of the entire network each month (12 per year), renewal rate of pipes and connections is limited to 5% (5 times the actual rate) per year considering an average lifespan for asset of 50 years (ambitious), and the renewal rate of meters is limited to 10% (two times the actual rate) which corresponds to a lifespan for meters of 10 years on average.

The definition of NSGA II parameters requires expertise and should be driven by tests. Results show the relevance of comparing different operators of selection and crossover. For the current study, roulette wheel seems more pertinent than other methods. The size of the population and the number of generations are also an important parameter to fit. The followed procedure aims at driving the implementation of the approach by: (i) defining the first suitable operators for a fixed-size population ($P = 200$); (ii) test the range of values for the number of generations (5 to 25); and (iii) increase the size of population for a given number of generations from 200 to 1000. Even if we cannot generalize the obtained results, it seems that this procedure leads to improve the shape of the Pareto front and to make it less discontinuous and more uniform. It can be interpreted as an improvement of the consistency of the front as shown in Figure 8.

The variety of solutions offered by the Pareto front constitutes a set of potential actions to implement depending on the context, constraints, and objectives to reach. This constitutes a valuable mitigation tool for decision makers and stakeholders.

Another advantage provided by the prospecting model is its capacity to be coupled with NSGA II in order to guide the search for the most relevant solutions. Even if results are really encouraging, some aspects have to be investigated. The dynamics of the model are not actually addressed: how is it possible to improve the planning of actions from year to year by updating input data? Another important aspect concerns the effect of asset management actions in the long term; it appears that maintenance actions significantly improve the value of *WER* with a low total cost. This can be considered as relevant in the short term, but it is not supposed to encourage the non-investment actions. A risk can be faced by the water utility due to an under-investment, which is the deterioration of the asset and the delivered service. One possible improvement is to constrain the rate of asset renewal when the solutions are searched for in order to avoid an important asset aging due to disinvestment.

5. Conclusions

The actual research is considered an encouraging improvement of our model based on ANN for predicting KPI's. The proposed improvement confirms that it is possible to predict and optimize KPIs for water utilities by coupling ANN and NSGA II in the context of the lack of local data. Many aspects should be checked in relation to the characteristics and parameters of ANN like the number of hidden layers and the number of neurons and activation functions. For NSGA II, the set of population and type of selection, crossover, and mutation have to be fixed before implementing the prediction model. All these aspects can render the model difficult to implement by the water utility because it requires specific skills. The actual model should be improved to gain simplicity for easy implementation by water utilities.

The absence of local data is encountered by the use of a national mandatory database and Monte Carlo simulations; this can be useful in the short term. We demonstrate to the water utility managers the usefulness of using data for prediction; this should encourage the water utility manager to improve their IS and converge to a smart water system in order to catch real-time data for supporting the decision making. The interpretation of results should take into account the context of the water utility. The preference of implementing maintenance actions versus renewal actions can be relevant when the value of *WER* is high. That means that the condition of the asset is good and does not require renewal. This can be acceptable in certain conditions but not adequate when the asset has deteriorated. For example, if assets are in a good condition and the water system is young, it is not necessary to check the network by leak detection. The context and condition of the network have to be considered when the boundaries of decision variables are set. Their range of variation may consider as low boundaries thresholds different from 0 to avoid the aging of assets in the long term.

The variation of *WER* and cost shown by the Pareto front seem realistic and offers a variety of potential solutions to the decision maker which is valuable.

Further research will explore the reproducibility of the developed approach for other KPIs by defining the set of input variables and how the ANN model and NSGA II can help to predict them. For example, the SISPEA database contains 25 additional KPIs that merit to be predicted in the same way as *WER*. We intend to explore the possibility to adapt the current model and make a general methodology for predicting water utility KPIs.

Author Contributions: A.N. conceived the methodology, analyzed and interpreted results, contributed to the paper writing, reviewing and editing. J.B. implemented the methodology, performed data processing, models fitting and participated to the paper writing.

Funding: The work presented is part of the French project "SPHEREAU", grant number AAP FUI n° 22. It was funded by "bpifrance", the basin water agency "Agence de l'Eau Rhin Meuse", French regional authorities "Région Centre Val de Loire" and "la region Grand Est".

Acknowledgments: We are grateful to the manager of the utility who helped us in our research and provided us with data, information and advice.

Conflicts of Interest: The authors declare no conflict of interest.

References

1. Alegre, H.; Hirnir, W.; Baptisia, J.M.; Parena, R. *Performance Indicators for Water Supply Services. Manual of Best Practice Series*; IWA Publishing: London, UK, 2000; p. 146.
2. Mutchek, M.; Williams, E. Moving Towards Sustainable and Resilient Smart Water Grids. *Challenges* **2014**, *5*, 123–137. [CrossRef]
3. Walski, T.M.; Pelliccia, A. Economic analysis of water main breaks. *J. Am. Water Work. Assoc.* **1982**, *74*, 3140–3147. [CrossRef]
4. Skipworth, P.J. *Whole Life Costing for Water Distribution Network Management*; Thomas Telford: London, UK, 2002.
5. Park, S.W.; Loganathan, G.V. Methodology for economically optimal replacement of pipes in water distribution systems: 1. Theory. *KSCE J. Civ. Eng.* **2002**, *6*, 539–543. [CrossRef]
6. Shamir, U.; Howard, C.D.D. Analytical approach to scheduling pipe replacement. *J. AWWA* **1979**, *71*, 248–258. [CrossRef]
7. Kleiner, Y.; Adams, B.J.; Rogers, J.S. Selection and scheduling of rehabilitation alternatives for water distribution systems. *Water Resour. Res.* **1998**, *34*, 2053–2061. [CrossRef]
8. Halhal, D.; Walters, G.A.; Ouazar, D.; Savic, D.A. Water Network Rehabilitation with Structured Messy Genetic Algorithm. *J. Water Resour. Plan. Manag.* **1997**, *123*, 137–146. [CrossRef]
9. Kim, K.; Seo, J.; Hyung, J.; Kim, T.; Kim, J.; Koo, J. Economic-based approach for predicting optimal water pipe renewal period based on risk and failure rate. *Environ. Eng. Res.* **2018**, *24*, 63–73. [CrossRef]
10. Xu, Q.; Chen, Q.; Ma, J.; Blanckaert, K. Optimal pipe replacement strategy based on break rate prediction through genetic programming for water distribution network. *J. Hydro Environ. Res.* **2013**, *7*, 134–140. [CrossRef]
11. O'Reilly, G.; Bezuidenhout, C.; Bezuidenhout, J.J. Artificial neural networks: Applications in the drinking water sector. *Water Sci. Technol. Water Supply* **2018**, *18*, 1869–1887. [CrossRef]
12. Cuesta Cordoba, G.A.; Tuhovcak, L.; Taus, M. Using Artificial Neural Network Models to Assess Water Quality in Water Distribution Networks. *Procedia Eng.* **2014**, *70*, 399–408. [CrossRef]
13. Mounce, S.R.; Machell, J. Burst detection using hydraulic data from water distribution systems with artificial neural networks. *Urban Water J.* **2006**, *3*, 21–31. [CrossRef]
14. Dongwoo, J.H.P.; Gyewoon, C. Estimation of Leakage Ratio Using Principal Component Analysis and Artificial Neural Network in Water Distribution Systems. *Sustainability* **2018**, *10*, 750. [CrossRef]
15. Caputo, A.C.; Pelagagge, P.M. An inverse approach for piping networks monitoring. *J. Loss Prev. Process. Ind.* **2002**, *15*, 497–505. [CrossRef]
16. Nafi, A.; Brans, J. Prediction of Water Utility Performance: The Case of the Water Efficiency Rate. *Water* **2018**, *10*, 1443. [CrossRef]

17. Deb, K.; Pratap, A.; Agarwal, S.; Meyarivan, T. A fast and elitist multiobjective genetic algorithm: NSGA-II. *IEEE Trans. Evolut. Comput.* **2002**, *6*, 182–197. [CrossRef]

18. Legifrance. Available online: https://www.legifrance.gouv.fr/affichTexte.do?cidTexte=ORFTEXT000000274838URL (accessed on 12 June 2019).

19. Kamínski, K.; Kami´nski, W.; Mizerski, T. Application of Artificial Neural Networks to the Technical Condition Assessment of Water Supply Systems. *Ecol. Chem. Eng. S* **2017**, *24*, 31–40. [CrossRef]

20. Duchi, J.; Hazan, E.; Singer, Y. Adaptive Sub-gradient Methods for Online Learning and Stochastic Optimization. *J. Mach. Learn. Res.* **2011**, *12*, 2121–2159.

21. Ding, L.; Zeng, S.; Kang, L. A fast algorithm on finding the non-dominated set in multi-objective optimization. In Proceedings of the Congress on Evolutionary Computation (CEC'03), Canberra, ACT, Australia, 8–12 December 2003.

22. Yang, L.; Guan, Y.; Sheng, W. A novel dynamic crowding distance based diversity maintenance strategy for MOEAs. In Proceedings of the International Conference on Machine Learning and Cybernetics (ICMLC), Ningbo, China, 9–12 July 2017; IEEE: Piscataway, NJ, USA, 2017; Volume 1, pp. 211–216.

23. Luo, B.; Zheng, J.; Xie, J.; Wu, J. Dynamic crowding distance—A new diversity maintenance strategy for MOEAs. In Proceedings of the Fourth International Conference on Natural Computation, Jinan, China, 18–20 October 2008; IEEE: Piscataway, NJ, USA, 2008; Volume 1, pp. 580–585.

24. Deb, K. *Multi-Objective Optimization Using Evolutionary Algorithms*; John Wiley & Sons: Hoboken, NJ, USA, 2001.

25. Lim, S.M.; Sultan, A.B.; Sulaiman, N.; Mustapha, A.; Leong, K.Y. Crossover and Mutation Operators of Genetic Algorithms. *Int. J. Mach. Learn. Comput.* **2017**, *7*, 9–12. [CrossRef]

26. Mendes, J.A. Comparative study of crossover operators for genetic algorithms to solve the job shop scheduling problem. *WSEAS Trans. Comput.* **2013**, *12*, 164–173.

27. Liagkouras, K.; Metaxiotis, K. An Elitist Polynomial Mutation Operator for Improved Performance of MOEAs in Computer Networks. In Proceedings of the 22nd International Conference on Computer Communication and Networks (ICCCN), Nassau, Bahamas, 30 July–2 August 2013; pp. 1–5.

28. Wolpert, D.H.; Macready, W.G. No Free Lunch Theorems for Optimization. *IEEE Trans. Evolut. Comput.* **1997**, *1*, 67–82. [CrossRef]

29. Sharapov, R.R. Genetic Algorithms: Basic Ideas, Variants and Analysis. In *Vision Systems: Segmentation and Pattern Recognition*; Goro, O., Ashish, D., Eds.; Springer: Berlin, Germany, 2009; pp. 407–422.

30. Deb, K.; Deb, D. Analysing mutation schemes for real-parameter genetic algorithms. *Int. J. Artif. Intell. Soft Comput.* **2014**, *4*, 1–28. [CrossRef]

31. Schott, R.J. Fault Tolerant Design Using Single and Multicriteria Genetic Algorithm Optimization. Master's Thesis, Massachusetts Institute of Technology, Cambridge, MA, USA, 1995.

Article

Nonlinear Dynamic Modeling of Urban Water Consumption Using Chaotic Approach (Case Study: City of Kelowna)

Peyman Yousefi [1], Gregory Courtice [1], Gholamreza Naser [2,*] and Hadi Mohammadi [1]

[1] School of Engineering, The University of British Columbia, Kelowna, BC V1V1V7, Canada; pe.yousefi@alumni.ubc.ca (P.Y.); greg.courtice@alumni.ubc.ca (G.C.); hadi.mohammadi@ubc.ca (H.M.)

[2] Associate Professor, School of Engineering, Shippensburg University of Pennsylvania, Shippensburg, PA 17257, USA

* Correspondence: gnaser@ship.edu; Tel.: +1-717-477-1359

Received: 11 November 2019; Accepted: 11 February 2020; Published: 9 March 2020

Abstract: This study investigated urban water consumption complexity using chaos theory to improve forecasting performance to help optimize system management, reduce costs and improve reliability. The objectives of this study were to (1) investigate urban water distribution consumption complexity and its role in forecasting technique performance, (2) evaluate forecasting models by periodicity and lead time, and (3) propose a suitable forecasting technique based on operator applications and performance through various time scales. An urban consumption dataset obtained from the City of Kelowna (British Columbia, Canada) was used as a test case to forecast future consumption values using varying lead times under different temporal scales to identify models which may improve forecasting performance. Chaos theory techniques were employed to inform model optimization. This study attempted to address the paucity of studies on chaos theory applications in water consumption forecasting. This was accomplished by applying non-linear approximation, dynamic investigation, and phase space reconstruction for input variables, to improve the accuracy in various periodicity and lead time. To reconstruct the phase space, lag time was calculated using average mutual information for daily resolution as 17 days to reconstruct the phase space. The optimum embedding dimension and correlation exponent for the phase space were 18 and 3.5, respectively. Comparing the results, the non-linear local approximation model provided the best performance. The forecasting horizon for the models was 122 days. Moreover, phase space reconstruction improved the accuracy of the models for the different lead times. The findings of this study may improve forecasting performance and provide evidence to support further investigation of the chaotic behaviour of water consumption values over different time scales.

Keywords: water consumption; chaos theory; local approximation; Kelowna; gene expression programming

---

1. Introduction

Global water scarcity concerns are increasing due to climate change, urban development, population growth, industrial development, economic expansion, and the cost of drinking water [1]. It is imperative that governments invest in integrated management plans that address consequences of water problems such as scarcity of available water resources, sufficient distribution and pipeline maintenance. Conflicts resulting from water scarcity are becoming more prevalent and severe than other natural resource scarcity due to rapid increases in population without access to reliable sources of fresh water in recent decades [1,2]. There were 153 water-related conflicts reported in the 19th century while 279 conflicts have been reported within the last decade [3]. Therefore, we highlight the necessity of a management plan that improves water consumption efficiency.

A robust operation of urban drinking water supply systems requires future water consumption values to inform the development of an efficient water consumption management plan to mitigate anticipated stressors on the system. The reliability of water distribution systems may be improved through the accurate simulation of hydraulic conditions in pipeline systems based on future water consumption forecasting. In other words, water consumption forecasting provides public suppliers with the necessary future consumption information to ensure consumption needs can be met [4,5]. Water consumption forecasting is a dynamic process as predictions are essential for the optimum operation and sustainable growth and development of urban water supply [6]. The reason for categorizing water consumption forecasting as a dynamic process is because of the influential variables in forecasting that are not stable and can change under varying conditions. Therefore, considering reliable forecasting horizon can be beneficial in any management plan. Moreover, forecasting consumption values in short-, mid-, and long-term (i.e., less than a week, a week to a month, a month to a year or more, respectively) time periods play a crucial role in water distribution systems' (WDS) daily operation basis by informing important factors such as optimized pumping, pipeline maintenance, minimizing energy and water supply cost, improving system reliability and the quality of allocated water [7–9]. Recent studies have improved the understanding of the nonlinearity and complexity of water consumption factors; however, more study on these concepts is required. The available studies related to these factors are limited and require (1) accurate estimation and forecasting of water consumption and (2) determination of the degree of nonlinearity among the influential variables in water consumption. Accurate estimation and forecasting data are influenced by the availability of high-resolution temporal scale datasets (e.g., daily and hourly). Additionally, data availability for other influential factors, such as holidays, humidity, peak consumption duration in a 24-h period, air temperature, population growth, and consumers' income, is important to accurately forecast consumption data.

Over the past three decades, two thoughts of deterministic and probabilistic methods have been proposed to forecast urban water consumption. The deterministic approach is solely based on input variables and their initial conditions, whereas a probabilistic model relies on modeling uncertainties and randomness of input variables. Researchers have confirmed the ability of nonlinear deterministic approaches in modeling the events with complexity in effective variables [10,11]. Individual consumption habits are an important factor in modeling and forecasting future consumption. However, there are a few studies that aimed to forecast water consumption at the individual household scale which is a technique to analyze consumption habits [12]. The authors are not aware of any recent studies that considered consumers' habits in total recorded consumption values when modeling urban water consumption in large scale of an urban district.

However, an individual's habits will influence consumption and thus, future values in the system will be highly influenced by individual consumer habits, resulting in a highly complex system. In a complex system, "the behavior of the integral part of the system is simple, but the behavior of the overall system is complex" [13]. Given the significant challenges and complexity of probabilistic methods, deterministic methods can provide a useful approximation to their probabilistic counterparts. Therefore, investigation of the complexity of effective variables that are used as input variables of water consumption models can provide information towards employing a reliable modeling technique. Therefore, this study investigates both the dynamic characteristics, and dynamic explanatory variables in water consumption within the City of Kelowna (British Columbia, Canada). Furthermore, this research focuses on a deterministic approach to forecast short-term and mid-term lead times under different temporal scales. To calibrate the forecasting models, urban dimension of the test data is considered (i.e., the whole urban scale, rather than neighborhoods or single buildings). This decision may have implications on usability of the model information for consumption and supply management.

1.1. Background

Information about future water consumption helps related authorities develop an integrated, efficient plan that reduces long-term supplier and consumer water stress. Accurate estimation of drinking water consumption and availability of resources is considered one of the solutions to reduce water stress [9,14–16]. Moreover, having an accurate estimation of future consumption values improves sustainability for planning purposes. Unlike forecasting for other water resources-related parameters of interest (e.g., river discharge, sedimentation, rainfall, etc.), where influential factors in forecasting the goal values are mostly correlated (e.g., rainfall factors in modeling and simulating of river discharge), influential variables in forecasting water consumption are more complex with a weaker correlation. Moreover, applying large datasets with many input variables can lead to overfitting issues, which can misrepresent model performance through overly confident forecasting model predictions in addition to adding model complexity without improving accuracy. Hence, consumption forecasting requires a balance of optimizing the quantity of input variables with improvements in forecasting accuracy. It can improve the accuracy of the models while reducing the number of input variables by defining the model's input with active variables. Input variable selection techniques (e.g., Garson Equation, Influential correlation) are employed to highlight the active variables in a system [17].

There are different types of variables affecting consumption values such as climatic (e.g., temperature, relative humidity, rainfall, etc.) and socioeconomic (e.g., population and income) [18]. Commonly climatic variables are used in short-term and mid-term forecasting while socioeconomic variables are used to forecast long-term consumption values [19–21]. Common climatic factors considered in recent studies are temperature, precipitation, and previously recorded consumption values [20,22,23]. The number of studies on short- and mid-term forecasting of water consumption is considerable; however, limited studies have specifically investigated impact of climatic factors on consumption forecasting in the same period [24–26].

The literature enlists implementation of different statistical and probabilistic approaches for forecasting consumption values. Generally, conventional techniques were reportedly prevalent for a better understanding of chosen variables in the modeling of water consumption [27–29], which consider linear relationships between functional variables and the value of water consumption, while observation indicates that these relationships exhibit nonlinear behavior. The literature is mostly categorized into physical-based and black box models. Physical-based models approximate the general internal sub-process and physical mechanism by fundamental laws of mass, energy, and momentum. Black box models implement artificial intelligence, fuzzy based and nonlinear deterministic methods (e.g., artificial neural networks, genetic programming, support vector machine, nonlinear local approximation, etc.) to ascertain the relationship between the input and output variables. The physical-based and black box models include conventional regression models [9], artificial neural networks (ANN) [14,20,30,31], feedforward neural networks (FNN) [22,32], general regression neural networks (GRNNs) [33,34], deep belief neural network (DBNN) [35], support vector machines (SVMs) [16,18,36–38], gene expression programming (GEP) [39,40], adaptive neural fuzzy inference system (ANFIS) [41], Fourier analysis [7], hybrid models (e.g., combined wavelet) [23,42,43], fuzzy regression [44], fuzzy cognitive map learning method [45], epidemiology-based forecasting framework [46], temporal disaggregation [47], harmonic analysis [48], and wavelet de-noising [49].

Influential factors in complex water consumption systems may benefit from chaos theory techniques to improve forecasting methods. As mentioned above, characteristics of influential factors in water consumption are characterized as complex systems. Chaos theory techniques may improve the application of forecasting models to estimate future values in complex systems and improve the prediction of nonlinearity within dynamic systems [13,50]. Descriptions of chaotic behavior have been used in various engineering applications. Chaotic systems are defined and characterized by large changes in behavior resulting from small changes in initial conditions of the system [51]. Using the 'extent of complexity' of a chaotic system (defined primarily in the context of the variability of relevant data), chaotic systems are classified as low-, medium- or high-dimensional [52]. Moreover, the availability of noise in time series increases the complexity of the data and results in a high embedding dimension [53]. In water distribution systems, factors such as temperature, humidity, precipitation, the economic condition of consumers, population size, holidays, etc., have an impact on water consumption.

By aggregating data at increasing temporal scales, the effects of scaling time series on deterministic chaos can be found [54,55] and any missing data can be generated [56,57]. This approach has been used in solving problems in various fields of study such as river discharge [58–60], sedimentation [61–63], climate [64], lake level variability [65,66], rainfall [67–70], traffic speed [71], finance [72], image processing [73] and ship motion prediction [74]. However, there is a paucity of study on the application of chaos theory on water consumption forecasting methods. Oshima [13] developed a real-time forecasting system for water consumption based on the chaos theory. To support operations with labor saving, facility maintenance and efficient use of energy, he introduced "information integration type chaos theory-based consumption forecasting". The report showed the application of chaos-based techniques in improving the accuracy of the results.

Evolutionary techniques became popular in modeling and optimizing fields. Genetic programming (GP) and gene expression programming (GEP) place on heuristic algorithms that are found in Darwin's evolution theory [66]. Evolutionary techniques adjust the population of a specific solution. In each stage, individuals are selected randomly from the current population. Then, the chosen individual plays the parent's role to be reproduced for another population for the next generation of solutions. This reproduction goes toward an optimal answer which has been defined as the goal values. The evolutionary techniques are used for optimization problems where standard methods are not suitable, such as discontinuity, nondifferentiable, stochastic, or highly nonlinear objective functions. Among evolutionary techniques, GP and GEP are used for modeling problems with the same abilities as GA such as informing data gaps and forecasting time series [18,75–77]. Aytek and Kishi [78] used the GP technique to model suspended sediment load in the Tongue River (United States) and reported better performance from GP compared to sediment rating curves and multiple linear regressions (MLR). Since GEP provides a tree-structure scheme, it makes GEP more convenient to interpret the results in comparison with GP. Moreover, GEP presents mathematical equations that clarify the relationship between input and output variables by a factor of 100-10000 [63,79,80]. The superiority of GEP and the advantages of this technique interested researchers to develop more sophisticated models with hybrid methods such as combining the extended Kalman filter [81], clustering the consumption values [82], Wavelet decomposition [39], and phase space reconstructed GEP (PSR-GEP) [42] in forecasting urban drinking water consumption. The results showed that GEP models are highly sensitive to wavelet decomposition when attempting to improve the performance of the models.

1.2. Problem Statement

Growth in peak water consumption, which overburdens urban water resources [30], requires an efficient management plan. This importance motivated the application of soft-computing techniques into urban water consumption forecasting methods to develop methods that are applicable in forecasting problems. Many techniques have been proposed to forecast water consumption under differing time scales; however, there has been limited investigation on performance comparisons between models to inform model selection under various conditions [5]. Moreover, non-stationary, non-linear, and inherent stochasticity of water consumption data makes forecasting problems more challenging in this field [83]. Donkor et al. [84] reported that periodicity and forecasting horizon influence the performance of the methods in short-term and long-term forecasting, such as artificial neural networks and econometric models, respectively. Therefore, understanding the forecasting horizon, which is dependent on the dataset considered, will help to categorize the performance of the developed models. It can be considered as a classification for short-, mid-, and long-term forecasting models, which help operators select the appropriate model to implement considering the available dataset [9,20,85]. Forecasting horizon can improve the reliability of consumption forecasting methods by informing which method is most useful for consumption forecasting under specified time frames; however, in general, there is currently no acknowledgement of time frame for these forecasting horizons [84]. This study focuses on the forecasting horizon for the considered models to inform performance under various periodicity (daily, 2-day, 4-day, weekly, bi-weekly, and monthly) and lead time.

Based on previous studies, common influential variables in water consumption forecasting include observed consumption values (e.g., historical recorded consumption values over various periodicity), climatic variables (e.g., temperature, humidity, levels of snow and rainfall) and socioeconomic variables (e.g., population growth rate, economic factors such as income and water cost) [5]. Although these variables are used in the literature, there are fewer studies that used socioeconomic variables in long-term forecasting than studies that employed the above-mentioned variables. Moreover, extrapolation models in water consumption forecasting are based on the recorded data and its error terms [84]. The limitation of these models is based on their dependence of past trends that will likely be observed in the future, but do not consider the role socioeconomic variables and consumer habits may play in influencing future consumption values. Recently published studies investigating these factors are limited and require (1) accurate estimation and forecasting of water consumption considering different periodicity and forecasting horizons to classify the models based on their performance, and (2) determination of the degree of nonlinearity among the influential variables to consider exogenous variables (e.g., socioeconomic). The role of these factors in the value of water consumption in the future is highlighted. Therefore, investigating the complexity of input data can be helpful in employing different techniques with various abilities to model forecasting water consumption problems.

1.3. Objective

The techniques presented in the literature are often considered for short-term forecasting, which is based on one-day-ahead lead time. In light of the previous discussion on model limitations, it has been reported that the weaknesses of the forecasting models are related to the lack of consideration for the complex behavior of water consumption datasets. Much of the existing research ignores the importance of dynamic behavior of consumption datasets although many pre-processing techniques have been introduced to improve model accuracy. This study investigates the dynamics of a dataset to detect chaotic behavior of water consumption time series. Then, a non-linear approximation technique is applied to forecast water demand consumption. The objective of this study was to (1) investigate system complexity and its role in forecasting technique performance, (2) evaluate selected forecasting models by periodicity and lead time, and (3) propose a suitable forecasting technique based on operator applications and performance through different time scales.

This study improves the understanding of nonlinear local approximation (NLA) performance in comparison with previously introduced methods in water consumption forecasting. Moreover, forecasting horizon is introduced as a factor to evaluate the performance of forecasting models. This study attempts to address gaps in previous studies by applying NLA, dynamic investigation, phase space reconstruction for input variables, and improving the accuracy in various periodicity and lead times.

2. Materials and Methods

Figure 1 provides a schematic description of the research methodology. Average mutual information (AMI) is used to calculate the lag time. Then, the existence of chaotic behavior in the test case is investigated by the correlation dimension method. Four different methods are employed to forecast the short- and mid-term consumption values. Non-Linear local approximation method is considered as the forecasting method in the condition of the test data that has chaotic behavior. The performance of selected models is evaluated in different lag and lead times.

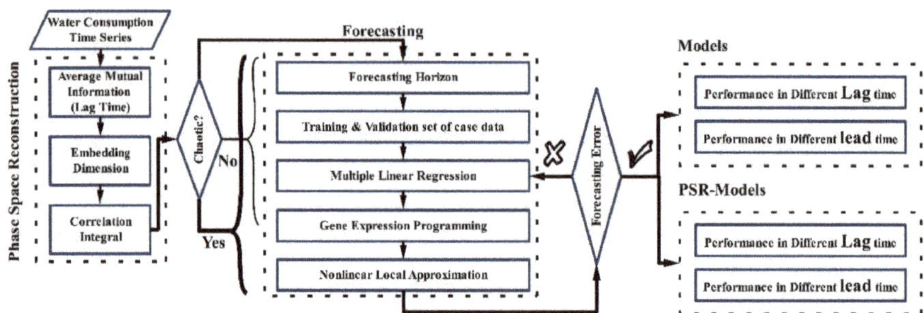

Figure 1. Research process scheme and methodology.

2.1. Phase Space Reconstruction

This research employed the phase-space concept to better understand the dynamic nature of a municipal water consumption dataset for the City of Kelowna. The dynamics of a water consumption system are represented by data points along a trajectory, whereby each position in time represents a system state. The lag-embedding technique can be used on deterministic, dynamic systems such as the present water consumption dataset to reconstruct phase-space from time series. The fundamental dynamics of a system can be studied by reconstructing an *m*-dimensional phase-space of *Dt* that is defined by [86,87]:

$$D_t = \left\{ d_t, d_{t-\tau}, d_{t-2\tau}, \dots, d_{t-(m-1)\tau} \right\}, \ t = 1, 2, 3, \dots, N \tag{1}$$

where D_t is a vector of the consumption data of $\{d_t\}_{t=1,\dots,N}$, N is the number of recorded consumption data points, τ is the lag time and m is the number of embedding dimension that generally varies from 1 to 10 or 1 to 20 [61,66,88,89]. In the case when m is greater than the minimum embedding dimension, the trajectory of reconstructed vectors can display the true state of the chaotic system. Indeed, the lag time (τ) is arbitrary as the data are often assumed to have infinite precision. The lag time should not be too small given the difference between various elements of the delay vectors and, it should not be too large as this can result in low coordinate correlation [90]. If the dynamics of the system can be reduced to a set of deterministic laws, trajectories will converge towards a subset of the phase-space with a fractional dimension called the attractor [89]. The lag-embedding method is sensitive to both embedding parameters of τ and m. Average mutual information (AMI) is a well-known method for estimating the lag time [91,92]. This research employed AMI to estimate the lag time [91]:

$$I(d_t, d_{t+\tau}) = \sum_{d_t} \sum_{d_{t+\tau}} P(d_t, d_{t+\tau}) \log_2 \frac{P(d_t, \ d_{t+\tau})}{P(d_t)P(d_{t+\tau})} \tag{2}$$

where the sum is extended over the total number of samples in the time series, $P(d_t)$ and $P(d_{t+\tau})$ are the marginal probabilities for measurements d_t and $d_{t+\tau}$ and $P(d_t, d_{t+\tau})$ is their joint probability. The optimal τ minimizes the value of the function $I(d_t, d_{t+\tau})$ for $t = \tau$. AMI considers the first local minimum as the lag time [91,93].

2.2. Correlation Dimension

The dimension of a system informs its complexity and indicates the number of required variables that specify a deterministic system. Kermani [59] classified different dimensions in a system including topological, Hausdorf, box counting, point-wise, and correlation dimensions. Additionally, the correlation dimension is used as an indicator of a deterministic or stochastic process [59,94]. The function below calculates the correlation integral value [95]:

$$C_m(r) = \frac{2}{N(N-1)} \sum_{j=1}^{N} \sum_{i=1}^{N} H\left(r - \|X_i - X_j\|\right) \tag{3}$$

where N is the number of data points, H is the Heaviside step function ($H(u) = 1$ for $u > 0$, $H(u) = 0$ for $u \le 0$ and $u = r - \|X_i - X_j\|$, X_i is the ith state vector, and r is the radius of a sphere with the content of X_i or X_j as the center. $C_m(r)$ is proportional to r for stochastic time series, whereas for chaotic time series, it scales with r as

$$C_m(r) \propto r^{c_e} \tag{4}$$

where c_e is the correlation exponent defined by approximating the slope of $\log C_m(r)$ versus $\log(r)$ in logarithmic scale. If the calculated c_e is unchanged by increasing the number of embedding dimensions, C_e can be considered as the correlation dimension of the attractor in the system. But, if c_e is not stable as a function of embedding dimensions, this system can be considered non-chaotic [54,96,97].

2.3. Nonlinear Local Approximation

Nonlinear local approximation (NLA) can forecast a system's future without development of an analytical model [51]. This research applied NLA to (1) test the chaotic nature of consumption data and (2) forecast the consumption in the case study. This was done by reconstructing the phase space of the dataset. It is critical in a chaos analysis to reconstruct the multi-dimensional phase space that provides a conceptual pattern for the time series. The presence of attractors in phase space indicates the possibility of chaos in the dataset. A phase-space reconstruction in a dimension m facilitates an interpretation of the underlying dynamics in the form of an m-dimensional map, f_T, by

$$X_{j+T} = f_T\left(X_j\right) \tag{5}$$

where X_j is the vector of dimension m at the current state of the system at time j and X_{j+T} is the vector of dimension m at the future state of the system at time $j + T$. NLA entails the subdivision of the f_T domain into many subsets. In other words, the dynamics of the system were described step-by-step locally in the phase-space [98]. In an m-dimensional space, estimating the change of trajectory with time would lead to forecasting. Considering the relation between two states $\vec{X}_t$ and $\vec{X}_{t+p}$, the behaviour at a future time (p) on the attractor was forecasted by the mapping $\vec{F}$ as [66]:

$$\vec{X}_{t+p} \cong \vec{F}\left(\vec{X}_t\right) \tag{6}$$

where the evolving dynamic of state $\vec{X}_t$ is influenced by nearby states. The future state $\vec{X}_{t+p}$ was determined by the first-order polynomial mapping $\vec{f}$ [99]:

$$\vec{X}_{t+p} \cong \vec{F}\left(\vec{X}_t\right) = \vec{a} + \vec{f}\left(\vec{X}_t, \vec{X}_{t-\tau}, \ldots, \vec{X}_{t-(m-1)\tau}\right) \tag{7}$$

While the mapping $\vec{f}$ is linear, the forecasted value is nonlinear [100]. This is because every state on the trajectory belongs to a different subset defined by different expressions for $\vec{F}$.

2.4. Largest Lyapunov Exponent

The rate of convergence or divergence in different dimensions is measured using the Lyapunov exponent. A deterministic system contains at least one positive Lyapunov exponent [101]. This research employed the approach proposed by Rosenstein et al. [101] to extract the value of Lyapunov exponent for the test case. After reconstruction of phase-space, a point Y_{n0} was selected and all the points in the neighborhood of Y_n, with the closer distance of r to that point were found. r is the radius of a sphere with the content of Y_n as the center. The procedure was repeated for N points in the route to find a stretch factor S:

$$S = \frac{1}{N} \sum_{n_0=1}^{N} \ln\left[\frac{1}{|u_{Y_{n0}}|} \sum |Y_{n0} - Y_n|\right] \tag{8}$$

where $|u_{Y_{n0}}|$ is the number of neighbors to point Y_{n0}. The plot of S verses N consists of linear and nonlinear components. The literature provides two methods for calculating the largest Lyapunov exponent (λ_{max}). Shang et al. [102] determined λ_{max} as the slope of the linear part of the curve, while Rosenstein et al. [101] determined λ_{max} as the average of the slope of the first part and that of the second part [89,102]. This research applied the second approach to study different temporal scales. The prediction horizon (Δt) is a time in which the consumption dataset sustains its dynamics in the most accurate forecasting; Δt was obtained as the inverse of the largest Lyapunov exponent:

$$\Delta t = \frac{1}{\lambda_{max}} \tag{9}$$

2.5. Gene Expression Programming

Evolutionary soft computing techniques may be applied to solve engineering problems. Among these evolutionary techniques, genetic algorithm (GA), genetic programming (GP), and gene expression programming (GEP) are considered, which were originally inspired from Darwin's theory of evolution [79,80,103,104]. GA and GP work with a string of numbers and a specific length that is known as a "chromosome." GEP defines an equation that shows the relationship between input and output values. Moreover, the process of the algorithm's learning starts with the generation of chromosomes for a given random raw dataset that works with chromosomes and expression trees. Expression trees demonstrate the relationship among variables connecting with arithmetic operators. Chromosomes contain genes that provide a series of symbols of two parts, head and tail, which have functions, terminals, and only terminals, respectively. GEP follows the genome restructuring process by mutation, recombination, transposition, and gene duplication randomly. This random reputation will deliver the best model to be selected. This cycle will be continued by the reproduction of randomly generated chromosomes to reach the model with satisfying results based on defined evaluation criteria. In this research, 30 chromosomes with the head size of 8 and 3 genes are reproduced with the linking functions of $\{+, -, \times, x, x^2, \sqrt{x}, \log x\}$. The fitness functions used as the selection criteria are root mean square error (RMSE), correlation coefficient (CC) and mean absolute error (MAE). All models were selected with a fair assessment by considering stopping condition of 5000 generations. Figure 2 shows the scheme of the process with GEP method.

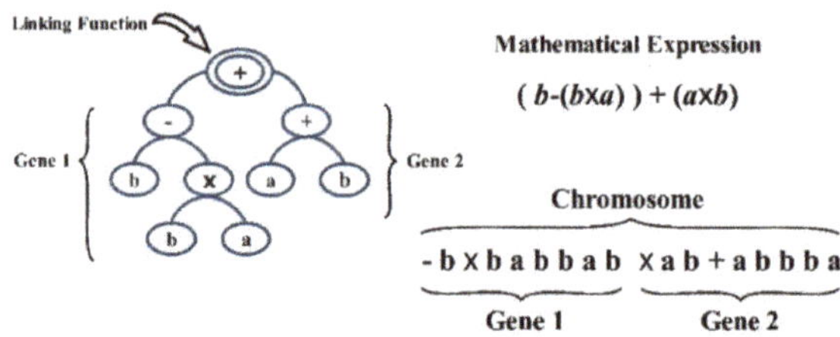

Figure 2. Expression tree and mathematical function of two gene chromosome.

2.6. Multiple Linear Regression

The common application of the linear regression (LR) modeling method is to model the goal variable (*Y*) with a single input variable (*X*). Multiple linear regression (MLR) describes the relationship between the goal value (*Y*) and the number of input variables (X_1, X_2, ... , X_n). In other words, MLR finds a relationship with a linear combination of independent explanatory variables (e.g., reconstructed phase space with a different number of embedding dimensions for recorded values of water consumption) and response variable (time ahead values of water consumption) by

$$Y_i = \alpha_0 + \alpha_1 X_{i1} + \alpha_2 X_{i2} + \cdots + \alpha_p X_{ip} + \epsilon \tag{10}$$

For *i* = *n* recorded values. Y_i is the output consumption value, X_i is the explanatory input variables, dependent upon the number of embedding dimensions, where α_0 is constant, α_p are slope coefficients for each input variables and ϵ is the unexplained residual value.

2.7. Models Selection Criteria

Correlation Coefficient (CC), root mean square error (RMSE) and mean absolute error (MAE) are fitness functions considered in this study to evaluate model performance and accuracy.

$$CC = \frac{\sum_{i=1}^{N_t}\left(R_i - \overline{R}\right)\left(F_i - \overline{F}\right)}{\sqrt{\sum_{i}^{N_t}\left(R_i - \overline{R}\right)^2}\ \sqrt{\sum_{i}^{N_t}\left(F_i - \overline{F}\right)^2}} \tag{11}$$

$$RMSE = \sqrt{\frac{\sum_{i=1}^{N_t}\left(R_i - F_i\right)^2}{N_t}} \tag{12}$$

$$MAE = \frac{1}{n}\sum_{i=1}^{n}|R_i - F_i| \tag{13}$$

Variables in the above equations include the number of values, N_t, and the recorded and forecasted values, *R* and *F*, respectively. $\overline{R}$ and $\overline{F}$ are the mean values of the recorded and forecasted water consumption, respectively. The range of CC is −1 and 1, where larger positive value of CC and smaller value for RMSE and MAE indicates better model performance in forecasting accuracy.

2.8. Test Case

Water consumption data from the City of Kelowna (British Columbia, Canada) were considered as a test case for this study. The water supply for the region comes from various resources, which include Lake Okanagan, Mission Creek, Mill Creek, Scotty Creek, Hydraulic Creek and numerous wells [105]. There is one primary distribution system that services 99% of the population via Poplar Point, Eldorado (seasonal intake uses only utilized during peak consumption) and Cedar Creek pump stations [105] plus Swick road pump station which services approximately 300 residents. This study considered Poplar Point, Eldorado and Cedar Creek stations. The SCADA system obtains information remotely from sensors installed at the intake locations. The software platform facilitates tracking of historical system performance for auditing and future decision making to optimize the system [105]. This study used six temporal scales of the consumption, including daily, 2-, 4-, 7-, 14- and 30 days from 1st of January 2010 to 30th December 2016. The test data are the whole scale of consumption values as the spatial scale for City of Kelowna. Table 1 shows the characteristics of the dataset in the test case.

Table 1. Characteristics of the temporal water consumption values in test case in whole urban scale.

Property	Daily	2-Day	4-Day	7-Day	14-Day	Monthly
Number of Data	2186	1092	552	312	156	72
Max. value (m³)	114,597.2	210,740.3	410,428.3	656,173.6	1,255,211	2,475,026
Min. value (m³)	14124	31,477.3	69,655.5	124,112.9	252,704.1	557,066.8
Average (m³)	43,046.4	86,102	170,332.4	301,357.3	602,714.7	1,291,944
Standard deviation (m³)	20,074.5	39,897	79,304.3	136,626.8	268,733.2	552,701.5
Coefficient of variation	0.46	0.46	0.46	0.45	0.44	0.42
Skew	0.73	0.71	0.72	0.66	0.63	0.54
Kurtosis	−0.38	−0.45	−0.51	−0.63	−0.79	−0.91

3. Results

Lag Time: Figure 3a,b presents the time variation of water consumption dataset for a six-year period (from January 2011 to December 2016) and the average 24-h consumption based on City of Kelowna's Utility report. The data were parsed into hourly maximum, minimum, and mean values for a 24-h period (Figure 3b). Figure 3b also presents a boxplot of the total mean, minimum, and maximum consumption value distributions for the six years period. Figure 3b indicates that the average consumption and minimum values have low frequency, demonstrating the highly deterministic behavior of the data. To calibrate the models, the data splinted into two folds; (1) 80% for training period (1st January 2016 to 31st December 2015); (2) 20% for test period (1st January 2016 to 31st December 2016). Water consumption estimation and forecasting of the average and minimum values are not as complex as compared to the peak consumption values. However, the maximum values exhibit a non-linear behavior which does not follow a specific pattern (i.e., appears to be random) unlike average and minimum values. Further, the maximum consumption values in a water distribution system are very important. This is because of estimation of peak consumption to supply customers' consumption and optimize WDS pipeline to make WDS more reliable, such as managing pipeline failures, improving peak pressure, reducing leakage, etc.

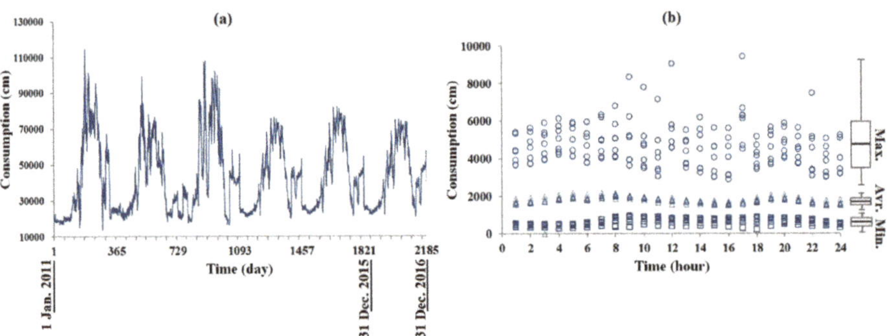

Figure 3. Time series plot of (a) daily water consumption; (b) 6-years consumption pattern within a 24-h period.

Average mutual information (AMI) was used to identify the proper lag time in this study. Figure 4a presents the first local minimum lag time of $\tau = 17$ days for daily time series as a function of lag time. The first minimum values of AMI were at lag times of 17, 12, 10, 6, 3 and 2 for daily, 2-, 4-, 7-, 14- and 30-days (monthly) data series of the water consumption, respectively. Table 2 summarizes the results for all other timescales. By using the lag time of $\tau = 17$ days, the phase-space was reconstructed, as presented in Figure 4b for daily scale. The vertical axis shows the time series and the other two axes show the same time series delayed by 17-(τ) and 34-days (2τ). This figure shows reconstructions in three dimensions; the projection attractor on the plane $\{x_i,\ x_{i+\tau}, x_{i+2\tau}\}$ with the lag time of 17 days. When comparing the two PSR graphs in Figure 4b, the presence of attractors becomes clear for $\tau = 17$ days (black line) comparing to $\tau = 1$ (blue dots). Indeed, the phase space is more spread for 1-day lag time and more concentrated on attractors for 17-days lag time. Based on the figure, PSR can make a consistent set of inputs for the model that makes better results in comparison with other sets having less consistency (e.g., peak consumption represented in Figure 3b).

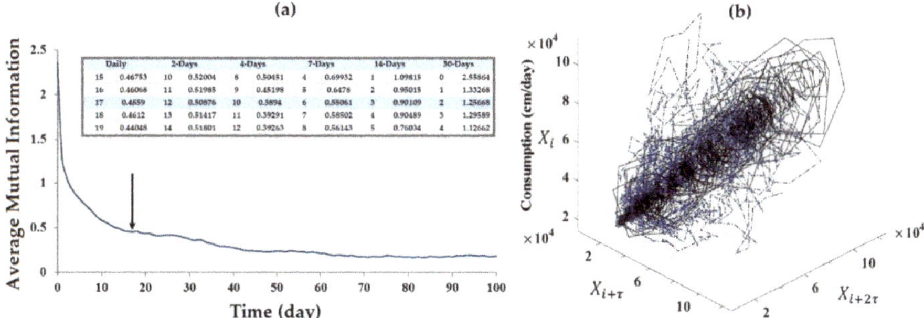

Figure 4. (**a**) Average mutual information for the daily value ($\tau = 17$); (**b**) reconstructed phase space by (τ and 2τ-day lag time).

Table 2. Average mutual information and correlation exponent values in different temporal resolutions.

Time Scale	AMI	C_e	$2LogN > C_e$
Daily	17	3.50	6.67
2-Day	12	3.37	6.07
4-Day	10	3.74	5.48
7-Day	6	3.94	4.98
14-Day	3	3.83	4.38
30-Day	2	3.49	3.71

Correlation Dimension: Figure 5a plots the results for correlation integral versus log (r) for different dimensions (m) in the range of 1 to 20 for daily consumption time series. The correlation exponents, C_e (m), were determined and the results were plotted in Figure 5b to evaluate the presence of chaos in the dataset. As demonstrated in Figure 5b, the correlation exponent increases with the embedding dimension up to a certain value and then remains steady. The figure reveals that the slope of larger embedding would become constant for all temporal scales.

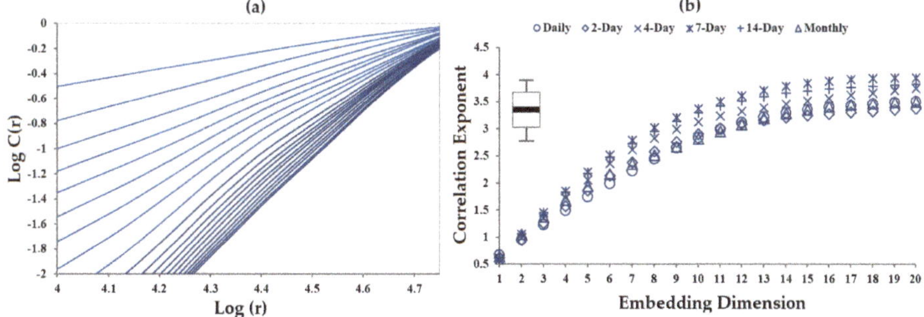

Figure 5. (**a**) The relation between correlation integral $C(r)$ and r by different embedding dimensions for daily temporal scale, (**b**) saturation of correlation dimension C_e (m) with embedding dimension m for different temporal scales.

The embedding dimension of 17 appears to explain the dynamics of the system (1- and 2-day consumption). The correlation exponents of 3.5, 3.37, 3.74, 3.94, 3.83 and 3.49 were determined for daily, 2-, 4-, 7-, 14- and 30-day, respectively. Regarding the concept of PSR, rounding up the correlation exponent to the nearest integer informs the number of dominant variables. Thus, it can be interpreted as the number of variables that dominantly govern the temporal dynamics of water consumption, which is approximately 4 for all temporal scales. Table 2 presents the results of lag time and correlation exponent for all temporal scales. Moreover, the condition of $C_e < 2 \log N$ (with N as the number of data point) was satisfied for the chaotic temporal scales [106].

Nonlinear Local Approximation and Largest Lyapunov Exponent: The first objective in this research was to investigate the impact of PSR in the accuracy of forecasted values. Forecasted values were evaluated for the embedding dimensions (ranging from 2 to 10) at lag times 1 and 17 days to forecast 1-day-ahead and highlight the effect of embedding dimension and performance of models. For $m = 2$, for a lag time of 1 day, two variables D_t and D_{t-1}, and for $m = 2$, for a lag time of 17 days, two variables D_t and D_{t-17}, were used as input variables to forecast 1-day ahead (D_{t+1}). For $m = 3$, for the lag time of 1 and 17 days, three variables D_t, D_{t-1}, D_{t-2} and D_t, D_{t-17}, D_{t-34} were used as input variables to predict 1-day-ahead. Moreover, phase space was reconstructed for $m > 3$. Table 3 presents a summary of the 1-day ahead forecasted values that reconstructed phase-space in dimensions ranging from 2 to 20 for $\tau = 1$ and 17 days. The overall average of fitness values for all embedding dimensions CC > 0.96, RMSE < 4200 (m3/day) (8% of average daily consumption) and MAE < 49. These results can be considered as reasonable performance of the model. Table 3 reveals the optimum embedding dimension for the most accurate forecasted value in bold. Using CC, RMSE, and MAE, the optimum embedding dimension (m_{opt}) was found to be 18 and 19 for the lag time of 1 and 17 days, respectively. The results of nonlinear local approximation (NLA) identify the closest results of correlation exponent in optimum embedding dimension ($m = 17$ for correlation dimension of daily value). The similar values for the embedding dimension of daily values imply that $m = 17$ or 18 is the optimum dimension of the system to be used as the models' input. The second objective was to evaluate the performance of selected m and τ. The optimum models for $m = 18$, $\tau = 1$ and $m = 19$, $\tau = 17$ have been applied to forecast 2-, 4-, 7-, 14-, 30- and 60-day lead time water consumption. Figure 6 compares the NLA with observed data using bolded values in Table 3.

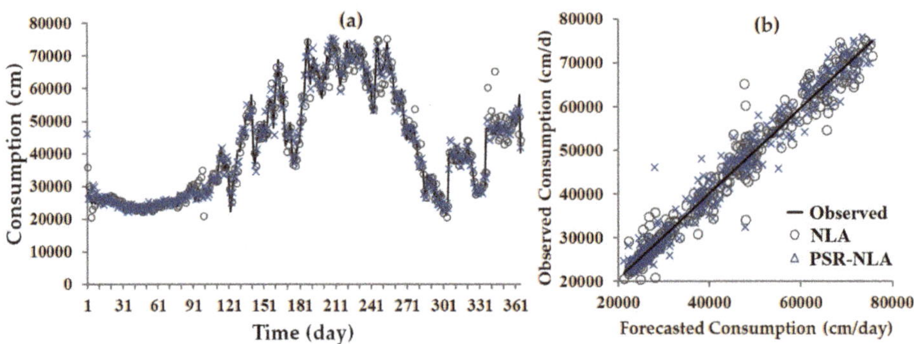

Figure 6. Forecasted values for water consumption by NLA (nonlinear local approximation) and PSR-NLA (phase space reconstructed-NLA) methods in comparison with observed values (a) consumption values time series (b) scatter pilot.

Figure 7a plots the stretching factor (S) versus the number of points ($N = 100$). The figure shows an overall increase in S by increasing N. Furthermore, the figure reveals two components: the first part reveals a sudden increase in S, while the second part (after $N = 25$) shows a more gradual increase in S. Figure 7a also indicates the best line (dashed line) fitted to the second part. Following Rosenstein et al. [101], the present study revealed λ_{max} of 0.014, 0.0102 and 0.0082 for $m = 17$, 18 and 19, respectively.

Three different values of λ_{max} are considered because of the embedding dimensions range that were determined by CD, FNN, NLA and PSR-NLA, which were between 17 and 19. Note that $m = 18$ and 19 were determined by NLA and PSR-NLA, respectively, which gave a higher accuracy than the other embedding dimensions (see Table 3). The forecasting horizons ($\Delta t = 1/\lambda_{max}$) were 72, 98 and 122 days for $m = 17$, 18 and 19, respectively. Moreover, $m = 17$ with forecasting horizon value of 72 was determined by CD. Figure 7b shows the frequency of the data is smooth from day 1 to 72. Even with the high data frequency between days 98 and 122, PSR-NLA increased the forecasting horizon from 98 days (NLA) to 122 days. Considering the results of CD, NLA, PSR-NLA, and forecasting horizon, it is reasonable to select $m = 17$ as the system's optimum dimension.

Table 3. Fitness values for NLA and PSR-NLA methods in different embedding dimension lag time and lead time.

	NLA, $\tau = 1$, T = 1				PSR-NLA, $\tau = 17$, T = 1				$\tau = 1$, $m = 18$		
m	CC	RMSE *	MAE	m	CC	RMSE *	MAE	T	CC	RMSE *	MAE
1								1	0.9842	2855.6	43.07
2	0.9759	3532.1	47.85	2	0.9751	3581.7	49.09	2	0.9783	3351.6	48.02
3	0.9771	3424.8	47.34	3	0.9773	3425.3	48.01	4	0.9331	5883.5	63.03
4	0.9762	3495.5	47.74	4	0.9789	3302.9	47.30	7	0.8932	7445.8	72.49
5	0.9785	3332.0	47.08	5	0.9785	3331.7	47.10	14	0.7877	10555.0	87.76
6	0.9795	3248.6	46.13	6	0.9795	3257.8	46.79	30	0.6735	13307.6	100.44
7	0.9802	3187.3	45.49	7	0.9829	2967.1	45.80	60	0.2523	20189.1	129.71
8	0.9805	3176.4	45.65	8	0.9838	2887.9	45.22		$\tau = 17$, $m = 19$		
9	0.9806	3164.1	45.65	9	0.9849	2792.8	43.95	T	CC	RMSE *	MAE
10	0.9803	3193.6	45.09	10	0.9846	2828.2	44.52	1	0.9852	2772.8	43.83
11	0.9813	3098.7	44.56	11	0.9850	2792.2	43.95	2	0.9898	2295.7	39.59
12	0.9763	3495.1	48.11	12	0.9804	3189.4	46.69	4	0.9415	5504.2	61.47
13	0.9752	3578.0	48.11	13	0.9768	3457.4	48.32	7	0.9002	7211.2	71.46
14	0.9779	3378.9	47.14	14	0.9788	3303.6	47.65	14	0.8048	10147.5	86.61
15	0.9806	3169.7	46.06	15	0.9790	3290.2	47.23	30	0.6776	13265.1	95.31
16	0.9765	3491.0	47.24	16	0.9796	3250.2	46.70	60	0.4363	17784.9	118.53
17	0.9810	3139.6	45.21	17	0.9825	2995.5	45.95				
18	0.9842	2855.6	43.07	18	0.9838	2894.0	45.21				
19	0.9685	4088.5	44.69	19	0.9852	2772.8	43.83				
20	0.9661	4209.2	45.29	20	0.9846	2833.1	44.43			* m^3	
Tot	0.9775	3394.2	46.30	Tot	0.9807	3142.9	46.34				
Best	0.9842	2855.6	43.07	Best	0.9852	2772.8	43.83				
EM	18	18	18	EM	19	19	19				

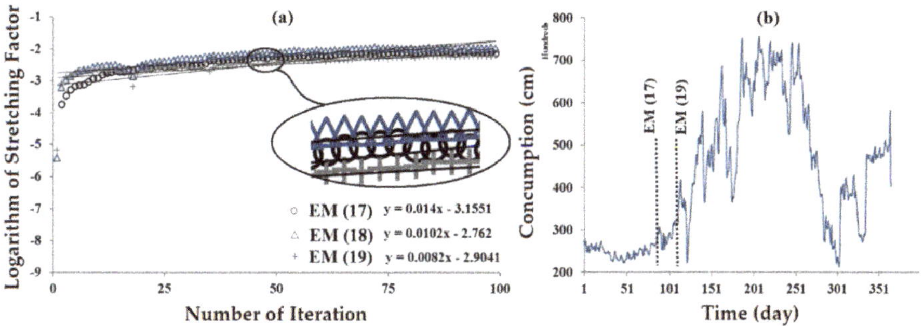

Figure 7. Estimation of largest Lyapunov exponent for daily consumption (**a**) for embedding dimension of 17, 18 and 19; (**b**) Observed values of water consumption in the test period.

Phase Space Reconstructed GEP (PSR-GEP) and Multiple Linear Regression (MLR) Models: One of the most important steps in developing an accurate model is the selection of the input variables. Input variable selection challenges the effect of the number of inputs in models' performance. In other words, there are diminishing returns on performance based on the number of input variables selected [42]. The combinations were selected in a way that included daily consumption data with lag times of $\tau = 1$ and 17 days. Different combinations of the time series of daily consumption were used to structure a policy for input dataset criteria. Combinations of $D_t, D_{t-1}, D_{t-2}, \ldots, D_{t-20}$ variables were used as input data with D_{t+1} (1-day time ahead) as output of the GEP and MLR model and combinations of $D_t, D_{t-\tau}, D_{t-2\tau}, \ldots, D_{t-20\tau}$ variables were used as input data with D_{t+1} as output of the PSR-GEP and PSR-MLR model (forecasting of 1 day time ahead). The combinations of arithmetic functions of $\{+, -, \times, x, x^2, \sqrt{x}, \log x\}$ and $\{+, -, \times,\}$ were used for PSR-GEP and PSR-MLR models, respectively. Ultimately, the best combination was selected using the criteria of CC, RMSE, and MAE. Table 4 reveals the 20 combinations of inputs and their performance with GEP and PSR-GEP models. In the table, the three criteria indicate the fourth combination of inputs ($m = 4$) with the lag time of 1-day, and ($m = 8$) with the lag time of 17 days as the best combinations of input data (reconstructed phase space) for GEP and PSR-GEP. The study reveals the following relationship for both GEP and PSR-GEP, respectively:

$$D_t = D_{t-1} + \log[\log(\log(9.58 D_{t-4})) \times (D_{t-2} D_{t-3})] + 8.253 \tag{14}$$

$$D_t = D_{t-\tau} - 0.04 D_{t-4\tau} + 8.03 \sqrt{D_{t-4\tau}} + 0.623 \tag{15}$$

Table 4. Fitness values for GEP (gene expression programming) and PSR-GEP (phase space reconstructed GEP) methods in different embedding dimension lag time and lead time.

GEP, $\tau = 1$, T = 1				PSR-GEP, $\tau = 17$, T = 1				$\tau = 1$, $m = 4$		
m	CC	RMSE *	MAE	m	CC	RMSE *	MAE	T	CC	RMSE *	MAE
1								1	0.9764	3486.6	47.83
2	0.9757	3543.7	48.14	2	0.9789	3636.9	48.57	2	0.9494	5112.2	57.92
3	0.9761	3517.8	47.91	3	0.9788	3644.6	48.59	4	0.9130	6716.4	67.48
4	0.9764	3486.6	47.83	4	0.9789	3647.1	48.56	7	0.8652	8376.4	76.57
5	0.9760	3519.0	47.95	5	0.9789	3635.5	48.68	14	0.7810	10,734.8	88.95
6	0.9760	3520.5	48.37	6	0.9788	3649.5	48.71	30	0.6548	13,649.0	97.03
7	0.9760	3500.9	47.91	7	0.9788	3649.0	48.70	60	0.2345	20411.3	130.23
8	0.9760	3521.4	47.97	8	0.9789	3631.6	48.62	$\tau = 17$, $m = 8$			
9	0.9760	3511.9	47.89	9	0.9788	3653.9	48.63	T	CC	RMSE *	MAE
10	0.9760	3514.7	47.89	10	0.9788	3650.6	48.73	1	0.9789	3631.6	48.62
11	0.9760	3514.6	47.88	11	0.9788	3656.6	48.65	2	0.9553	5267.3	58.52
12	0.9760	3514.7	47.89	12	0.9787	3657.4	48.69	4	0.9227	6894.5	68.47
13	0.9760	3510.1	47.91	13	0.9789	3645.4	48.55	7	0.8713	8848.2	78.11
14	0.9760	3516.6	47.91	14	0.9787	3655.7	48.69	14	0.7782	11,571.8	91.84
15	0.9760	3510.4	47.86	15	0.9788	3650.5	48.61	30	0.6334	14,631.5	105.98
16	0.9760	3498.7	47.88	16	0.9789	3644.4	48.54	60	0.3864	18,670.2	126.20
17	0.9759	3515.1	47.89	17	0.9789	3638.9	48.56				
18	0.9760	3509.7	47.85	18	0.9789	3646.6	48.57				
19	0.9759	3514.8	47.93	19	0.9787	3650.6	48.74				
20	0.9759	3514.2	47.90	20	0.9789	3642.5	48.51			* m^3	
Tot	0.9759	3519.0	47.96	Tot	0.9788	3646.5	48.62				
Best	0.9764	3486.6	47.83	Best	0.9789	3631.6	48.51				
EM	4	4	4	EM	8	8	20				

The results were not found to be more sensitive to lag time calculated by AMI (17 days) than to the 1-day lag time. However, PSR made the performance of the model better than $m = 4$ with 1-day lag time. Equation (14) used all four variables made by $m = 4$ ($D_{t-1}, D_{t-2}, D_{t-3}, D_{t-4}$), while equation (15) only used two of the variables out of 8 variables ($m = 8$; $D_{t-\tau}$, $D_{t-4\tau}$). Table 4 compares further details for GEP and PSR-GEP models for different lead times forecasted values. However, the results of both GEP and PSR-GEP are satisfactory, although they do not forecast as accurately as NLA and PSR-NLA. Table 5 presents the results for the MLR model that was used as a test to compare the results of the techniques studied. The application of PSR in improving accuracy is shown in the table. Surprisingly, the best embedding dimension for MLR is 17, which was identified by CD and NLA. Additionally, as expected, PSR-MLR results are more accurate than MLR when forecasting for different lead times.

Table 5. Fitness values for MLR and PSR-MLR methods in different embedding dimension lag time and lead time.

MLR, $\tau = 1$, T $= 1$				PSR-MLR, $\tau = 17$, T $= 1$				$\tau = 1$, $m = 17$			
m	CC	RMSE *	MAE	m	CC	RMSE *	MAE	T	CC	RMSE *	MAE
1								1	0.9789	3638.9	48.56
2	0.7825	11658.6	92.37	2	0.9758	3763.4	50.48	2	0.9494	5139.4	58.65
3	0.9790	3762.3	49.89	3	0.9809	3336.0	48.17	4	0.9130	6701.6	68.26
4	0.9790	3792.3	50.19	4	0.9811	3443.9	49.41	7	0.8595	8511.9	77.68
5	0.9790	3814.1	50.42	5	0.9766	3905.5	52.04	14	0.7595	11204.4	91.66
6	0.9790	3958.0	52.11	6	0.9148	22846.5	145.73	30	0.6447	13707.5	101.02
7	0.9790	4133.6	54.24	7	0.9765	3568.2	48.60	60	0.2282	20211.2	129.01
8	0.9790	4187.1	54.85	8	0.9764	3811.2	51.08	$\tau = 17$, $m = 3$			
9	0.9791	4432.7	57.65	9	0.9766	3568.2	48.71	T	CC	RMSE *	MAE
10	0.9792	5024.0	63.46	10	0.9766	3680.4	49.81	1	0.9809	3336.0	48.17
11	0.9792	5576.0	68.14	11	0.9767	3610.6	49.13	2	0.9555	5336.5	59.61
12	0.9792	5921.4	70.92	12	0.9767	3603.9	49.04	4	0.9232	6889.3	69.17
13	0.9793	6276.8	73.59	13	0.9766	3601.4	49.13	7	0.8723	8841.9	78.12
14	0.9793	7267.0	80.61	14	0.9767	3584.5	48.92	14	0.7790	11549.8	91.70
15	0.9794	9128.5	92.30	15	0.9769	3560.6	48.79	30	0.6344	14504.2	104.84
16	0.9794	10115.3	97.81	16	0.9769	3550.4	48.73	60	0.3859	18351.2	125.12
17	0.9789	3638.9	48.56	17	0.9769	3550.5	48.73				
18	0.9794	10114.2	97.80	18	0.9769	3560.4	48.81				
19	0.9794	10115.3	97.81	19	0.9768	3561.6	48.87				
20	0.9795	9618.8	95.07	20	0.9769	3610.5	49.39		* m^3		
Tot	0.9595	6709.6	72.00	Tot	0.9738	4579.2	54.20				
Best	0.9795	3638.8	48.56	Best	0.9811	3336.0	48.17				
EM	20	17	17	EM	4	3	3				

4. Discussion

A steady value of the correlation exponent in a certain embedding dimension is an indication of chaotic behavior in all six temporal scales. The correlation exponent became steady after an embedding dimension of 17 for the temporal scales. The correlation exponent results show that the 17th embedding dimension may be considered the optimum embedding dimension for this system; however, we conducted an additional investigation to better understand the optimum embedding dimension for the test case. Moreover, the saturation and constant slope of correlation exponent for all temporal scales revealed a highly chaotic behaviour exhibited by the whole system. Many studies that investigated the chaotic behaviour of the system concluded that high-resolution timescales demonstrate chaotic behaviour while the low-resolution timescale of the same time series does not exhibit chaotic behaviour.

Regarding the fractal law, it remains uncertain whether a chaotic system should have chaotic behaviour in all temporal scales; we speculate that this is an important factor to consider under this application. Based on the presented results, this study suggests that to investigate the availability of chaos in any dataset, investigation for different temporal scales is needed to indicate if the system has a chaotic behaviour, as this conclusion supports the findings presented by other researchers [54]. The present investigation of chaotic behaviour in different time scales using the test case provides enough information to evaluate the objectives of this research. However, additional study is required to make definitive conclusions regarding the findings of this study and more generally, the chaotic behaviour of consumption data. Nevertheless, the study findings provide encouraging evidence to further investigate the chaotic behaviour of water consumption values over different time scales of a dataset.

The results indicate that forecasting was more sensitive to optimum embedding dimension and lag time calculated by AMI than to the reconstructed phase space by 1-day lag time. However, the difference between the results of NLA and PSR-NLA was not considerable; the performance of PSR-NLA was better than NLA to forecast consumption values in different lead time. Moreover, $m_{opt} = 19$ provided more accurate predictions than $m_{opt} = 18$ (Table 3). Figure 8 compares observed values to the forecasted results, which highlights the negligible error in the results of PSR-NLA and GEP during the test period. Regarding the average fitness results for all dimensions, the performance of GEP is better than PSR-GEP, while the application of PSR is shown in forecasting with equal dimension and different lead times. Figure 8a shows the greatest accuracy from the results shown by PSR-NLA compared to the other models. Figure 8b shows the value of residual for forecasted consumption by PSR-NLA in two forecasting horizons. The performance of PSR-NLA in forecasting high-frequency values is shown in Figure 8b. The figure reveals that the optimum embedding dimension of 19 was more acceptable than the embedding dimension of 17 determined by the correlation function (Figure 5b).

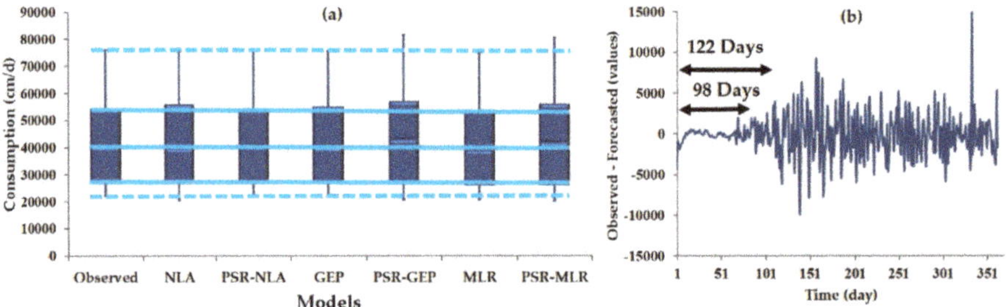

Figure 8. Performance of the models (**a**) box-plot of forecasted values in comparison with observed values; (**b**) Residual of PSR-NLA (best performance) in two forecasting horizons.

The results in Figure 8b cannot reject the optimum embedding dimension that was determined from other methods in this study. It can be concluded that the results of the 17th dimension are more reliable than those of the 19th dimension. However, the 19th dimension will offer a longer forecasting horizon. Therefore, based upon the primary goal of modeling, either dimension may be a candidate for optimum forecast modeling. The importance of embedding dimension is demonstrated when considering the small difference among the forecasted values by different embedding dimensions in all models. Therefore, it is recommended to test the optimum embedding dimension by various methods. However, the model results are quite similar; the overview of approximation for all the techniques for multiple dimensions concluded that the optimum embedding dimension is between $m = 17$ and $m = 19$, demonstrating reliability of the methods. To compare the models, Table 6 shows the comparison of statistics of each forecasted value with observed data. The check marks belong to the models where the results are closer to the observed values' statistics. The results reveal the PSR technique had a positive impact on the accuracy of all models at all temporal scales, especially in high forecasting horizons. As Figure 9 shows, the performance of NLA and PSR-NLA was approximately similar for daily and 2-days ahead forecasting, but when considering the lead time with longer period, PSR-NLA was more accurate than NLA in lead time of 1- and 2-months ahead. The value of correlation coefficient for NLA and PSR-NLA shows 43% improvement. Following CC, the values of RMSE and MAE showed the performance of PSR-NLA was better than NLA in forecasting for longer lead time. Nevertheless, the slope of the fitted line for the fitness functions shows that PSR-NLA can forecast more accurate values in comparison with NLA. The results of this application of PSR may also serve useful for application in long-term forecasting, which was previously identified in the literature as a topic requiring further study in drinking water consumption modeling.

Table 6. Statistics comparison of observed and forecasted consumption in test period by the selected models.

Property	Observed	NLA $\tau = 1, m = 18$	PSR-NLA $\tau = 17, m = 19$	GEP $\tau = 1, m = 4$	PSR-GEP $\tau = 17, m = 8$	MLR $\tau = 1, m = 17$	PSR-MLR $\tau = 17, m = 3$
Max. value	75,620.26		√				
Min. value	21,313.72				√		
Average	42,500.82		√				
Standard deviation	16,117.34				√		
Coefficient of variation	0.38	√					
Skew	0.43	√					
Kurtosis	−1.13	√					

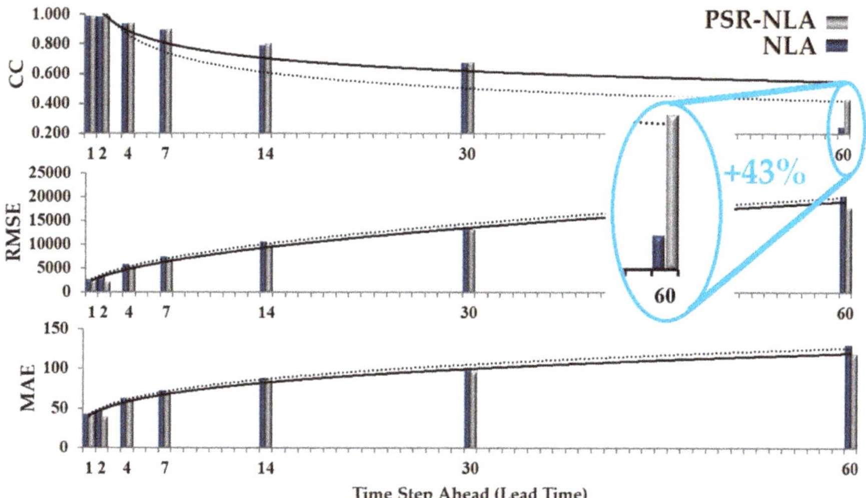

Figure 9. Performance of NLA and PSR-NLA in time ahead forecasting by the fitness functions of Correlation Coefficient (CC); Root Mean Square Error (RMSE); Mean Absolute Error (MAE) for lead time of 1-, 2-, 4-, 7-, 14-, 30- and 60-days.

5. Conclusions

This study presented a novel approach to improve the accuracy of models in forecasting residential daily water consumption values. The residential water consumption dataset from the City of Kelowna (British Columbia, Canada) was used as a test case to forecast future consumption values using varying lead times under different temporal scales to identify models which may improve forecasting performance. A chaos approach based nonlinear local approximation was compared with previously studied forecasting methods to assess and consider the applicability of the chaotic behavior of urban consumption values to better improve forecasting performance. To investigate the existence of chaos in test data, different temporal scales were considered. The results showed all temporal scales, daily at 2-, 4-, 7-, 14- and 30-days, have chaotic behavior with similar correlation exponent values for all scales. The value of the largest Lyapunov Exponent showed the reliable time period for the models as a forecasting horizon. Dynamic investigation of explanatory variables and phase space reconstruction of the input variables based on optimum correlation dimension was considered to improve the accuracy in various periodicity and lead-times. The findings suggest that considering the chaotic behavior of consumption values by taking the forecasting horizon into account, will give insight about the behavior of data. It was recommended to investigate the performance of the models in forecasting future values with different lead-times. Considering the fitness values for the models by different lead-times within the forecasting horizon, substantial unstudied criteria remain to organize different techniques based on the behavior of the models input data. Moreover, the high results of the optimum embedding dimension in reconstructed phase space revealed that to improve the models, considering a high number of input variables is not beneficial. Therefore, this research suggested investigating the dynamic behavior of explanatory variables in modeling and forecasting problems to improve the reliability and accuracy of the water consumption models.

Author Contributions: P.Y. Analyzed the data, defined the models, investigated the accuracy of the models and wrote the first draft of the manuscript; G.C. contributed in writing and analyzing; conceived and designed the experiments; G.N. contributed in writing and editing, analyzing the data and models; H.M. contributed in writing and editing. All authors have read and agree to the published version of the manuscript.

Funding: This research received no external funding.

Acknowledgments: The authors would like to thank the Okanagan Basin Water Board and the City of Kelowna are thanked for providing test case information.

Conflicts of Interest: The authors declare no conflict of interest.

References

1. Yousefi, P. Integrated Management Plan of Water Distribution Systems: Forecasting Approach. Ph.D. Thesis, University of British Columbia, Kelowna, BC, Canada, 2020. [CrossRef]
2. Xenochristou, M.; Kapelan, Z.; Hutton, C. Using Smart Demand-Metering Data and Customer Characteristics to Investigate Influence of Weather on Water Consumption in the UK. *J. Water Resour. Plan. Manag.* **2020**, *146*, 4019073. [CrossRef]
3. Water Conflict—World's Water. Available online: https://www.worldwater.org/water-conflict/ (accessed on 14 February 2019).
4. Billings, R.B.; Jones, C.V. *Forecasting Urban Water Demand*; American Water Works Association: Denver, CO, USA, 2008.
5. Ghalehkhondabi, I.; Ardjmand, E.; Young, W.A.; Weckman, G.R. Water demand forecasting: Review of soft computing methods. *Environ. Monit. Assess.* **2017**, *189*, 313. [CrossRef]
6. Sastri, T.; Valdes, J.B. Rainfall Intervention Analysis for On-Line Applications. *J. Water Resour. Plan. Manag.* **2008**, *115*, 397–415. [CrossRef]
7. Odan, F.K.; Reis, L.F.R. Hybrid Water Demand Forecasting Model Associating Artificial Neural Network with Fourier Series. *J. Water Resour. Plan. Manag.* **2012**, *138*, 245–256. [CrossRef]
8. Iwanek, M.; Kowalska, B.; Hawryluk, E.; Kondraciuk, K. Distance and time of water effluence on soil surface after failure of buried water pipe. Laboratory investigations and statistical analysis. *Eksploat. I Niezawodn. Maint. Reliab.* **2016**, *18*, 278–284. [CrossRef]
9. Ghiassi, M.; Zimbra, D.K.; Saidane, H. Urban Water Demand Forecasting with a Dynamic Artificial Neural Network Model. *J. Water Resour. Plan. Manag.* **2008**, *134*, 138–146. [CrossRef]
10. Jayawardena, A.W.; Gurung, A.B. Noise reduction and prediction of hydrometeorological time series: Dynamical systems approach vs. stochastic approach. *J. Hydrol.* **2000**, *228*, 242–264. [CrossRef]
11. Lisi, F.; Villi, V. CHAOTIC FORECASTING OF DISCHARGE TIME SERIES: A CASE STUDY. *J. Am. Water Resour. Assoc.* **2001**, *37*, 271–279. [CrossRef]
12. Cominola, A.; Giuliani, M.; Piga, D.; Castelletti, A.; Rizzoli, A.E. Benefits and challenges of using smart meters for advancing residential water demand modeling and management: A review. *Environ. Model. Softw.* **2015**, *72*, 198–214. [CrossRef]
13. Oshima, N. Information Integration Type Chaos Theory-Based Demand Forecasting for Predictive Control of Waterworks. *Water Purify Technol.* **2015**, *164*, 6–12.
14. Jain, A.; Ormsbee, L.E. Short-term water demand forecast modeling techniques—Conventional methods versus AI. *J. Am. Water Work Assoc.* **2002**, *94*, 64–72. [CrossRef]

15. Kame'enui, A.E. Water Demand Forecasting in the Puget Sound Region: Short and long-Term Models. 2003, pp. 1–97. Available online: http://citeseerx.ist.psu.edu/viewdoc/download?doi=10.1.1.461.405&rep=rep1&type=pdf (accessed on 18 February 2019).

16. Herrera, M.; Torgo, L.; Izquierdo, J.; Pérez-García, R. Predictive models for forecasting hourly urban water demand. *J. Hydrol.* **2010**, *387*, 141–150. [CrossRef]

17. Yousefi, P.; Naser, G.; Mohammadi, H. Surface Water Quality Model: Impacts of Influential Variables. *J. Water Resour. Plan. Manag.* **2018**, *144*, 4018015. [CrossRef]

18. Shabani, S.; Yousefi, P.; Adamowski, J.; Naser, G. Intelligent Soft Computing Models in Water Demand Forecasting. In *Water Stress in Plants*; IntechOpen: London, UK, 2016. [CrossRef]

19. Miaou, S.-P. A stepwise time series regression procedure for water demand model identification. *Water Resour. Res.* **1990**, *26*, 1887–1897. [CrossRef]

20. Jain, A.; Kumar Varshney, A.; Chandra Joshi, U. Short-Term Water Demand Forecast Modelling at IIT Kanpur Using Artificial Neural Networks. *Water Resour. Manag.* **2001**, *15*, 299–321. [CrossRef]

21. Gato, S.; Jayasuriya, N.; Roberts, P. Temperature and rainfall thresholds for base use urban water demand modelling. *J. Hydrol.* **2007**, *337*, 364–376. [CrossRef]

22. Bougadis, J.; Adamowski, K.; Diduch, R. Short-term municipal water demand forecasting. *Hydrol. Process.* **2005**, *19*, 137–148. [CrossRef]

23. Adamowski, J.; Fung Chan, H.; Prasher, S.O.; Ozga-Zielinski, B.; Sliusarieva, A. Comparison of multiple linear and nonlinear regression, autoregressive integrated moving average, artificial neural network, and wavelet artificial neural network methods for urban water demand forecasting in Montreal, Canada. *Water Resour. Res.* **2012**, *48*. [CrossRef]

24. Zhou, S.L.; McMahon, T.A.; Walton, A.; Lewis, J. Forecasting daily urban water demand: A case study of Melbourne. *J. Hydrol.* **2000**, *236*, 153–164. [CrossRef]

25. Mukhopadhyay, A.; Akber, A.; Al-Awadi, E. Analysis of freshwater consumption patterns in the private residences of Kuwait. *Urban. Water.* **2001**, *3*, 53–62. [CrossRef]

26. Dos Santos, C.C.; Pereira Filho, A.J. Water Demand Forecasting Model for the Metropolitan Area of São Paulo, Brazil. *Water Resour. Manag.* **2014**, *28*, 4401–4414. [CrossRef]

27. Brekke, L.; Larsen, M.D.; Ausburn, M.; Takaichi, L. Suburban Water Demand Modeling Using Stepwise Regression. *J. Am. Water Works Assoc.* **2002**, *94*, 65–75. [CrossRef]

28. Polebitski, A.S.; Palmer, R.N. Seasonal Residential Water Demand Forecasting for Census Tracts. *J. Water Resour. Plan. Manag.* **2010**, *136*, 27–36. [CrossRef]

29. Lee, S.-J.; Wentz, E.A.; Gober, P. Space–time forecasting using soft geostatistics: A case study in forecasting municipal water demand for Phoenix, Arizona. *Stoch. Environ. Res. Risk Assess.* **2010**, *24*, 283–295. [CrossRef]

30. Adamowski, J.; Karapataki, C. Comparison of Multivariate Regression and Artificial Neural Networks for Peak Urban Water-Demand Forecasting: Evaluation of Different ANN Learning Algorithms. *J. Hydrol. Eng.* **2010**, *15*, 729–743. [CrossRef]

31. Cutore, P.; Campisano, A.; Kapelan, Z.; Modica, C.; Savic, D. Probabilistic prediction of urban water consumption using the SCEM-UA algorithm. *Urban. Water J.* **2008**, *5*, 125–132. [CrossRef]

32. Adamowski, J.F. Peak Daily Water Demand Forecast Modeling Using Artificial Neural Networks. *J. Water Resour. Plan. Manag.* **2008**, *134*, 119–128. [CrossRef]

33. Zhou, T.; Wang, F.; Yang, Z. Comparative Analysis of ANN and SVM Models Combined with Wavelet Preprocess for Groundwater Depth Prediction. *Water* **2017**, *9*, 781. [CrossRef]

34. Firat, M.; Yurdusev, M.A.; Turan, M.E. Evaluation of Artificial Neural Network Techniques for Municipal Water Consumption Modeling. *Water Resour. Manag.* **2009**, *23*, 617–632. [CrossRef]

35. Xu, Y.; Zhang, J.; Long, Z.; Chen, Y. A Novel Dual-Scale Deep Belief Network Method for Daily Urban Water Demand Forecasting. *Energies* **2018**, *11*, 1068. [CrossRef]

36. Msiza, I.S.; Nelwamondo, F.V.; Marwala, T. Artificial neural networks and support vector machines for water demand time series forecasting. In Proceedings of the IEEE International Conference on Systems, Man and Cybernetics, Montreal, QC, Canada, 7–10 October 2007; pp. 638–643. [CrossRef]

37. Msiza, I.S.; Nelwamondo, F.V.; Marwala, T. Water demand prediction using artificial neural networks and support vector regression. *J. Comput.* **2008**, *3*, 1–8. [CrossRef]

38. Shabani, S.; Yousefi, P.; Naser, G. Support Vector Machines in Urban Water Demand Forecasting Using Phase Space Reconstruction. *Procedia Eng.* **2017**, *186*, 537–543. [CrossRef]

39. Yousefi, P.; Shabani, S.; Mohammadi, H.; Naser, G. Gene Expression Programing in Long Term Water Demand Forecasts Using Wavelet Decomposition. *Procedia Eng.* **2017**, *186*, 544–550. [CrossRef]

40. Shabani, S. Water Demand Forecasting: A Flexible Approach. Ph.D. Thesis, University of British Columbia, Kelowna, BC, Canada, 2018. [CrossRef]

41. Ambrosio, J.K.; Brentan, B.M.; Herrera, M.; Luvizotto, E.; Ribeiro, L.; Izquierdo, J. Committee Machines for Hourly Water Demand Forecasting in Water Supply Systems. *Math. Probl. Eng.* **2019**, *2019*, 1–11. [CrossRef]

42. Yousefi, P.; Naser, G.; Mohammadi, H. Application of Wavelet Decomposition and Phase Space Reconstruction in Urban Water Consumption Forecasting: Chaotic Approach (Case Study). In *Wavelet Theory and Its Applications*; IntechOpen: London, UK, 2018. [CrossRef]

43. Yousefi, P.; Naser, G.; Mohammadi, H. Hybrid Wavelet and Local Approximation Method for Urban Water Demand Forecasting—Chaotic Approach. In Proceedings of the WDSA Conference, Kingstone, ON, Canada, 23–25 July 2018.

44. Azadeh, A.; Neshat, N.; Hamidipour, H. Hybrid Fuzzy Regression–Artificial Neural Network for Improvement of Short-Term Water Consumption Estimation and Forecasting in Uncertain and Complex Environments: Case of a Large Metropolitan City. *J. Water Resour. Plan. Manag.* **2011**, *138*, 71–75. [CrossRef]

45. Ahmadi, S.; Alizadeh, S.; Forouzideh, N.; Yeh, C.H.; Martin, R.; Papageorgiou, E. ICLA imperialist competitive learning algorithm for fuzzy cognitive map: Application to water demand forecasting. In Proceedings of the IEEE International Conference on Fuzzy Systems, Beijing, China, 6–11 July 2014; pp. 1041–1048. [CrossRef]

46. Navarrete-López, C.; Herrera, M.; Brentan, B.; Luvizotto, E.; Izquierdo, J. Enhanced Water Demand Analysis via Symbolic Approximation within an Epidemiology-Based Forecasting Framework. *Water.* **2019**, *11*, 246. [CrossRef]

47. Yousefi, P.; Naser, G.; Mohammadi, H. Estimating High Resolution Temporal Scale of Water Demand Time Series—Disaggregation Approach (Case Study). In Proceedings of the 13th International Conference on Hydroinformatics (HIC 2018), Palermo, Italy, 1–6 July 2018; Volume 3, pp. 2408–2416. [CrossRef]

48. Kozłowski, E.; Kowalska, B.; Kowalski, D.; Mazurkiewicz, D. Water demand forecasting by trend and harmonic analysis. *Arch. Civ. Mech. Eng.* **2018**, *18*, 140–148. [CrossRef]

49. Campisi-Pinto, S.; Adamowski, J.; Oron, G. Forecasting Urban Water Demand Via Wavelet-Denoising and Neural Network Models. Case Study: City of Syracuse, Italy. *Water Resour. Manag.* **2012**, *26*, 3539–3558. [CrossRef]

50. Casdagli, M. Chaos and Deterministic *Versus* Stochastic Non-Linear Modelling. *J. R Stat. Soc. Ser. B* **1992**, *54*, 303–328. [CrossRef]

51. Lorenz, E.N. Atmospheric Predictability as Revealed by Naturally Occurring Analogues. *J. Atmos. Sci.* **2004**, *26*, 636–646. [CrossRef]

52. Sivakumar, B.; Jayawardena, A.W.; Li, W.K. Hydrologic complexity and classification: A simple data reconstruction approach. *Hydrol. Process.* **2007**, *21*, 2713–2728. [CrossRef]

53. Ng, W.W.; Panu, U.S.; Lennox, W.C. Chaos based Analytical techniques for daily extreme hydrological observations. *J. Hydrol.* **2007**, *342*, 17–41. [CrossRef]

54. Regonda, S.K.; Sivakumar, B.; Jain, A. Temporal scaling in river flow: Can it be chaotic? *Hydrol. Sci. J.* **2004**, *49*, 373–385. [CrossRef]

55. Salas, J.D.; Kim, H.S.; Eykholt, R.; Burlando, P.; Green, T.R. Aggregation and Sampling in Deterministic Chaos: Implications for Chaos Identification in Hydrological Processes. June 2005. Available online: https://hal.archives-ouvertes.fr/hal-00302625/ (accessed on 29 July 2019).

56. Elshorbagy, A.; Simonovic, S.P.; Panu, U.S. Estimation of missing streamflow data using principles of chaos theory. *J. Hydrol.* **2002**, *255*, 123–133. [CrossRef]

57. Elshorbagy, A.; Simonovic, S.P.; Panu, U.S. Noise reduction in chaotic hydrologic time series: Facts and doubts. *J. Hydrol.* **2002**, *256*, 147–165. [CrossRef]

58. Sivakumar, B.; Wallender, W.W. Predictability of river flow and suspended sediment transport in the Mississippi River basin: A non-linear deterministic approach. *Earth Surf. Process. Landforms.* **2005**, *30*, 665–677. [CrossRef]

59. Zounemat-Kermani, M. Investigating Chaos and Nonlinear Forecasting in Short Term and Mid-term River Discharge. *Water Resour. Manag.* **2016**, *30*, 1851–1865. [CrossRef]

60. Ghorbani, M.A.; Khatibi, R.; Danandeh Mehr, A.; Asadi, H. Chaos-based multigene genetic programming: A new hybrid strategy for river flow forecasting. *J. Hydrol.* **2018**, *562*, 455–467. [CrossRef]

61. Sivakumar, B. A phase-space reconstruction approach to prediction of suspended sediment concentration in rivers. *J. Hydrol.* **2002**, *258*, 149–162. [CrossRef]
62. Sivakumar, B.; Jayawardena, A.W. An investigation of the presence of low-dimensional chaotic behaviour in the sediment transport phenomenon. *Hydrol. Sci. J.* **2002**, *47*, 405–416. [CrossRef]
63. Ghorbani, M.; Khatibi, R.; Asadi, H.; Yousefi, P. Inter-Comparison of an Evolutionary Programming Model of Suspended Sediment Time-Series with Other Local Models. In *Genetic Programming—New Approaches and Successful Applications*; IntechOpen: London, UK, 2012. [CrossRef]
64. Petkov, B.H.; Vitale, V.; Mazzola, M.; Lanconelli, C.; Lupi, A. Chaotic behaviour of the short-term variations in ozone column observed in Arctic. *Commun. Nonlinear Sci. Numer. Simul.* **2015**, *26*, 238–249. [CrossRef]
65. Ghorbani, M.A.; Kisi, O.; Aalinezhad, M. A probe into the chaotic nature of daily streamflow time series by correlation dimension and largest Lyapunov methods. *Appl. Math. Model.* **2010**, *34*, 4050–4057. [CrossRef]
66. Khatibi, R.; Ghorbani, M.A.; Aalami, M.T.; Kocak, K.; Makarynskyy, O. Dynamics of hourly sea level at Hillarys Boat Harbour, Western Australia: A chaos theory perspective. *Ocean Dyn.* **2011**, *61*, 1797–1807. [CrossRef]
67. Rodriguez-Iturbe, I.; Febres De Power, B.; Sharifi, M.B.; Georgakakos, K.P. Chaos in rainfall. *Water Resour. Res.* **1989**, *25*, 1667–1675. [CrossRef]
68. Jayawardena, A.W.; Lai, F. Analysis and prediction of chaos in rainfall and stream flow time series. *J. Hydrol.* **1994**, *153*, 23–52. [CrossRef]
69. Sivakumar, B.; Berndtsson, R.; Olsson, J.; Jinno, K.; Kawamura, A. Dynamics of monthly rainfall-runoff process at the Gota basin: A search for chaos. *Hydrol. Earth Syst. Sci.* **2000**, *4*, 407–417. [CrossRef]
70. Maskey, M.L.; Puente, C.E.; Sivakumar, B. Temporal downscaling rainfall and streamflow records through a deterministic fractal geometric approach. *J. Hydrol.* **2019**, *568*, 447–461. [CrossRef]
71. Wang, J.; Shi, Q. Short-term traffic speed forecasting hybrid model based on Chaos–Wavelet Analysis-Support Vector Machine theory. *Transp. Res. Part. C Emerg. Technol.* **2013**, *27*, 219–232. [CrossRef]
72. Ravi, V.; Pradeepkumar, D.; Deb, K. Financial time series prediction using hybrids of chaos theory, multi-layer perceptron and multi-objective evolutionary algorithms. *Swarm Evol. Comput.* **2017**, *36*, 136–149. [CrossRef]
73. Abdechiri, M.; Faez, K.; Amindavar, H.; Bilotta, E. The chaotic dynamics of high-dimensional systems. *Nonlinear Dyn.* **2017**, *87*, 2597–2610. [CrossRef]
74. Li, M.W.; Geng, J.; Han, D.F.; Zheng, T.J. Ship motion prediction using dynamic seasonal RvSVR with phase space reconstruction and the chaos adaptive efficient FOA. *Neurocomputing* **2016**, *174*, 661–680. [CrossRef]
75. Kalra, R.; Deo, M.C. Genetic programming for retrieving missing information in wave records along the west coast of India. *Appl. Ocean. Res.* **2007**, *29*, 99–111. [CrossRef]
76. Ustoorikar, K.; Deo, M.C. Filling up gaps in wave data with genetic programming. *Mar. Struct.* **2008**, *21*, 177–195. [CrossRef]
77. Gaur, S.; Deo, M.C. Real-time wave forecasting using genetic programming. *Ocean. Eng.* **2008**, *35*, 1166–1172. [CrossRef]
78. Aytek, A.; Kişi, Ö. A genetic programming approach to suspended sediment modelling. *J. Hydrol.* **2008**, *351*, 288–298. [CrossRef]
79. Ferreira, C. Gene Expression Programming in Problem Solving. In *Soft Computing and Industry*; Springer: London, UK, 2002; pp. 635–653. [CrossRef]
80. Ferreira, C. Function Finding and the Creation of Numerical Constants in Gene Expression Programming. In *Advances in Soft Computing*; Springer: London, UK, 2003; pp. 257–265. [CrossRef]
81. Nasseri, M.; Moeini, A.; Tabesh, M. Forecasting monthly urban water demand using Extended Kalman Filter and Genetic Programming. *Expert Syst. Appl.* **2011**, *38*, 7387–7395. [CrossRef]
82. Shabani, S.; Candelieri, A.; Archetti, F.; Naser, G. Gene Expression Programming Coupled with Unsupervised Learning: A Two-Stage Learning Process in Multi-Scale, Short-Term Water Demand Forecasts. *Water* **2018**, *10*, 142. [CrossRef]
83. Gutzler, D.S.; Nims, J.S. Interannual Variability of Water Demand and Summer Climate in Albuquerque, New Mexico. *J. Appl. Meteorol.* **2006**, *44*, 1777–1787. [CrossRef]
84. Donkor, E.A.; Mazzuchi, T.A.; Soyer, R.; Alan Roberson, J. Urban Water Demand Forecasting: Review of Methods and Models. *J. Water Resour. Plan. Manag.* **2012**, *140*, 146–159. [CrossRef]
85. Alvisi, S.; Franchini, M.; Marinelli, A. A short-term, pattern-based model for water-demand forecasting. *J. Hydroinformatics* **2006**, *9*, 39–50. [CrossRef]

86. Sivakumar, B.; Berndtsson, R.; Olsson, J.; Jinno, K. Evidence of chaos in the rainfall-runoff process. *Hydrol. Sci. J.* **2001**, *46*, 131–145. [CrossRef]

87. Takens, F. Detecting strange attractors in turbulence. In *Dynamical Systems and Turbulence, Warwick*; Springer: Berlin/Heidelberg, Germany, 1981; pp. 366–381. [CrossRef]

88. Sivakumar, B. Forecasting monthly flow dynamics in the western united states: A nonlinear dynamical approach. *J. Environ. Model. Softw.* **2003**, *17*, 721–728. [CrossRef]

89. Khatibi, R.; Sivakumar, B.; Ghorbani, M.A.; Kisi, O.; Koçak, K.; Farsadi Zadeh, D. Investigating chaos in river stage and discharge time series. *J. Hydrol.* **2012**, *414–415*, 108–117. [CrossRef]

90. Meng, Q.; Peng, Y. A new local linear prediction model for chaotic time series. *Phys. Lett. Sect. A Gen. At. Solid State Phys.* **2007**, *370*, 465–470. [CrossRef]

91. Fraser, A.M.; Swinney, H.L. Independent coordinates for strange attractors from mutual information. *Phys. Rev. A* **1986**, *33*, 1134–1140. [CrossRef] [PubMed]

92. Holzfuss, J.; Mayer-Kress, G. An Approach to Error-Estimation in the Application of Dimension Algorithms. In *Dimensions and Entropies in Chaotic Systems*; Springer: Berlin/Heidelberg, Germany, 2011; pp. 114–122. [CrossRef]

93. Hegger, R.; Kantz, H.; Schreiber, T. Practical implementation of nonlinear time series methods: The TISEAN package. *Chaos Interdiscip. J. Nonlinear Sci.* **1999**, *9*, 413–435. [CrossRef] [PubMed]

94. Zounemat-Kermani, M.; Kisi, O. Time series analysis on marine wind-wave characteristics using chaos theory. *Ocean. Eng.* **2015**, *100*, 46–53. [CrossRef]

95. Grassberger, P.; Procaccia, I. Measuring the strangeness of strange attractors. *Phys. D Nonlinear Phenom.* **1983**, *9*, 189–208. [CrossRef]

96. Islam, M.N.; Sivakumar, B. Characterization and prediction of runoff dynamics: A nonlinear dynamical view. *Adv. Water Resour.* **2002**, *25*, 179–190. [CrossRef]

97. Tongal, H.; Berndtsson, R. Impact of complexity on daily and multi-step forecasting of streamflow with chaotic, stochastic, and black-box models. *Stoch. Environ. Res. Risk Assess.* **2017**, *31*, 661–682. [CrossRef]

98. Farmer, J.D.; Sidorowich, J.J. Predicting chaotic time series. *Phys. Rev. Lett.* **1987**, *59*, 845–848. [CrossRef] [PubMed]

99. Itoh, K.-I. A method for predicting chaotic time-series with outliers. *Electron. Commun. Jpn. Part III Fundam Electron. Sci.* **1995**, *78*, 44–53. [CrossRef]

100. Porporato, A.; Ridolfi, L. Nonlinear analysis of river flow time sequences. *Water Resour. Res.* **1997**, *33*, 1353–1367. [CrossRef]

101. Rosenstein, M.T.; Collins, J.J.; De Luca, C.J. A practical method for calculating largest Lyapunov exponents from small data sets. *Phys. D Nonlinear Phenom.* **1993**, *65*, 117–134. [CrossRef]

102. Shang, P.; Li, X.; Kamae, S. Chaotic analysis of traffic time series. *Chaos Solitons Fractals* **2005**, *25*, 121–128. [CrossRef]

103. Holland, J.H. Genetic algorithms and the optimal allocation of trials. In *Evolutionary Computation: The Fossil Record*; Society for Industrial and Applied Mathematics: Philadelphia, PA, USA, 1998; Volume 2, pp. 443–460. [CrossRef]

104. Goldberg, D.E.; Holland, J.H. Genetic Algorithms and Machine Learning. *Mach. Learn.* **1988**, *3*, 95–99. [CrossRef]

105. Strategic Value Solution. Kelowna Integrated Water Supply Plan. Kelowna. 2017. Available online: https://www.kelowna.ca/city-services/water-wastewater/ (accessed on 26 February 2020).

106. Ruelle, D. The Claude Bernard Lecture, 1989. Deterministic Chaos: The Science and the Fiction. *Proc. R Soc. A Math. Phys. Eng. Sci.* **1990**, *427*, 241–248. [CrossRef]

Article

Hydraulic Simulation and Analysis of an Urban Center's Aqueducts Using Emergency Scenarios for Network Operation: The Case of Thessaloniki City in Greece

Alexandros Mentes [1], Panagiota Galiatsatou [2,*], Dimitrios Spyrou [3], Achilleas Samaras [4] and Panagiota Stournara [2]

[1]   Department of Strategic Planning, Hydraulic Works & Development, Thessaloniki Water Supply and Sewerage Company S.A. (EYATH S.A.), Tsimiski Str. 98, 54622 Thessaloniki, Greece; amentes@eyath.gr
[2]   Division of Geoinformatics, Topography & Hydraulic Modeling, Department of Strategic Planning, Hydraulic Works & Development, EYATH S.A., 54622 Thessaloniki, Greece; pstournara@eyath.gr
[3]   Division of Research and Development, Department of Strategic Planning, Hydraulic Works & Development, EYATH S.A., 54622 Thessaloniki, Greece; dspyrou@eyath.gr
[4]   Division of Hydraulic Studies & Tenders, Department of Strategic Planning, Hydraulic Works & Development, EYATH S.A., 54622 Thessaloniki, Greece; achilleas.samaras@gmail.com
*   Correspondence: pgaliatsatou@eyath.gr; Tel.: +30-2310-966917

Received: 31 March 2020; Accepted: 27 May 2020; Published: 6 June 2020

Abstract: The present work aims at developing a hydraulic simulation model for the aqueducts of Thessaloniki city in Greece to model the current operating state of the network, as well as its response to emergency conditions resulting from failure in one of them. Hydraulic simulations performed using WaterGEMS software in an extended period simulation (EPS) mode entail estimating water demand in all areas of the conurbation and calibrating the model under both normal and abnormal conditions. Calibration parameters set include the pipes' roughness coefficients and head loss characteristics of throttle control valves (TCVs). Failure in the city's aqueducts is confronted with the development and hydraulic simulation of five emergency scenarios of network operation, two of which consider possible interconnections of the studied aqueducts. These scenarios, which include appropriately defined intermittent water supply schedules for the aqueducts, are created on the basis of fair and equitable management of water among the different areas of the city, also assuming a small number of interventions/operations during the crisis. The simulations performed reveal quite a satisfactory compliance of the system's operation with the defined schedules, and an improved management of limited water reserves in some areas of the network when considering interconnections of the city's aqueducts.

Keywords: hydraulic simulation; water demand; emergency scenario; intermittent water supply; water management; WaterGEMS software

---

1. Introduction

The Water Supply and Sewerage Company of Thessaloniki (EYATH S.A.) in Greece maintains and manages a large number of installations and water distribution systems, including pumping stations, water tanks, and large diameter pipeline, aiming to transport water to all areas of the Thessaloniki conurbation under its responsibility, and to satisfy the water needs of its numerous consumers. The city of Thessaloniki, the second largest city in Greece, has a dual and quite complex water supply system, partially covering its water needs with groundwater (aqueduct of Aravissos) and surface water (aqueduct of Aliakmonas river). The aqueducts of Thessaloniki city have been studied and constructed over a long period depending on the water needs and available resources, while nowadays the whole system operates on a daily basis to meet the water demand needs of the city.

1.1. Aim and Objectives

The present work primarily aims at the development and implementation of a detailed hydraulic model for the aqueducts of the Thessaloniki conurbation, which will be able to simulate and assess the hydraulic behavior of the system's elements and critical infrastructure for its current operational status, as well as its response to different management strategies in case of emergency conditions. To accomplish this main goal, a thorough attempt is made to: (1) produce reliable estimates of water demand in the different areas of Thessaloniki conurbation, (2) collect and update available topographical, geometrical and hydraulic data regarding the current status of the water supply system, and (3) adjust the calibration parameters of the model accordingly based on the aforementioned data. In this work, hydraulic simulation of the water distribution system (WDS) of Thessaloniki in its current operational status is also considered with the development, analysis, and simulation of alternative emergency scenarios for distributing water to the different areas of the city in case of a failure in one of its aqueducts. The former emergency scenarios also incorporate the utilisation of existing interconnections between the aqueducts, not deployed under normal operating conditions of the system.

The hydraulic simulation of Thessaloniki's aqueducts under normal and emergency conditions, such as failure in any of the city's aqueducts causing water shortage for quite an elongated period of time and discomfort to the city's residents, will significantly assist understanding the aqueducts' operating rules and restrictions. Assessing the hydraulic behavior of the studied system under abnormal and sometimes critical conditions targets the management of available water reserves fairly and equally among the different areas of the city, with the fewest possible interventions and operations by the managing company's staff during the crisis. It also contributes to finding optimum ways of managing available water resources, as well as formulating operating rules for existing pumping stations and water tanks of the network, in order to handle and manage a crisis during the time needed for repair works. This work was highly motivated by a severe structural failure which occurred on the main pipeline of one of the city's aqueducts (aqueduct of Aravissos) in 2018, resulting in water shortage for almost one week in the entire city and causing emergency conditions for both the managing company, EYATH S.A., and the city itself. This serious crisis reinforced the need for a detailed and up-to-date hydraulic simulation of the city's aqueducts, which incorporates, explores, and simulates alternative scenarios for distributing the available resources among the different areas of the network in a fair and equitable way, in case of a severe failure in one of the city's aqueducts. A flow diagram of the proposed methodology is presented in Figure 1.

The methodology introduced in this work is quite simple, comprehensive, and practical, mainly on the basis of the simulation procedures used to cope with water shortage conditions in the aqueducts of quite a large and complex WDS, with a limited existing SCADA (supervisory control and data acquisition) system. For emergency conditions resembling those described in this work, a modified hydraulic analysis is currently applied in the literature. This analysis does not necessarily consider the required demands of the WDS, but examines water availability based on pressure conditions at its nodes. Therefore, unlike the demand-driven analysis (DDA), demand at a certain node of the network is herein considered a function of pressure, therefore flow is set equal to zero if the nodal head is below a critical minimum value and increases up to a desirable threshold of the nodal head until an adequate flow is satisfied (nodal demand). However, in this work, an analysis based on pressure-dependent demands is avoided. The necessity to deviate from such an analysis results mainly from difficulties in extracting typical pressure-dependent demand curves for the different areas of the WDS considering that the system is poorly instrumented, as well as in defining thresholds for both the minimum and desirable head of the different demand nodes. The approach introduced in this work is beneficial in the sense that it considers a demand satisfaction ratio, defined as the ratio of the flow that is available to the flow that is required, for each area of the WDS equal to unit during water supply periods, which assists in fulfilling the main objectives of water distribution management under abnormal conditions of the system.

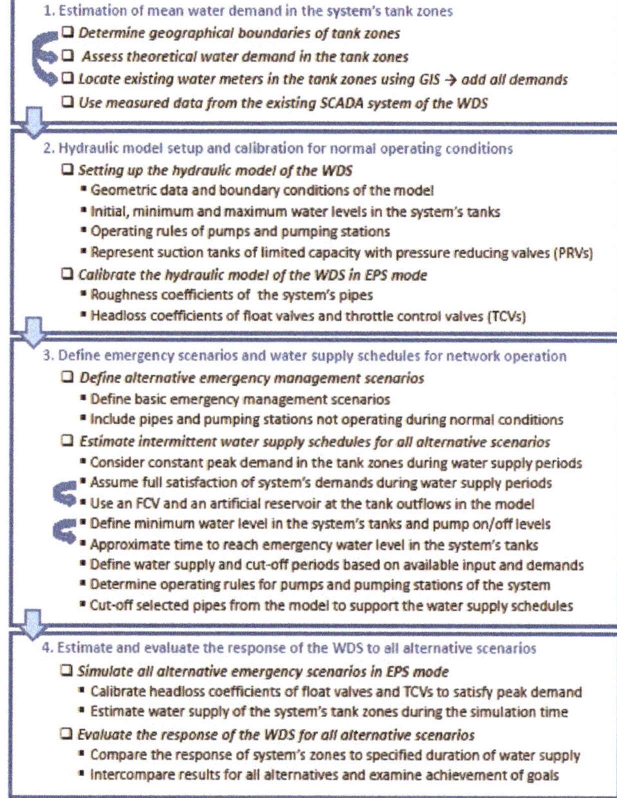

Figure 1. Flow diagram of the proposed methodology.

The methodology presented also includes different aspects that could possibly lead to more reliable and feasible hydraulic simulations of WDSs, especially for those that are poorly instrumented, under both normal and abnormal conditions. Assessing water demand in the different areas of the WDS is one such aspect. In this work, water demand estimation in the different areas of the city combines: (a) design/theoretical methods, (b) measurements from the existing limited SCADA system, and (c) measured data of water consumption from water meters installed in the different areas of the WDS. After assessing water demand in the different areas of the network, the hydraulic model is calibrated for both pipe roughness characteristics and the hydraulic properties of float valves and TCVs in EPS mode. To accomplish this, the system is first divided into smaller and distinct parts, and an iterative procedure is followed for certain hydraulic parameters to converge to measured quantities or satisfy defined constraints.

This work introduces and simulates different scenarios of system operation in the case of failure in the aqueducts of a real case study, following a real emergency event which happened in the city Thessaloniki in 2018, analysing and comparing the response of the system to each of these scenarios and examining the fulfilment of requirements and objectives set by the managing authority. Within this framework, different alternative scenarios of system operation in the case of emergency are considered.

These scenarios do not only include different intermittent water supply schedules, but also consider completely cutting off water supply in entire areas after some time, or utilise existing pipes or/and pumping stations not used under normal operating conditions of the WDS.

1.2. Literature Review

Hydraulic simulation of pressurised pipe systems is quite a crucial step for managing large water distribution networks (WDNs), as well as for decision making under both normal and abnormal conditions. EPANET is a freely distributed modeling software package by EPA (United States Environmental Protection Agency) which is used, among other things, to design new WDNs and water infrastructure, as well as to analyse and manage existing WDNs, optimise the operations of tanks and pumps, and investigate water quality problems [1]. OpenFlows WaterGEMS software [2], based on the hydraulic engine of EPANET, is a flexible modeling environment to design, analyse and optimise WDSs, provided with advanced hydraulic features, geospatial model-building, optimisation and asset management tools. Yu-Kun et al. [3] present a modeling approach of the WDN of Zhengzhou city in China using WaterGEMS software, based on information acquired from a GIS (Geographic Information System) database. Peaking factor and pipe roughness coefficient of the studied network are assessed by means of a sensitivity analysis. Avesani et al. [4] describe modifications to EPANET software for modeling WDNs with variable tank heads in unsteady flow conditions. Jiang et al. [5] introduce a detailed analysis of a water supply network in WaterGEMS software, emphasising on pipe modeling processes, as well as on collection of flow and pressure SCADA data used for calibrating the model. Elsheikh et al. [6] model the aging WDN of Tanta city in Egypt using WaterCAD Haestad software to consider optimal interventions and possible extensions of the network. Steady state measurements, fire flow tests, and extended period measurements of pressure and flow are used to assess the roughness coefficient of pipes of the studied network. Okeya et al. [7] introduce an online modeling approach of a WDS combing a hydraulic model with SCADA and data assimilation techniques, to improve the real-time estimation of water demand and to predict reliable future system states. Alves et al. [8] present an EPANET-based approach to modeling and calibrating a WDN presenting pressure problems and characterised by a scarce data inventory. The model is run multiple times, finally achieving a good representation of the simulated system and a satisfactory level of accuracy. Sonaje and Joshi [9] review available software used for modeling WDNs, intercomparing them in terms of application. Kara et al. [10] present an EPANET-based hydraulic model of a touristic area in Antalya, Turkey, characterised by large temporal and spatial variations of water consumption profiles and rates throughout the highly varying topographic levels of the area. Pressure and flow measurements from a SCADA station located at the entrance of the DMA (District Metered Area), pressure measurements at different elevations, monthly water consumption rates of the population served, and hourly water consumption rates of representative end users of the service, are used to calibrate the hydraulic model. Świtnicka et al. [11] use WaterGEMS software to optimise the operation of a WDS by maximising flow velocity in the system's pipes, regulating pressure head, and minimising water retention time in the network and pump energy consumption. Optimal solution for the network, including pipe diameters, hydraulic parameters of pumps, as well as locating new pressure reducing valves (PRVs) and pipe connections, is found using metaheuristic methods, such as genetic algorithms. Chatzivasili et al. [12] introduce a hybrid, two-stage approach for modeling large scale WDNs and evaluate the proposed approach through WaterGEMS software, obtaining very good results. The proposed approach uses the Geometric Partitioning Method to divide the WDN into smaller areas, followed by the application of Student's t-mixture model to find optimal locations for isolation valves and to separate DMAs.

Under pressure-deficient conditions of a WDS, demand can only be partially satisfied until the system fully recovers. Under such emergency conditions, the actual water amount available at a demand node depends on available pressure conditions by means of establishing a node head-flow relationship (NHFR). Analysis based on NHFR has been used, among others, in parameter calibration, leakage management, and vulnerability and reliability analysis of WDSs [13–19]. Ingeduld et al. [20]

adjust the EPANET source code to model intermittent water supply and present a simplified model where node demands depend on respective pressure values. Wu and Walski [21] develop and present an approach to simulate abnormal events in WDNs based on pressure-dependent demand modeling. Fiorini Morosini et al. [22] propose a pressure-driven analysis (PDA) based approach to efficiently manage a WDN during emergency conditions. Pressure-deficient conditions are usually simulated by a NHFR acting to reduce or even eliminate the nodal demand when the pressure drops below certain appropriately defined thresholds [23,24]. Such conditions are also simulated by incorporating artificial elements (i.e., reservoirs, pipes, flow control valves, emitters) in the WDS [24–27]. Two principal schemes of this category can be summarised in the inclusion of artificial reservoirs coupled with flow control valves (FCVs), check valves, and artificial pipes with suitable resistance properties [27] and emitters connected to demand nodes coupled with FCVs and check valves [24].

Lee et al. [28] propose modification techniques in the applied hydraulic models to resolve issues of negative pressure DDA or of total head reverse in PDA, arising during abnormal conditions of the network caused by pipe damages or rapid increases in demand. Nardo et al. [29] use fractal and topological metrics to assess the resilience of a WDN to a possible pipe failure and test the proposed methodology on two real WDNs in southern Italy. Batish [30] introduces a methodology to design intermittent water supply networks to adjust to financial constraints in the operation of drinking water systems in developing countries. Fontanazza et al. [31] propose a methodology for analyzing an intermittent water supply system under water scarcity conditions, resulting in significant alterations to both water availability for the users, as well as to the performance of the network itself. Vairavamoorthy et al. [32] analyse guidelines for the design and control of intermittent WDSs under water scarcity conditions in developing countries, and incorporate all required modifications in a network analysis simulation tool coupled with an optimal design tool. Soltanjalili et al. [33] propose hedging or intermittent water supply as an appropriate means to confront with water shortage conditions. Their methodology considers available water, consumption fluctuations, and consumer satisfaction, among other factors, by linking the EPANET 2 hydraulic simulator to the honey-bee mating optimisation (HBMO) algorithm, and calculating WDN performance criteria. However, Agathokleous and Christodoulou [34] highlight that intermittent water supply practices and operations negatively affect the vulnerability of WDNs, estimating an increase in the number of water loss incidents during and immediately after such periods.

2. Study Area and Available Data

To meet the water demand needs of the conurbation of Thessaloniki, EYATH S.A. supplies water to around 500,000 water meters in the area. The water needs of the city are currently covered by three resources: (1) Aliakmonas river (surface water), (2) Aravissos springs (source water), and (3) water supply boreholes in the plain of Thessaloniki (groundwater). The abovementioned water sources provide drinking water to the conurbation of Thessaloniki via the aqueduct of Aliakmonas, which conveys water from Aliakmonas river to the Thessaloniki Water Treatment Plant (TWTP), and via the aqueduct of Aravissos, which transports water from Aravissos springs and water supply boreholes of the broader area of Thessaloniki to the central pumping station of Dendropotamos at the southwestern edge of the city. Within the conurbation of Thessaloniki, there is a wide network of water tanks and pumping stations that receive water from one of the two aqueducts, or from both aqueducts at the same time. The aqueducts of Thessaloniki extend over 460 km, while the entire WDS consists of water supply pipes with a total length exceeding 2690 km.

Operation of the aqueduct of Aliakmonas started in 2003, transporting water to central tanks of high-altitude areas in the conurbation of Thessaloniki, while in the low-altitude areas of the city, water was still supplied by the aqueduct of Aravissos (the oldest aqueduct used to supply the entire city before 2003). The aqueduct of Aliakmonas brought significant changes to both the operation and management of pumping stations, tanks, and WDNs of Thessaloniki, effectively transferring the "gravity center" of the water supply system. However, nowadays there are still tanks in the water system of the city which are supplied simultaneously by both aqueducts. During the last decades, the inclusion of new areas characterised by increased urban development (residential and commercial), together with an apparent degradation of the water quality of some of Thessaloniki's water supply boreholes, which renders them inappropriate for use, resulted in a marginal coverage of the city's total water needs, especially during the summer months. Considering the aforementioned problem, a main priority for EYATH S.A. included in the Water Resources Management Plan of Thessaloniki [35] is to increase the capacity of the TWTP to an output of 300,000 m^3/day (2nd stage of the TWTP), considering that its current capacity can reach 150,000 m^3/day (1st stage of the TWTP).

For Aravissos aqueduct, two pressure pipes (DN800 and DN1000) start from the central pumping station-suction tank of Dendropotamos (P-ST35) transporting water to the pumping station suction tank of Evangelistria (P-ST17), pumping group of DN800, and pumping stations-suction tanks of Kalithea (P-ST12) and Kassandrou (P-ST14), pumping group of DN1000. From P-ST17, water is pumped to the tanks of Agios Pavlos (T18) and Saranta Eklisies (T19). P-ST12 pumps water to the tanks of Neapolis (T10), Kafkasou (T11), and Sykies (T13), while water is pumped from T13 to the high-altitude tank of Eptapirgio (T15). P-ST14 supplies water to tanks of Vlatades (T16) and Toumpa (T20). From T20, water is transferred to the tank of Kalamaria (T27), and pumped to the tank of Pylaia (T21), which in turn pumps water to a series of tanks in the settlement of Panorama in eastern Thessaloniki (T22—Lykoi, T23—Gymnasio, T61—Tompoudes, T24—Analipsi, T25—Toumpitsa). For Aliakmonas river aqueduct, a DN1700 steel pipe starts from the outlet of the TWTP and ends up to the pumping station of Ionia (P-3), after suppling the industrial area of Thessaloniki and the tank of Diavata (T36) with water. T36, which also collects water from boreholes in the areas of Axios and Eleousa, supplies water to the petroleum industry EKO and pumps water to the tank of Kato Evosmos (T6), in the western part of the city. From P-3, water is transferred to the regulating tank of Oraiokastro (T4) and from there to a number of tanks located in moderate to high-altitude areas of the city. More specifically, the aqueduct of Aliakmonas provides water to the tanks of Ano Evosmos (T5), Efkarpia (T63), Meteora (T9), Polichniotisa (T7), Androutsou (T48), Zefiron (T59), Akropolis (T64), as well as tanks in the area of Oraiokastro (T44, T45, T47, T50, T52, T69) and partly to the tanks of Neapolis (T10), Vlatadon (T16), Toumpa (T20) and Kalamaria (T27). Figure 2 presents a basic layout of Thessaloniki's aqueducts within the conurbation. Blue, red, purple, and orange colors represent water supplied by the aqueduct of Aravissos, Aliakmonas, Aravissos, and Aliakmonas, or Aliakmonas and water supply boreholes, respectively. Figure 2 also includes information on material and diameter of the system's main pipes.

From Figure 2, it is quite obvious that the majority of the system's tanks is supplied by pumping stations, while there are some tanks supplied by another tank by means of gravity flow in pressurised pipes. For the latter, mostly observed at the pipeline of Aliakmonas aqueduct, water supply is controlled using TCVs or float valves. TCVs are also installed on pipes entering P-ST12 and P-ST14 (pumping group DN1000) because both installations are supplied from P-ST35 through the main pressure pipe DN1000. Field trips were conducted for the purposes of this study, focusing on data collection for TCVs (butterfly valves) of the city's aqueducts. The opening and closing angles of the system's TCVs were measured in the field and transformed to a minor loss coefficient for each valve by means of the minor loss versus valve opening angle curve provided by the manufacturing company. Such estimates are used as initial values of TCV head loss coefficient within the hydraulic model calibration process described in Section 3.

To set up the hydraulic model of the city's water supply system, dimensions and main geometrical and hydraulic properties of tanks and pumping stations were collected. For the system's water tanks the following data was gathered from existing drawings, the company's GIS system, as well as field measurements: (a) base elevation, (b) minimum elevation of water in the tank, usually considered equal to the elevation of the outlet pipe's top, (c) maximum elevation of water in the tank, usually considered equal to the elevation of the tank's overflow pipe, (d) shape of the tank's cross section, and (e) dimensions of the tank's cross section. For all pumps and pumping stations of the system, pump curves or nominal flow and head data were collected where available.

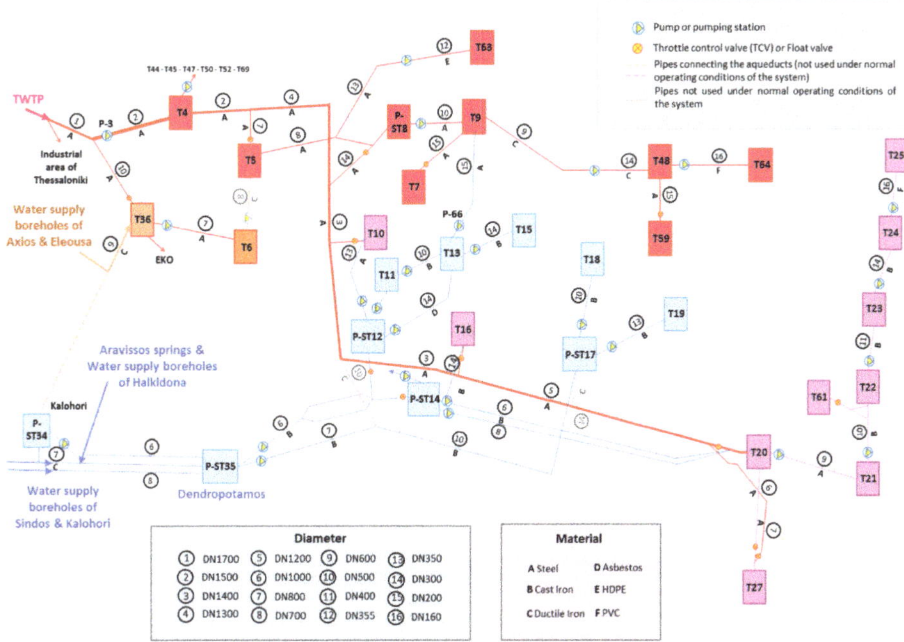

Figure 2. Aqueducts of Thessaloniki within the conurbation. Colors blue, red, purple and orange represent water supplied by the aqueducts of Aravissos, Aliakmonas, Aravissos and Aliakmonas, or Aliakmonas and water supply boreholes, respectively.

Inflows to the WDS of Thessaloniki conurbation were calculated based on available data on daily water quantities transferred from Aravissos springs, Sindos, Halkidona, Axios, and Eleousa water supply boreholes, as well as from the TWTP, provided on a monthly basis for the year 2018. Daily water quantities were also available for the outflows of P-ST35, pumping station P-3 and tank T36, as well as for supplying the industrial area of the city and EKO industry. Using simple averaging of water quantities in all months, and considering the balance of inflows and outflows in both aqueducts of the system, inflows were estimated at 5052 m^3/h for the TWTP, 4256 m^3/h for Aravissos springs and water supply boreholes of Halkidona, 731 m^3/h for the water supply boreholes of Sindos and 469 m^3/h for the water supply boreholes of Axios and Eleousa. The average daily water consumption of the conurbation was estimated at 252,217 m^3/day, while the minimum and maximum daily water consumption observed in 2018 is 216,964 m^3/day and 286,137 m^3/day, respectively. The total active volume of the system's tanks was estimated to correspond to 60.3% of the latter.

The severe structural failure of the Aravissos main pipeline in 2018, resulted in a reduction of about 100,000 m^3/day of water supplied to the city, leading to water shortage and causing emergency conditions for both the managing company, EYATH S.A., and the city itself. This serious crisis reinforced the need for a detailed and up-to-date hydraulic simulation of the city's aqueducts, creating an urgent need to formulate alternative emergency scenarios based on appropriately defined intermittent water supply schedules to distribute the available resources among the different areas of the network. Hydraulic simulation of the city's aqueducts in case of emergency conditions, such as the ones mentioned above, will contribute to an equitable management of available water quantities, but also to enquire optimal use of available water resources and operation of existing pumping stations and tanks of the WDS.

3. Materials and Methods

3.1. Estimation of Mean Water Demand in the City's Tank Zones

The hydraulic model of Thessaloniki's aqueducts includes a network of almost 136 km. The TWTP, and more specifically the tank at the outlet of the facility (T3), T36, and the facility of Kalohori (P-ST34) upstream of P-ST35 are used to define the hydraulic boundary conditions of the modeled water supply system. The hydraulic model created includes twenty-six tanks providing water to their areas of influence (tank zones), while there are seven smaller areas supplied directly by pressure pipes from pumping stations (see Figure 3). It should be noted that the city center of Thessaloniki is supplied directly from pressure pipe DN800 from P-ST35. The industrial area of Thessaloniki is supplied with almost 35,000 m^3/day, while a constant supply of 4250 m^3/day is also considered for EKO industry.

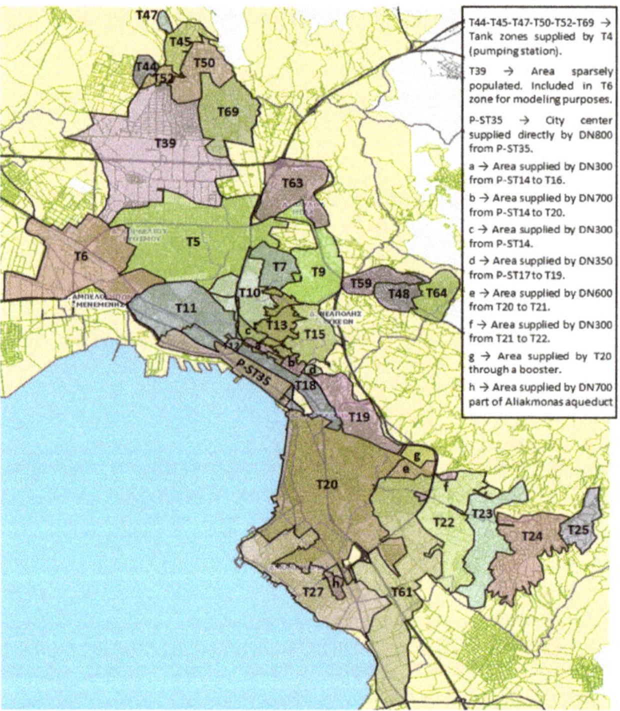

Figure 3. Tank influence areas (tank zones) in the conurbation of Thessaloniki, Greece.

To estimate water demand in all tank zones of the conurbation, both theoretical and measured data was considered in this work. The theoretical estimation of water demand relied upon using population data, as well as data on specific water consumption of the different urban planning units of Thessaloniki. Population data for all urban planning units of Thessaloniki, referring to year 2018, was obtained using information from 2011 census, as well as information from each municipality's general urban plan. The specific water consumption for each urban planning unit was assessed processing real data of water consumption, increased by about 25% to account for water losses of the network. The above resulted in estimates of specific water consumption in the conurbation of Thessaloniki ranging from 200 lit/c/day to 350 lit/c/day. The theoretical water demand in all tank zones was then assessed, after calculating the areas of all urban planning units located within the boundaries of each tank zone, estimating water demand within each such part, and finally adding up water demands of all parts of different urban planning units located in a tank zone.

To assist water demand estimation in the tank zones, water consumption data was also acquired for a total of more than 500,000 water meters installed in the conurbation of Thessaloniki. The datasets available include, among other information, the geographical coordinates of each water meter and data on its age and total water volume recorded. The water meters were first geo-allocated on the company's GIS system and mean annual water consumptions (divide total water volume recorded by age for each water meter) of all water meters belonging to the same tank zone were added. It should be noted that results of the abovementioned procedure should only be used with caution, due to the large number of uncertainties entering the estimation process, that mostly concentrate on inaccuracies in the coordinates of the water meters, location of the boundaries of the system's tank zones, water meter accuracy (especially for ageing water meters), inaccuracies in water meter age and volume records (especially for very old devices), estimation of mean water consumption ignoring variations throughout the years, and water losses in the WDS.

Final estimates of water demand in each tank zone, to be used in the hydraulic model, were assessed by also considering available information from the existing SCADA system of EYATH S.A. However, it should be noted that the aforementioned SCADA system records the operating state of the WDS only partly, due to the small number of pressure and flow meters installed, as well as the fact that water level is not monitored in all tanks of the system. For tank zones with installed flow meters at the tank's outlet pipe or downstream a pumping station, measured data were used to assess water demand (i.e., use daily flow data from the pumps' working schedule at P-ST17 to assess water demand in tank zone T19, or assume a constant rate of decay in water level of T59 to assess water demand in its tank zone). In the majority of tank zones in the study area, theoretical and measured data (i.e., daily water level variations in a tank, daily flow variations at a pumping station's pressure pipe, water balance of inflows and outflows at a subsystem of the network) were indeed combined to assess water demand, while in areas with no available information, theoretically estimated water demand was assigned to the respective tank zones.

Water demand for each tank zone, estimated as summarised above, was assigned in the hydraulic model as an outflow from each respective tank. For the sake of simplicity, water demand of some small areas supplied directly from pressure pipes (see Figure 3), was added to water demand of the upstream or downstream tank zone. Therefore, the water demand of areas d, e, and g of Figure 3 was added to water demand of tank zone T19 (zone d), and T20 (zones e and g), respectively. Tank T39 was not included in the hydraulic model of the system, due to the very low water consumption in its area of influence, while its demand was added to the one of tank zone T6. Water demands of tank zones T44, T45, T47, T50, T52, and T69 were simulated as water consumption from tank T4.

3.2. Calibration of the Hydraulic Model for the Existing Operating State of the System

The existing SCADA system provided available water level data in some of the system's tanks, as well as pressure and flow measurements downstream of the pumping stations, during a typical day. The data collected were used to calibrate the hydraulic model of the WDS, as well as to extract important information on its operating state. Hydraulic modeling of Thessaloniki's aqueducts was performed using Bentley OpenFlows WaterGEMS software [2]. Calibration of the system's hydraulic model was performed manually in EPS mode for an entire day, having as main calibration parameters the roughness coefficients of the system's pipes, and head loss characteristics of float valves and TCVs located upstream of some of the network's tanks. Hydraulic model calibration was performed dividing it in smaller and distinct parts and units, following an iterative procedure for certain hydraulic parameters to converge to measured quantities or satisfy defined constraints.

Initial, minimum, and maximum water levels in the system's tanks were acquired from EYATH S.A's installed SCADA system, where available. In tanks with no available water level measurements, minimum and maximum water levels were estimated from elevations of the tank's inlet and overflow pipes, while initial water levels were configured somehow lower than the respective maximum water levels. The basic hydraulic simulation was conducted for an entire day, starting from 6:00 am. At the starting time, water in all tanks was considered at its maximum observed level. However, in some of the key tanks of the system (P-ST17, T10, P-ST12, T20, T36), initial water level was considered lower to avoid overflow caused by the limited water demand of the early morning hours, combined with the constant supply of these tanks during the day. Operating rules for pumps and pumping stations of the system were adjusted using water levels of downstream tanks as reference points. For tanks supplied by more than one pumps, the last one turns on when water is at the minimum operating level of the tank. The latter was acquired for the majority of the system's tanks from EAYTH S.A.'s existing SCADA system. The rest of the pumps turn on 10 cm higher in a row. Correspondingly, the first pump of the pumping station turns off at the maximum operating level of the tank (provided by the SCADA system), while the others turn off 10 cm lower in a row. Therefore, each successive pump of the pumping station turns on when water demand of the respective tank zone augments, as water level in the downstream tank falls.

For the pumping group DN800 in Aravissos aqueduct, there is always one operating pump in P-ST35, while the second turns on conditional on water intake, as well as on water demand of tank zones supplied by P-ST17. Therefore, operation of the second pump of pumping group DN800 was adjusted conditional on water level in P-ST17. The pumping group DN1000 usually operates with 2–3 pumps supplying water to P-ST12 and P-ST14. At P-ST12, each of the pumping stations to T10, T11 and T13 includes two similar pumps, which operate following the general rules described above. As tank T13 is also supplied by T11, water level in T11 was also used to adjust the operation of both pumping stations to T13. At P-ST14, one pump constantly operates supplying water directly to a small area of Thessaloniki's conurbation (area c in Figure 3), a second one supplies water to T16 which is also accompanied with a throttling valve to tank T20 (the valve opens more when T16 is full), and a maximum of four other pumps transport water to T20, two of which operate constantly on a daily basis. From these four pumps to T20, two pumps were adjusted to operate conditional on the operation of the second and the third pump in P-ST35 (pumping group DN1000), respectively.

In the aqueduct of Aliakmonas river, water is pumped from T36 to T6 using two similar pumps, which operate following the general rules described above. At P-3 two similar pumps were adjusted to operate conditional on water level in T4. The regulating tank T4 transports water to the tanks of Aliakmonas aqueduct, as well as to P-ST8. Water is pumped from P-ST8 to T9 by means of two similar pumps which turn on in a row following the formulated operating rules. In times of high water demand in tank zone T9 (i.e., especially during the summer months) water could also be pumped from tank T13 through P-66 (water pumped from Aravissos aqueduct), with the existing pump serving as the last one to turn on in this sub-system (P-66 turns on when water level in T9 is minimum and turns off 20 cm below its maximum value).

Suction tanks of the city's WDS with a limited storage capacity (i.e., suction tank of P-ST14 and P-ST8) were simulated in the hydraulic model as pressure reducing valves (PRVs). Hydraulic grade setting of the valve was considered equal to water elevation in the suction tank. This transformation ensures the preservation of a constant water level in the suction tank, overcoming possible numerical instabilities during the simulation process.

Float valves (T9 to T7, T22 to T61, T20 to T27, P-ST14 to T16) were represented as isolation valves in the hydraulic model, accompanied with an additional adjustment of pipe closing or opening based on water level in the downstream tank. Head loss coefficients of isolation valves were calibrated to satisfy water demand of the respective tank zones. Opening or closing of the valve was connected to a minimum or a maximum operating level at the downstream tank, acquired from the installed SCADA system. Head loss coefficients of TCVs of Aliakmonas aqueduct were calibrated to satisfy the total daily demand of the respective tank zones. Initial values used in the calibration process were collected during field measurement campaigns (see Section 2). The main necessity to calibrate the head loss coefficients of the network's TCVs results from the fact that, if the initial values were kept, the tanks could empty or overflow after a number of days. Therefore, head loss coefficients of TCVs were calibrated, so that tank water levels at the end of the day are really close to the initial ones.

Roughness coefficients of pressure pipes were calibrated using available flow and pressure data from EYATH S.A.'s SCADA system downstream of pumps or pumping stations. For pipes with no available pressure measurements, roughness coefficients were adjusted based on pipe material and age information, conditional on satisfying daily demand of the different tank zones during the simulation interval. Hazen Williams coefficients for the system's pipes range between 60 and 150. For the majority of pipes, the Hazen Williams coefficient was fixed at 110.

3.3. Emergency Scenarios for Network Operation in Case of Failure in a Central Aqueduct

The assumption of failure in one of the central aqueducts of the studied WDS brought about different alternative scenarios to encounter problems in water distribution management. The fact that there is quite a large number of tanks currently supplied by both aqueducts (see Section 2), that could receive water from a single source during a possible crisis, significantly assisted to define water supply schedules for the system's tank zones. Therefore, a basic failure scenario was defined for each aqueduct, where the majority of tanks supplied by both aqueducts during normal operating conditions (few exceptions were allowed caused by network topology, and demand satisfaction balance between the two aqueducts) received water only from the aqueduct operating properly. High water consumption of the industrial area of Thessaloniki led to the formulation of an additional scenario that considers interrupting its water supply in case of failure in Aravissos aqueduct. It should be noted that in case of failure in Aliakmonas aqueduct, water supply of this area is automatically interrupted after the first hours of the crisis. The water supply capacity of the two aqueducts, elevation of the system's tanks, and network configuration with high and low elevation areas supplied by the aqueducts of Aliakmonas and Aravissos, respectively, and existing links between the two aqueducts not used under normal operating conditions, resulted in formulating two additional scenarios in case of failure in Aravissos aqueduct based on utilising existing interconnections.

Therefore, the majority of the alternative emergency scenarios formed relate with failure in Aravissos aqueduct. This is also attributed to the aqueduct's age (compared to Aliakmonas aqueduct), combined with its reinforced concrete material and the existence of quite a high water table in the area it crosses, as well as to the recent experience from the 2018 emergency in the city of Thessaloniki. The alternative scenarios of this work were formed to confront with: (1) a basic failure scenario at Aravissos aqueduct, (2) a scenario of failure at Aravissos aqueduct, interrupting water supply of the industrial area of Thessaloniki after verifying the failure, (3) a failure scenario at Aravissos aqueduct, activating the interconnection of the two aqueducts using an existing pipeline from T36 to P-ST34, (4) a scenario of failure at Aravissos aqueduct, activating a newly constructed DN500 pipeline connecting the two aqueducts, and (5) a basic failure scenario at Aliakmonas aqueduct cutting off water supply from the TWTP. The four failure scenarios at Aravissos aqueduct include cutting off water supply from Aravissos springs ($\approx$87,000 m^3/day), while preserving water quantities from Halkidona and Sindos boreholes ($\approx$33,000 m^3/day). Water production from the TWTP was increased to 135,000 m^3/day (from 121,000 m^3/day during normal conditions), while preserving water quantities from Axios and Eleousa boreholes entering T36. Failure at Aliakmonas aqueduct is simulated in the hydraulic model by cutting off water supply from the TWTP ($\approx$121,000 m^3/day).

Emergency scenarios for network operation in case of failure in one of the city's aqueducts were formulated to include intermittent water supply, depending on the available amount of water per aqueduct, ensuring a fair and equitable distribution of water reserves and periods of water supply interruption among all tank zones of the system. Achievement of the aforementioned goal was further assisted (see Section 4) by the development and application of scenarios (3) and (4). In all five scenarios examined, operating rules of pumping stations and tanks of both aqueducts were also investigated and adjusted accordingly to contribute to achieving the principal objectives of the study.

To formulate all alternative water supply scenarios, water demand in all different tank zones of the system was considered fixed and equal to the existing peak demand of each area (average water consumption in each tank zone multiplied by a peak factor). However, water demand during emergency conditions, like the ones described above, is thought to significantly increase, introducing high uncertainty in hydraulic simulations, as well as in the intermittent water supply schedules. In this work, the adjusted schedules were formed to keep a low number of necessary interventions in the system while managing the crisis, not only for the sake of simplicity, but also to reduce the probability of failure in other parts of the WDS during the time of repair. The basic characteristics of intermittent water supply schedules created rely heavily upon estimated water demand in each tank zone of the system, and are therefore subject to possible revisions and alterations in case of any significant change. The lack of detailed recording of the system's hydraulic behavior under normal operating conditions, coupled with a necessity to readjust the head loss characteristics of Aliakmonas' aqueduct TCVs to achieve the simulation objectives in case of emergency, increase the uncertainty of the extracted simulation results for all alternative scenarios.

During the initial hours of detected failure, and before implementing an intermittent water supply schedule, an attempt was made to use the volume of water stored in the system's tanks. Therefore, the hydraulic model of the system was adjusted to interrupt water supply of almost all tanks, until a minimum water level was attained in each of them. This minimum water level used for all emergency scenarios was considered at 0.80 m from each tank's outlet pipe, for all tanks supplied by pumps or pumping stations, except from tank T13, where a higher value of 1.95 m was preserved. For the rest of the system's tanks, the emergency water level was set at 0.50 m (unless the minimum water level during normal operating conditions was lower). For tanks supplied by more than one pump, the last one was assumed to turn on when water was at the emergency level, while the rest of the pumps were considered to turn on 10 cm higher in a row. All pumps were adjusted to turn off 20 cm higher (compared to their starting water level). Head loss characteristics of TCVs upstream of Aliakmonas' aqueduct tanks were readjusted to satisfy water demand in all studied tank zones during the water supply periods of the intermittent water supply schedule applied in each emergency scenario. All hydraulic simulations for the emergency scenarios were run in EPS mode of WaterGEMS software for a time period of 2.5 days (almost half a day to use the water stored in the system's tanks and two entire days to apply the intermittent water supply schedules).

Implementing an intermittent water supply schedule, primarily based on adjusted operating rules of the pumps, requires to control the flow downstream the system's tanks. This should be performed to ensure that water supply in each tank zone is interrupted when the respective tank empties, while when a minimum water level is assured, a peak demand of the respective tank zone will be satisfied. Since hydraulic modelling during emergency conditions is not typical in the sense of flows being controlled by demands, outflows of tanks to their tank zones were complimented with FCVs and ended up at reservoirs. The initial flow setting of each FCV was adjusted to the peak flow value of the respective tank zone. This was performed to meet the unit demand satisfaction ratio (defined as the ratio of the flow that is available to the flow that is required) of each tank zone during water supply periods. To better achieve the objectives of the simulation, FCVs of a limited number of critical tank zones were kept closed during the intervals of water supply interruption.

The main failure scenario at Aravissos aqueduct (scenario 1) was based on the assumption of water supply cut off from Aravissos springs, so that water only from water supply boreholes enters P-ST35. Emergency scenario 1 was based on an assumption of uninterrupted water supply to the industrial area of Thessaloniki, to EKO industry and to critical infrastructure, such as large hospitals in the east of the city. Tank zones supplied solely by Aravissos aqueduct under normal conditions were assumed to belong to the system of P-ST35 (tank zone T16 was also included), while Aliakmonas aqueduct was assumed to transport water to all other tank zones. Daily water demand of the two aqueducts was thus covered by almost 48% and 63%, respectively, used as a basis to form the intermittent water supply schedule. The system supplied by the DN800 pumping group was therefore subject to a six-hourly intermittent water supply schedule. Operation of the system supplied by the DN1000 pumping group was simulated to depend on the water level in P-ST35 suction tank, conditional on the necessary water quantity for the DN800 system being lower than the total inflow to P-ST35. During operating hours of the DN800 system, water supply to P-ST14 and T13 was interrupted. The latter was performed to ensure sufficient water quantities for the high elevation T15, during operating hours of the DN1000 system. During the first hours of the event, an attempt was made to use the volume of water stored in the system's tanks (water at the beginning of the simulation process was assumed to be at its lowest level observed during normal operating conditions). This time needed for the tanks of Aliakmonas aqueduct (tanks supplied by Aliakmonas aqueduct or by both aqueducts during normal conditions, except for T16) was considered fixed and equal to 12 h. After these initial 12 h interval, an intermittent water supply schedule was selected for the tanks of Aliakmonas aqueduct, cutting off water supply every 5 h for 3 h intervals. The water supply of T10 and T20 from Aravissos aqueduct, as well as the water supply of T27 from T20 were interrupted.

Emergency scenario 2 was also based on a failure detected in Aravissos aqueduct. Water supply to the industrial area of Thessaloniki was interrupted after the first 12 h of the detected failure. In this case daily water demand of the two aqueducts was covered by almost 48% and 84%, respectively. Therefore, this scenario considered an intermittent water supply schedule for Aravissos aqueduct, similar to the one presented for scenario 1. After the initial 12 h interval, an intermittent water supply schedule was also selected for the tanks of Aliakmonas aqueduct, cutting off water supply every 10 h for 2 h intervals.

Emergency scenario 3 was assumed to include an increase in water inflow to P-ST35 ($\approx$469 m^3/h) from the boreholes of Axios and Eleousa. A DN600 pipe connecting the two aqueducts was used to transport the additional water quantity to the system of P-ST35 (during normal operating conditions, water from Axios and Eleousa boreholes is transported to T36 of Aliakmonas aqueduct). Tank zones supplied solely by Aravissos aqueduct under normal conditions were assumed to belong to the system of P-ST35 (tank zones T16 and T10 were also included), while Aliakmonas aqueduct was assumed to transport water to all other tank zones. Daily water demand from both aqueducts was thus covered to almost 58%. For Aravissos aqueduct, the intermittent water supply schedule was simulated to start after the first 8 h of the crisis, including cutting off water supply almost every 4.5 h for time intervals of 3.5 h. During the time periods of continuous water supply, only one pump was set active for the pumping group DN800, while the pumps of DN1000 were simulated to turn on to preserve the water level in the suction tank of P-ST35 in the range of 1.40–2.00 m, even during the period of water supply cut off. During the periods of water supply interruption, the TCV upstream of P-ST14 was set to closed. P-ST14 was also simulated to pump water to P-ST17 (see Figure 2) through a DN300 pipe (inactive during normal operating conditions), while T20 was adjusted to receive water only from Aliakmonas aqueduct. Pumps from P-ST12 to T10 were adjusted to operate only during the periods of continuous water supply of the implemented schedule. For Aliakmonas aqueduct, the implemented intermittent water supply schedule was similar to the one followed in Aravissos aqueduct, while water supply of the industrial area of Thessaloniki, EKO industry, as well as critical infrastructure remained uninterrupted.

Emergency scenario 4 was assumed to include an interconnection of the city's aqueducts, performed with the newly constructed DN500 pipe (see Figure 4), which supplies the city center (supplied directly from DN800 of Aravissos aqueduct during normal operating conditions) with water directly from the TWTP. Tank zones supplied by the pumping group DN1000 of Aravissos aqueduct under normal conditions were assumed to belong to the system of P-ST35 (tank zone T10 and part of T16 were also included), while Aliakmonas aqueduct was assumed to transport water to all other tank zones. Daily water demand from both aqueducts was thus covered to almost 58%. For Aravissos aqueduct, the intermittent water supply schedule started after the first 12 h of the crisis, including cutting off water supply almost every 4.5 h for time intervals of 3.5 h. Adjustments similar to those of scenario 3 were performed in the hydraulic model for pumps of DN1000 in P-ST35, the TCV upstream of P-ST14, water supply of P-ST17 and T20, as well operation of pumps from P-ST12 to T10. It should be noted that for this scenario, T10 was assumed to receive water only from Aravissos aqueduct. For Aliakmonas aqueduct, most adjustments of this scenario were similar to those of scenario 3.

However, it should be emphasised that the pumping group of DN800 was supplied by both aqueducts of Aliakmonas (pipe DN500 connects the Aliakmonas main pipeline to DN800) and Aravissos (pipe DN300 from P-ST14 to P-ST17 was activated), following the applied intermittent water supply schedule for Aravissos aqueduct.

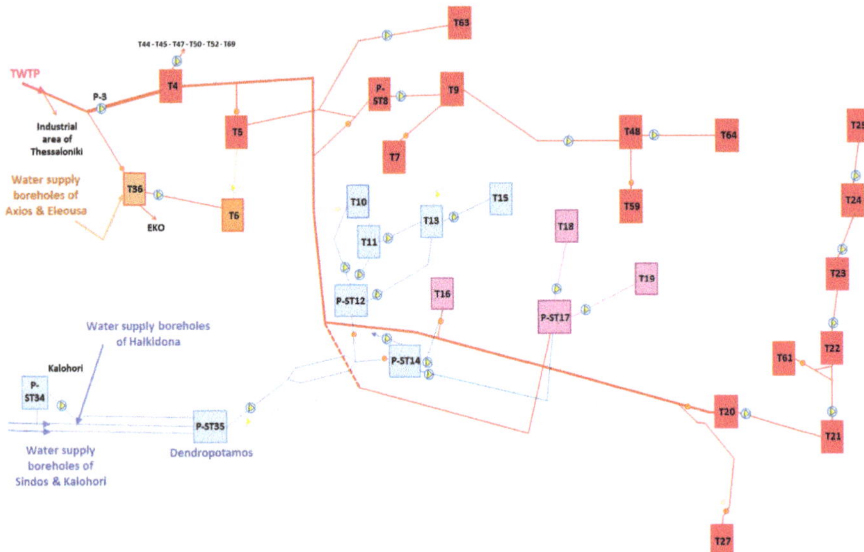

Figure 4. Schematization of Thessaloniki's aqueducts for emergency scenario 4. Colors blue, red, purple and orange represent water supplied by the aqueducts of Aravissos, Aliakmonas, Aravissos and Aliakmonas, or Aliakmonas and water supply boreholes, respectively. The red dashed line represents the newly constructed pipe DN500 activated in this scenario.

Emergency scenario 5 concerns a major failure at the aqueduct of Aliakmonas, interrupting water supply especially to high elevation areas of the city. Conditional on this failure scenario, the boreholes of Axios and Eleousa are the main water supply sources to the entire aqueduct of Aliakmonas. Water exchange between the city's aqueducts was mainly achieved, here, using an existing DN600 pipe to transport water from P-ST34 to T36 (this pipe was used in emergency scenario 3 for transporting water from Axios and Eleousa boreholes to P-ST34). The maximum capacity of the pipe reaches 600 m^3/h. It should be noted that the supply of the industrial area of Thessaloniki was assumed to continue until the water level in the TWTP downstream tank falls below 4.60 m, while when this tank empties, pumping of water to tank zones of T4 (T44, T45, T47, T50, T52, T69) stops. To create the intermittent water supply schedule for this emergency scenario, tank zones supplied solely by the aqueduct of Aliakmonas during normal conditions, were assumed to receive water through the aqueducts' central infrastructure (tank T36 and regulating tank T4), while the aqueduct of Aravissos was assumed to supply the rest of the tank zones. Considering the abovementioned, emergency management for this scenario was formed to include two distinct phases, one corresponding to the first hours of the failure (the first 19 h for the tank downstream of TWTP to empty) and one to the upcoming hours. During the first hours of the crisis, tank T36 was simulated to receive input from both the TWTP and water boreholes, supplying T6 with water. The pumping group of DN800 was simulated to operate with one or two pumps at P-ST35, to preserve the water level of P-ST17 to emergency standards. The pumping group of DN1000 was simulated to operate with 2–3 pumps in P-ST35, with the third one turning on or off conditional on the water level in the suction tank of P-ST35. During the second part of the scenario, an intermittent water supply schedule of 4.25 h continuous water supply followed by a 3.75 h cut off was implemented for both aqueducts, activating the pipe DN600 from P-ST34 to support the aqueduct of Aliakmonas, as well as a partial contribution from Aravissos aqueduct to T9 through the pump P-66. The pumping station of T6 was also activated during this stage to transport water to T5. The pumping group of DN800 was modelled to follow the intermittent water supply schedule, while the pumping group of DN1000 was adjusted to operate conditional on water level elevation in the suction tank of P-ST35. Water supply to T20 was assumed to be continuous and uninterrupted, while the pipe DN350 from T20 to T27 was kept closed. To prevent critical tanks from emptying during the water supply interruption intervals, consumption valves (FCVs in the hydraulic model) for tank zones T13 and T15 were kept closed during these intervals.

4. Results

Figure 5 presents final estimates of mean water demand assessed for all tank zones of Thessaloniki's WDS. These estimates are compared with the respective values assessed theoretically -design demands- (see Section 3.1). It should be noted that total water demands resulting from adding mean annual consumptions of all water meters located in each tank zone of the system (see Section 3.1) were assessed in the majority of cases lower than final tank zone demand estimates presented in Figure 5. For the most representative tank zones of the WDS these differences range in the interval 25–40%, possibly approximating water losses in the different areas of the network. However, the respective water demands are not included in Figure 5, mainly due to significant uncertainties in the estimation process.

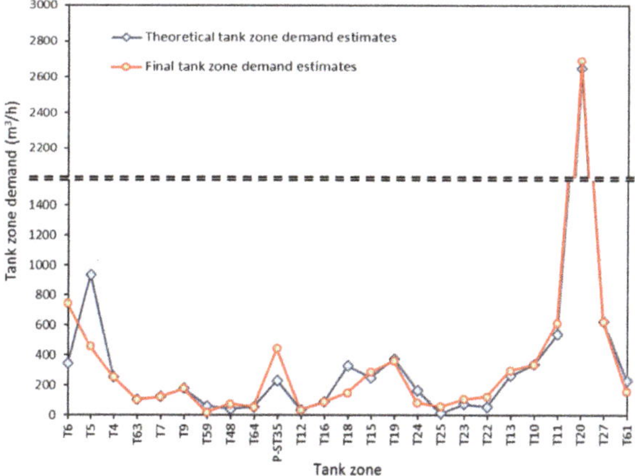

Figure 5. Final and theoretical mean water demands in tank zones of Thessaloniki conurbation.

Theoretical and final water demand estimates are really close in most of the tank zones shown in Figure 5. Higher differences are observed for tank zones T6 and T5 (in the western part of the city), for tank zone T18, as well as for the city center (tank zone P-ST35). Figure 6 presents daily variation of the coefficient of hourly water demand considered in this work. Separate variation schemes of the coefficient were considered for tank zones located in the city (majority of tank zones) and in its suburbs (i.e., tank zones T48, T59, T64, T23, T24 and T25). Daily variation of water demand for urban tank zones is based on analyzed measured data from a previous study in Thessaloniki city [36]. The selected scheme corresponds to tank zone T11, which serves quite a large area close to the city center. For the suburban tank zones, a daily variation scheme from existing literature [37–40] was used in this work.

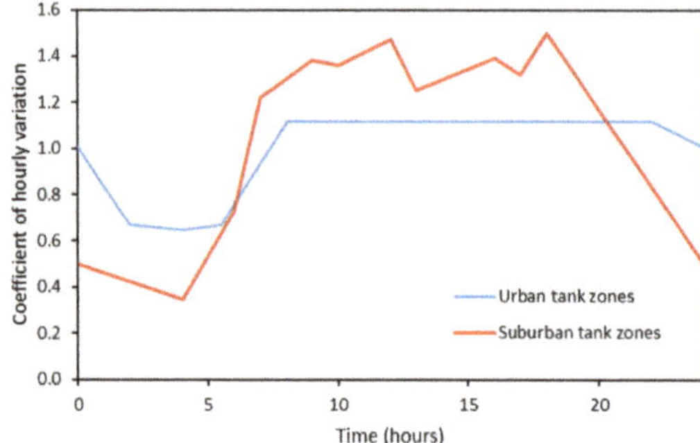

Figure 6. Coefficient of hourly water demand for tank zones of Thessaloniki conurbation.

Hydraulic simulation of the system was performed in OpenFlows WaterGEMS software [2] following the methodology presented in Section 3.2 Mean water demand in the city's tank zones (Figure 5), as well as hourly water demand variations within the day (Figure 6), were used in the hydraulic model to simulate the system's response to normal operating conditions. Figure 7 presents the daily operation time for all pumping stations or pumps of Thessaloniki's aqueducts, estimated from the hydraulic simulation of the system during normal operating conditions. Pump operation was adjusted following the methodology described in Section 3.2 of this work. The total working hours of the system's pumps exceed 650 h per day, while the total amount of water pumped is estimated at about 380,000 m^3/day.

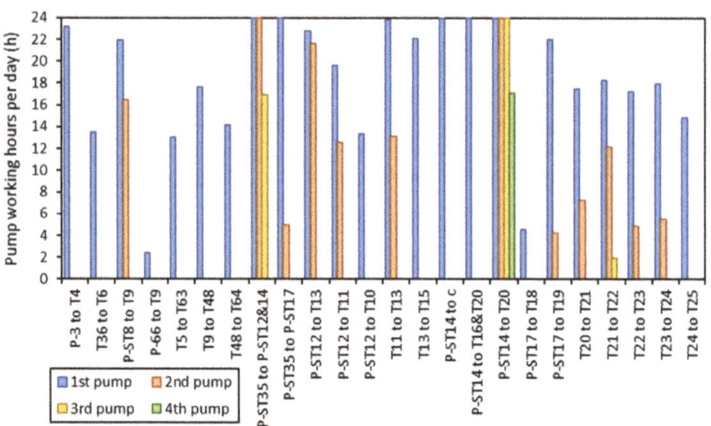

Figure 7. Daily working hours of pumps in the WDS of Thessaloniki city.

Results of the simulation reveal that the majority of the system's pumps operate at medium to quite high efficiencies. However, it should be noted that pump efficiency significantly changes throughout the years, and therefore the use of manufacturers' efficiency data for the system's pumps (most pumps of the system are really old) can sometimes lead to unreliable results especially in terms of energy consumption and greenhouse gas emissions (GHG) of the entire system. Daily energy usage is really high for the pump P-3 downstream of TWTP, the two pumps of pumping group DN1000 to P-ST12 and P-ST14 and the one pump of pumping group DN800 to P-ST17, with the abovementioned pumps using more than 45% of the energy of the entire system on a daily basis. Energy usage is also high for pumps from T13 to T15, T36 to T6, T8 to T9, P-ST14 to T20, and P-ST17 to T19. It should be noted that optimization of pump operation for the WDS of Thessaloniki to minimize operating costs and GHG emissions of the system is a subject for future research.

During emergency conditions of the system (see Section 3.3), water demand in all tank zones is assumed to attain its maximum value during the entire period of study. A peak factor of λ = 1.117 (see Figure 6) is used to estimate water demand in the tank zones of the system (both urban and suburban). This peak factor is only applied to tank zones with domestic consumers, while water demand in industrial areas of the system is kept constant. Considering the conservative case in which a failure at one of the city's aqueducts happens when water in the system's tanks is at its lowest level during normal operating conditions, Figure 8 presents the time needed for each tank to reach its emergency water level. From Figure 8, it is evident that water stored in some of the tanks of Aliakmonas aqueduct (i.e., T9, T48, T59, T64) could satisfy water demand in their tank zones for many hours during a water crisis period. In Aravissos aqueduct, water demand of tank zone T18 could also be satisfied for more than 20 h at the beginning of the emergency period.

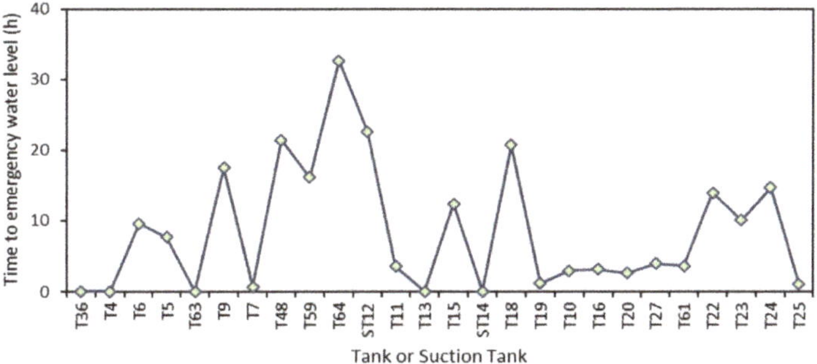

Figure 8. Time to reach emergency water level in the system's tanks after failure in one of the city's aqueducts.

For emergency scenario 1, different intermittent water supply schedules were created for the city's aqueducts (see Section 3.3). Figure 9 presents total flow at the two central pipes (DN800 and DN1000) of Aravissos aqueduct for the simulation period of 2.5 days. It can be noted that water supply of DN800 strictly follows the intermittent water supply schedule for Aravissos aqueduct (six-hourly water supply followed by six-hourly cut off). For the pumping group of DN800, one pump constantly operates at P-ST35 during water supply periods, while the second one turns on conditional on water elevation in P-ST17. Water supply of pipe DN1000 is continuously interrupted, due to adjustments performed in the hydraulic model (turning a pump on or off depending on water level in P-ST35).

Figure 10 presents water consumption from the pumping group of DN800, including water supply of the city's center, and water supply of tank zones T18 and T19. Water supply of the city center from pressure pipe DN800 justifies its direct response to the imposed intermittent schedule. For tank zones T18 and T19, water supply seems to continue during the interruption periods. Especially for tank zone T18, prolongation of water supply periods could be mainly attributed to large storage volume of T18, compared to water demand in the respective tank zone. For the pumping group of DN1000, not shown here for the sake of brevity, tank zone T11 seems to receive water during the assigned six-hourly periods with very short interruptions caused by water supply adjustments/controls in the hydraulic model, water supply of tank zone T13 strictly follows the imposed schedule due to adjustments performed directly at the FCV at the tank's outlet, while water supply of tank zone T15 seems to be problematic. The high-altitude tank T15 seems to supply its tank zone with water during the first 12 h of the failure until water level in the tank reaches its emergency value, while in the next two days water supply of the tank zone is limited. The latter is mainly attributed to its dependence to water level elevation in the upstream tank T13, as well as the capacity of the existing pumps (from T13 to T15), detected to be marginal compared to the tank zone's water needs.

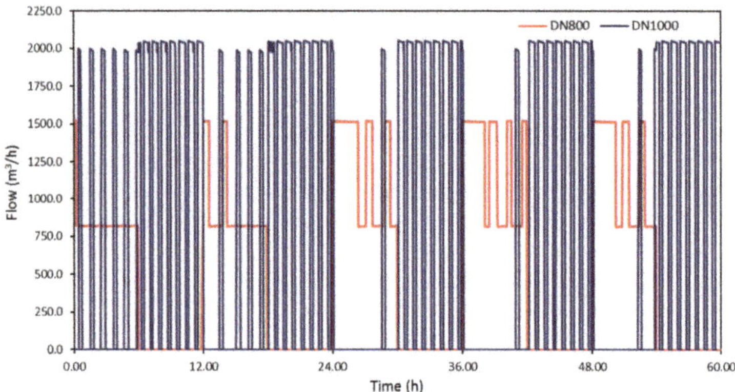

Figure 9. Total flow at pipes DN800 and DN1000 during the simulation period for emergency scenario 1.

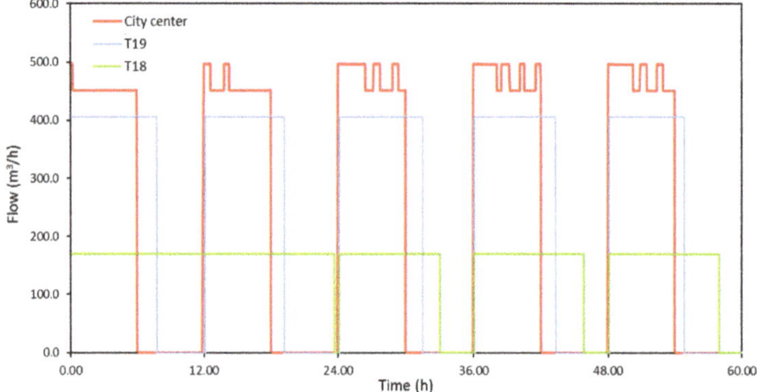

Figure 10. Water supply of the city center and tank zones T19 and T18 during the simulation period for emergency scenario 1.

Figure 11 presents water supply in some indicative tank zones of Aliakmonas aqueduct (T4, T5, T9, T22, T24, T25). For tank zones T4, T5 and T6 (not shown here), the objectives of the intermittent water supply schedule for Aliakmonas aqueduct are quite fully accomplished. However, in the majority of the aqueduct's tanks, it is rather difficult to strictly follow the designed schedule, mainly due to network topology issues and the relative comparison between the storage volume of each tank and water consumption/demand in its tank zone. For the series tanks T22, T61, T23, T24 in the eastern part of Thessaloniki city, it is quite evident that water supply periods are longer than the ones assigned in the schedule, mainly caused by pumping schedules applied in this scenario. However, the water supply of the high-altitude tank T25 seems to be highly problematic.

Figure 12 presents daily water supply (in hours and in %) for all tank zones of the system during a typical day (application of the intermittent water supply schedule has already started) for emergency scenario 1. Specified duration of water supply periods based on the applied intermittent water supply schedule for both aqueducts is also included as a straight line (blue and red color for the aqueducts of Aravissos and Aliakmonas, respectively). The green bars correspond to areas of the system, where water supply is considered uninterrupted (the industrial area of Thessaloniki, EKO industry, and critical infrastructure of the city).

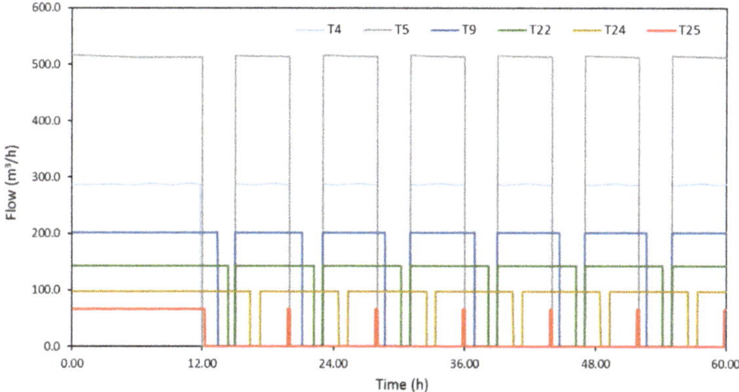

Figure 11. Water supply of tank zones T4, T5, T9, T22, T24, and T25 during the simulation period for emergency scenario 1.

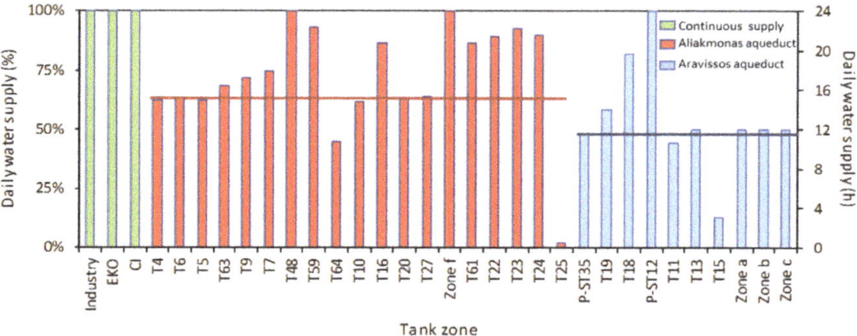

Figure 12. Daily water supply (in hours and %) for all tank zones of Thessaloniki's aqueducts for emergency scenario 1.

Results for emergency scenario 1 show: (a) a total inability of the system to supply the high-altitude tanks of the network (T15 and T25), which was also verified for the event of failure at the aqueduct of Aravissos in 2018; (b) good coverage of daily water needs for the majority of tanks supplied by both aqueducts, mainly the ones connected in series (mostly due to the configured operational rules of the pumps upstream of the tanks), as well as the majority of tanks in the suburbs except for the high-altitude tank of Acropolis (T64) supplied with water for less than 11 h a day; (c) partial coverage (60–75% of the daily intervals) of the rest of the tank zones in the aqueduct of Aliakmonas (tank zones fully or partially supplied by the aqueduct of Aliakmonas under normal operating conditions of the system i.e., T20 and T27); and (d) approximately 50% (48–58% of the daily intervals) coverage of the daily water demand for the aqueduct of Aravissos (excluding T18, which covers the daily water needs of its area/zone for more than 19 h, due to its high storage volume compared to the area's water demand).

For emergency scenario 2 (Figure 13), which considers interrupting water supply to the industrial area of Thessaloniki after the first 12 h of a failure in Aravissos aqueduct, an increase is estimated for water quantities and water supply time periods for areas of Aliakmonas aqueduct. However, no significant benefit is expected for the areas supplied by the aqueduct of Aravissos. In particular, this scenario achieves coverage of daily water demand of areas in the aqueduct of Aliakmonas to a proportion exceeding 83% (for the area supplied by the high-altitude T25 this proportion reaches almost 50%), while for the aqueduct of Aravissos, results do not differ from those extracted for emergency scenario 1. Figure 13 includes a comparison of daily water supply extracted from scenario 2 with the respective results for scenario 1.

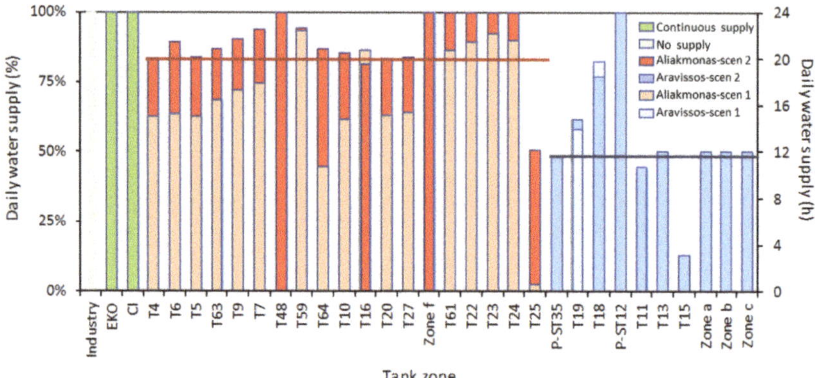

Figure 13. Daily water supply (in hours and %) for all tank zones of Thessaloniki's aqueducts for emergency scenario 2 compared to results produced for emergency scenario 1.

For emergency scenario 3, which activates an interconnection of the two aqueducts through P-ST34, a common intermittent water supply schedule was assigned to both aqueducts. Figure 14 presents total flow at the two central pipes (DN800 and DN1000) of Aravissos aqueduct for the simulation period of 56 h (two days plus 8 h to use water stored in the system's tanks). It can be noted that water supply of DN800 strictly follows the applied intermittent water supply schedule. For the pumping group of DN800, only one pump constantly operates at P-ST35 during water supply periods. Water supply of pipe DN1000 is continuously interrupted due to adjustments performed in the hydraulic model (turning a pump on or off depending on water level in P-ST35). Water supply of the city center from pressure pipe DN800 strictly follows the imposed intermittent schedule, tank zone T19 seems to not follow the implemented schedule with shorter periods of water supply after the first 1.5 days of the simulation period, while for tank zone T18 water supply periods present a time delay with respect to the implemented schedule (after the first 1.5 days where water supply of the tank zone is continuous). For the pumping group of DN1000, not shown here for the sake of brevity, tank zone T13 strictly follows the imposed schedule, while water supply of tank zones T10 and T11 seems to last longer than 4.5 h. The main difference in the system's response with respect to emergency scenario 1 has to do with water supply of the high-altitude tank zone T15, which seems to improve. Figure 15 presents water supply in some indicative tank zones of Aliakmonas aqueduct (T5, T6, T9, T22, T24, T25). The objectives of the intermittent water supply schedule for Aliakmonas aqueduct are quite fully accomplished for tank zones T5 and T6 (and for T4), while water supply periods are longer than those assigned in the schedule for tank zone T9, as well as tank zones in the eastern part of the city (e.g., T22, T23, and T24). Water supply of the high-altitude T25 remains highly problematic. Figure 16 presents daily water supply (in hours as well as %) for all tank zones of the system during a typical day (application of the intermittent water supply schedule has already started) for emergency scenario 3. Specified duration of water supply periods based on the applied intermittent water supply schedule for emergency scenario 3 is also included as a straight line. The industrial area of Thessaloniki, EKO industry, and critical infrastructure in the northeastern part of the city are considered to receive continuous water supply during the crisis. Figure 16 includes a comparison of daily water supply extracted from emergency scenario 3 (pink bars) with the respective results for emergency scenario 1 (blue bars).

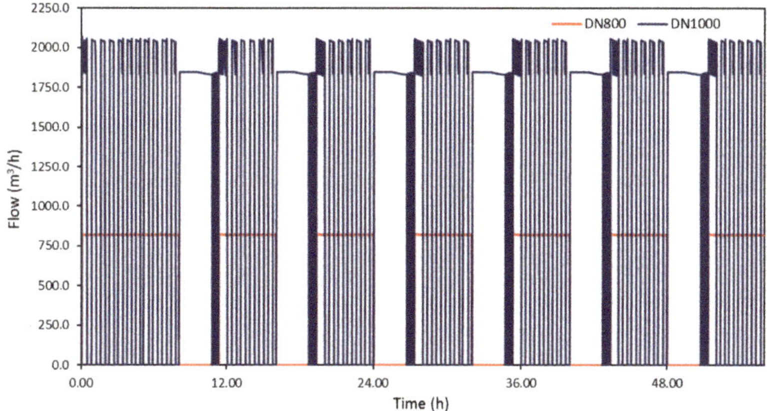

Figure 14. Total flow at pipes DN800 and DN1000 during the simulation period for emergency scenario 3.

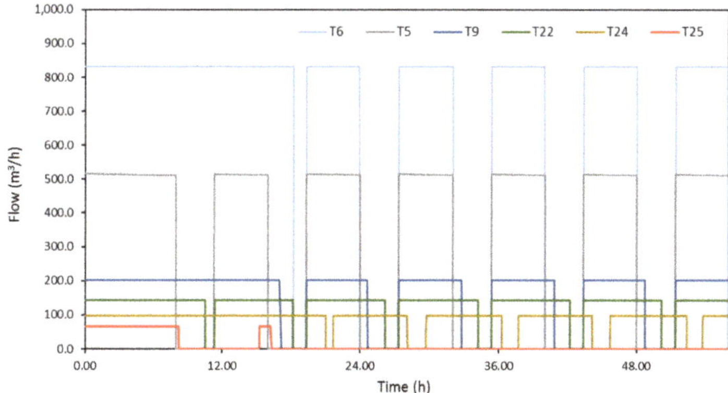

Figure 15. Water supply of tank zones T6, T5, T9, T22, T24, and T25 during the simulation period for emergency scenario 3.

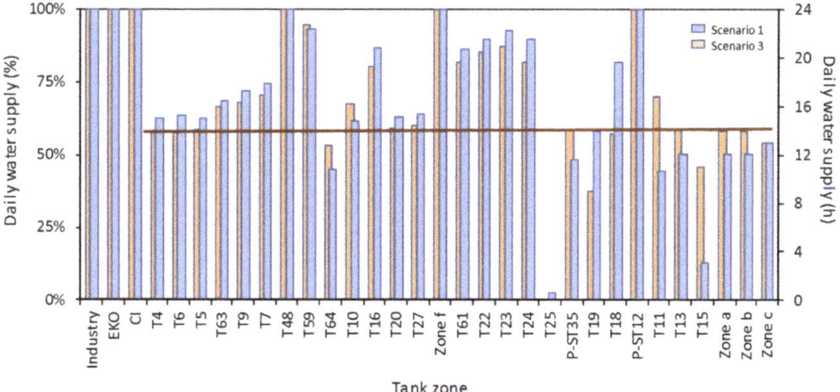

Figure 16. Daily water supply (in hours and %) for all tank zones of Thessaloniki's aqueducts for emergency scenarios 1 and 3. The straight line corresponds to assigned water supply duration for emergency scenario 3 following the designed intermittent water supply schedule.

Results for emergency scenario 3 show that, compared to emergency scenario 1, there is: (a) an increase in the duration of water supply for some tank zones of Aravissos aqueduct (this is not the case for tank zones T18 and T19 where a decrease of 30% and 36% respectively, is anticipated); (b) an increase in duration of water supply for the city center; (c) a significant improvement in water supply of high-altitude tank zone T15, while this is not the case for T25 in the eastern part of the city; and (d) a significant improvement in water supply of large tank zones, such as T11. For tank zones T11 and T13, the duration of daily water supply periods increases more than 56% and 16% respectively, while the respective increase for the city center reaches 21.5%. For the high-altitude tank zone T15, total duration of daily water supply increases more than 2.5 times.

For emergency scenario 4, where the two aqueducts are fully connected by means of a DN500 pipeline, a common intermittent water supply schedule, similar to the one applied for emergency scenario 3, was assigned for the entire system. Figure 17 presents total flow at the city center, supplied directly from Aliakmonas aqueduct through DN500, as well as water supply of tank zones T18 and T19 during the 2.5 days (two days plus 12 h to use storage volume of tanks) simulation period. Water supply of the city center strictly follows the applied intermittent water supply schedule, tank zone T19 is supplied for somehow longer periods than those defined, while water supply of tank zone T18 covers the most of the simulation period. The water supply of tank zone T11 lasts longer than assigned, tank zone T10 follows the intermittent water supply schedule with very short interruptions therein, tank zone T13 strictly follows the intermittent water supply schedule, while the performance of the high-altitude tank zone T15 is improved (compared to emergency scenario 1). Some tank zones in Aliakmonas aqueduct perform quite similar to emergency scenario 3, while some differences arise for tank zones T5, T6, T7, T9, T10, and T64. Figure 18 presents daily water supply (in hours as well as %) for all tank zones of the system during a typical day (application of the intermittent water supply schedule has already started) for emergency scenario 4 compared to those of emergency scenario 3. Specified duration of water supply periods based on the applied intermittent water supply schedule is also included as a straight line. Continuous water supply was assumed for the industrial area of Thessaloniki, EKO industry, and critical infrastructure in the northeastern part of the city.

Results for emergency scenario 4 show that compared to emergency scenario 3 there is: (a) an increase in duration of daily water supply for the high-altitude tank zone T15 (almost 15%), however it is still not possible to satisfy the objectives set by the implemented schedule; (b) a significant increase in daily water supply of tank zones T18 and T19, reaching almost 21 h/day for T18 (increase more than 50% for T18 and 65% for T19); (c) a significant decrease in daily water supply duration for tank zones T10 (almost 19%) and T64 (more than 50%); (d) a decrease in daily water supply of tank zones T5, T7, and T9 (up to 9%). Therefore, considering the current water production from TWTP, emergency scenario 4 offers benefits compared to emergency scenario 3, especially for tank zones of pumping group DN800, while water supply of the high-altitude tank zone T25 remains problematic. However, it should be noted that both scenarios 3 and 4, offer significant advantages compared to emergency scenario 1, where no connection of the two aqueducts is considered.

Emergency scenario 5 considers responding to a failure at the central aqueduct of Aliakmonas. Hydraulic simulation for this scenario was performed for a period of 67 h (two days plus 19 h to use water stored in system's tanks) and resulted in continuous water supply of the industrial area of Thessaloniki and pumping of water to tank zones of T4 (T44, T45, T47, T50, T52, T69) for the first 5.25 h and 18.15 h of the crisis, respectively. After this time, there is no alternative way to supply these areas with water. Tank zones T5 and T6 strictly follow the water supply schedule implemented, due to turning on or off FCVs downstream these areas' head tanks, while water supply of tank zones T9 and T63 lasts longer than assigned. Figure 19 presents total flow at the two central pipes (DN800 and DN1000) of Aravissos aqueduct for the simulation period. Water flow in DN800 strictly follows the intermittent water supply schedule applied, with one or two pumps operating in P-ST35. Water flow in DN1000 is continuous, using two or even three pumps, conditional on water level in the suction tank of P-ST35. Water supply in tank zone T11 lasts longer than assigned by the schedule, T10 presents interruptions during water supply periods, while tank zones T13 and T15 present difficulties in following the defined schedule.

Figure 20 presents total flow for the city center, as well as for tank zones T18 and T19 (pumping group DN800). The city center strictly follows the applied schedule, while water supply in tank zone T19 and especially in T18 exceeds the assigned duration. Tank zones in the eastern part of the city (T22, T23, T24, T61) do not follow the intermittent water supply schedule, while water is pumped to the high-altitude tank zone T25 only during the first 19 h of the crisis (Figure 21). Figure 22 presents daily water supply (in hours as well as %) for all tank zones of the system during a typical day (application of the intermittent water supply schedule has already started) for emergency scenario 5. Blue bars correspond to tank zones supplied directly from P-ST35, while red bars correspond to tank zones supplied by water transferred from Aravissos to Aliakmonas aqueduct through DN600 from P-ST34, as well as from T13 to T9 through P-66 (see Figure 2). The specified duration of daily water supply periods based on the applied intermittent water supply schedule is also included as a straight line.

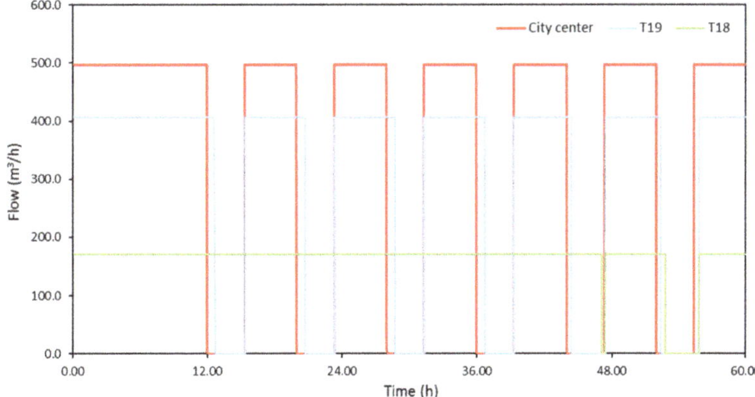

Figure 17. Water supply of the city center and tank zones T19 and T18 during the simulation period for emergency scenario 4.

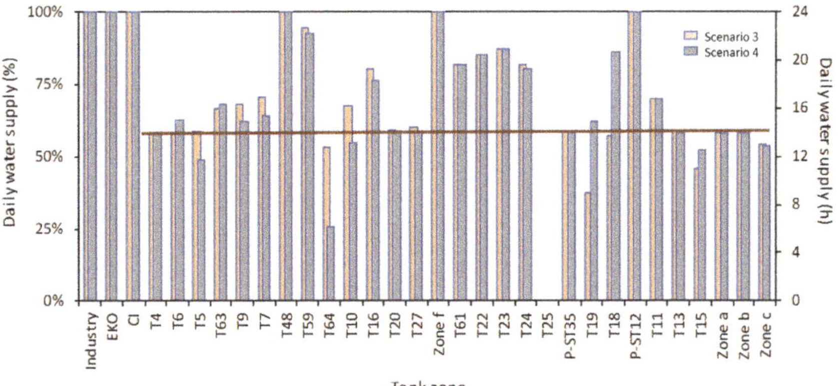

Figure 18. Daily water supply (in hours and %) for all tank zones of Thessaloniki's aqueducts for emergency scenarios 3 and 4. The straight line corresponds to assigned water supply duration for both emergency scenarios following their intermittent water supply schedules.

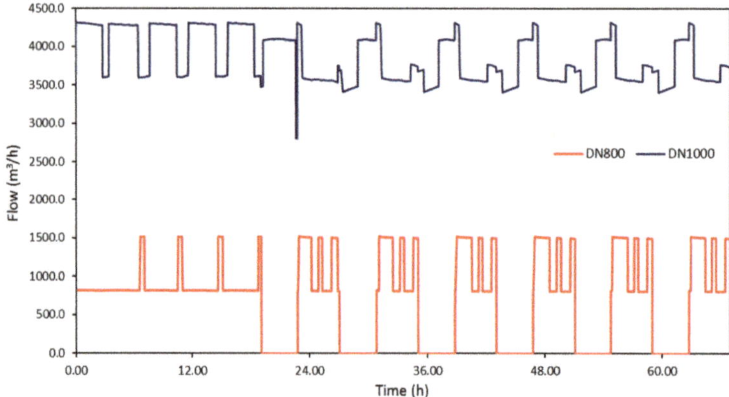

Figure 19. Total flow at pipes DN800 and DN1000 during the simulation period for emergency scenario 5.

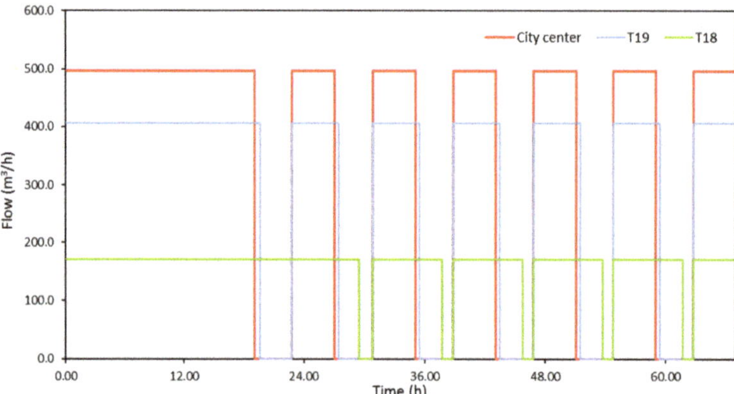

Figure 20. Water supply of the city center and tank zones T19 and T18 during the simulation period for emergency scenario 5.

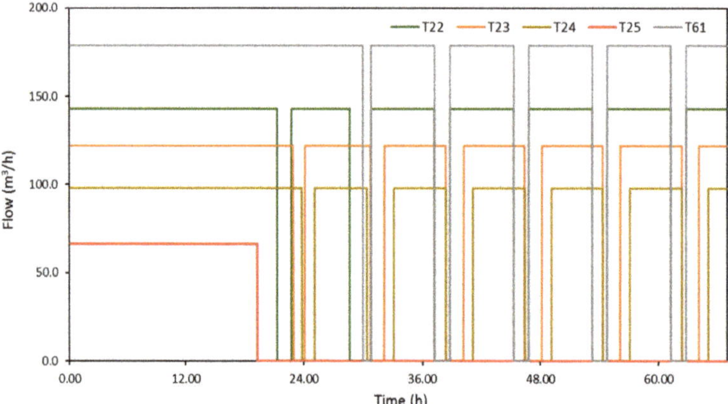

Figure 21. Water supply of tank zones T22, T23, T24, T25 and T61 during the simulation period for emergency scenario 5.

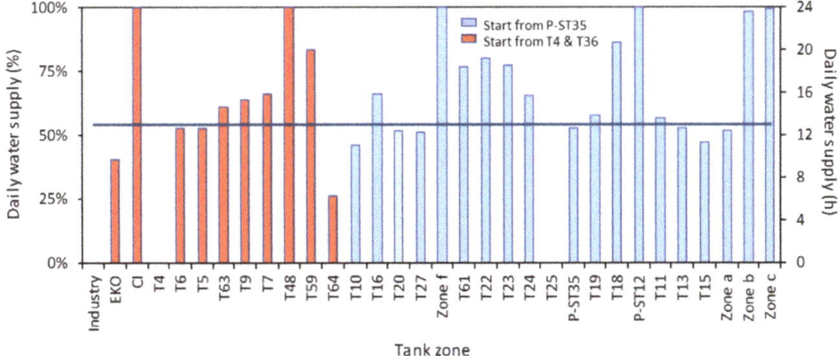

Figure 22. Daily water supply (in hours and %) for all tank zones of Thessaloniki's aqueducts for emergency scenario 5. The straight line corresponds to assigned water supply duration following the intermittent water supply schedule.

It is obvious from Figure 22 that the water supply of the industrial area of Thessaloniki, as well as of tank zones supplied by pumped water from T4, stops completely after the first hours of the crisis. Water can be transported to EKO industry for less than 12 h during a typical emergency day, while water supply of the city center stays close to estimates for emergency scenario 1 (almost 10% higher). Water supply of tank zone T25 is highly problematic, while for tank zones T10, T15, T20, T27 and T64, daily duration of water supply is estimated lower than the defined threshold. Based on results extracted for emergency scenarios 1 and 5, it can be concluded that a possible failure in Aliakmonas aqueduct is more difficult to manage, compared to a failure in Aravissos aqueduct. For the large majority of tank zones (apart from tank zones P-ST35, T11, T13, T15 and T18), duration of water supply periods is estimated higher for a tentative failure in Aravissos aqueduct (emergency scenario 1), compared to a respective failure in Aliakmonas aqueduct (emergency scenario 5), while in the latter an intermittent water supply schedule seems more difficult to follow.

5. Conclusions

In this work, a simple, comprehensive, and practical methodology was introduced to cope with water shortage conditions in the aqueducts of quite a large and complex WDS, with a limited existing SCADA system. Results of the simulations for both normal and abnormal conditions of the WDS suggest that the procedures followed are reliable and robust, while keeping the computational time low. The methodology presented assesses water demands in the different areas of the system combining both theoretical and measured data, while calibration of the WDS was performed in EPS mode following an iterative procedure for certain hydraulic parameters to converge to measured quantities or satisfy defined constraints. For emergency conditions of the network, the analysis conducted deviates from PDA, mainly because of large uncertainties in extracting typical pressure-dependent demand curves for the different tank zones and in defining thresholds for both the minimum and desirable head of the different demand nodes, ensuring a unit demand satisfaction ratio during water supply periods in each tank zone.

A hydraulic simulation model for the aqueducts of Thessaloniki city in Greece was developed to model the current operating state of the network, as well as its response to emergency conditions resulting from a major failure in one of them. Hydraulic simulations were performed using OpenFlows WaterGEMS software in EPS mode, including estimating water demand in all areas of the conurbation and calibrating the hydraulic model under both normal (pipe roughness coefficients, head loss characteristics of TCVs) and abnormal conditions (head loss characteristics of TCVs). Failure in the city's aqueducts was confronted with the development and hydraulic simulation of five emergency scenarios of network operation, two of which consider possible interconnections of the studied aqueducts. These scenarios, which include appropriately defined intermittent water supply schedules for the aqueducts, were created on a basis of a fair and equitable management of water among the different areas of the city, also assuming a small number of interventions/operations during the crisis. The simulations revealed quite a satisfactory compliance of the system's operation with the defined schedules, and an improved management of limited water reserves in some areas of the WDS when considering interconnections of the city's aqueducts.

Hydraulic simulations for all emergency scenarios showed that the main objectives set for managing such adverse conditions in the city of Thessaloniki were satisfactorily met. However, it is evident that it was not always possible to strictly comply with the designed intermittent water supply schedules. A better distribution of the available water resources in each such case would require lots of additional interventions in the city's water system performed during the crisis. The application of operating rules to consumer valves (FCVs at the outlet of tanks) following the intermittent water supply schedule of each emergency scenario (close the valve during water supply cut off periods) could significantly assist in fulfilling the simulation objectives and perfectly meeting the assigned requirements at the cost of a huge increase in the number of necessary operations and handlings from the water supply company's (EYATH S.A.) working staff, which could in turn significantly increase vulnerability of the system's elements to sudden and severe failures.

For emergency scenario 1, including response to a major failure in Aravissos aqueduct, the hydraulic simulation verified that the objective of applying a six-hourly intermittent water supply schedule for the tanks of Aravissos aqueduct was not always feasible. The water schedule was fully satisfied for the city center, while some tanks of the network downstream of pumping stations continued to supply their tank zones during water supply interruption periods. Water needs in tank zones fully or partially supplied by the aqueduct of Aliakmonas during normal operating conditions were satisfied up to almost 60–75% (higher proportions were observed for the series tanks in the eastern part of the city), while the water supply of high-altitude tank zones in both aqueducts was proven highly problematic. For emergency scenario 2, which considers interrupting water supply to the industrial area of Thessaloniki after the first 12 h of a major failure in the aqueduct of Aravissos, an increase was estimated for water quantities and water supply time periods for areas of Aliakmonas aqueduct, compared to emergency scenario 1, however without achieving any benefit for the areas supplied by the aqueduct of Aravissos.

Emergency scenarios 3 and 4 were formulated to include responses to a major failure in Aravissos aqueduct considering possible interconnections of the city's central aqueducts. Both scenarios improved the performance of the entire water system (compared to emergency scenarios 1 and 2), assisting in fulfilling the principal objectives of the hydraulic simulation under abnormal conditions of the network. Major improvements can be summarised to an increase in the duration of water supply for large tank zones of Aravissos aqueduct, including high-altitude tank zones of great historical interest.

For emergency scenario 5, including response to a major failure in Aliakmonas aqueduct, hydraulic simulation has proven that application of a defined intermittent water supply schedule in such a case is more difficult, compared to a tentative failure in Aravissos aqueduct (emergency scenarios 1–4), rendering such conditions more adverse to handle and to manage. For this scenario, hydraulic simulation has indicated interruption of water supply of the industrial area of Thessaloniki, as well as of some high-altitude areas of Aliakmonas aqueduct, after the first hours of the crisis, while for some large tank zones of the city daily duration of water supply was estimated lower than the one assigned by the intermittent water supply schedule.

In the present work, basic reasoning and rules for constructing emergency management scenarios for the operation of aqueducts in a large city characterised by a dual water supply system were introduced and analysed further. Such scenarios were defined to include intermittent water supply schedules for the different areas of the network, considering that it still operates with its normal topology and elements but based on a different distribution of available water reserves between the system's sources, completely interrupting water supply to certain areas of the network if necessary, or taking advantage of existing connections between the aqueducts and using system elements or infrastructure not utilised under normal operating conditions. The definition of case specific emergency water levels in the system's tanks, estimation of average time to reach them (for each aqueduct) when the emergency starts, definition of water supply and cut off periods based on available input and demands per aqueduct, determination of pump operating rules, and deactivation of network elements to support the defined schedules are some basic concepts and steps used when formulating the abovementioned emergency scenarios.

All emergency scenarios of this work have not been tested and validated in practice, and represent theoretical ones, considering the purely theoretical methodology followed and the basic assumptions (see Section 3) it was based on. The above emergency scenarios simulate the direct response of the network to the different operations and handlings performed by the water supply company, which differs from the actual response of the system due to inertia conditions, handling of valves, air-fixing in the network, and pipe conductivity, among others. It is also noted that it was not possible to simulate the exact response of the system's aqueducts in the week between 28 March and 1 April 2018, mainly due to the lack of very detailed information on hydraulic quantities of the entire system and on operations performed during the crisis, and also due to the very large number of changes in the implemented management measures during the same period. Although this work presents theoretical scenarios, the defined intermittent water supply schedules follow some of the basic directions and operations performed during the actual crisis management, resulting in simulated operating conditions resembling those that occurred during the real event.

It is worth noting that simulations for emergency scenarios considered in this work were based on an assumption of increased water consumption in all tank zones of the studied system. During such conditions, as well as in cases of planned or unplanned system maintenance or power failure at pump stations, the daily water demand is usually not based on the notions of diurnal variation, but on a maximum water quantity collected during the supply hours. This assumption was also verified during the event of failure in the Aravissos aqueduct in 2018. Therefore, in this work water demand for all emergency scenarios was considered constant during the assigned time intervals of water supply in each tank zone and equal to the peak demand during normal operating conditions of the system. However, it should be noted that estimation of water demand during emergency conditions introduces a large amount of uncertainty in the formulated intermittent water supply schedules and in the simulation results in general, and requires further investigation and analysis.

A strict compliance with the initially defined water supply schedules will be supported by installing a detailed SCADA system in the network for full monitoring of its hydraulic parameters, as well as by replacing manually controlled TCVs with modern electronic ones controlled by a central server. The above are also expected to significantly contribute to further development of the city's aqueducts' hydraulic model, enabling a more reliable dynamic calibration of the model thereby improving its ability to predict the system's behavior under different operating scenarios. They are also expected to improve the managing company's response to similar conditions, and to significantly contribute to the development of newly improved intermittent water supply schedules for available water reserves. Following the installation of a detailed SCADA system at the watermains and tanks of the WDS the aforementioned alternative scenarios are also expected to contribute to the development of a decision support system (DSS) for emergency management.

Once the installation of a fully equipped SCADA system is completed, it will be connected to the hydraulic model of the WDS to create a digital twin. Such an approach will allow to quickly adjust daily demands of the different areas/tank zones of the entire system and will assist in controlling and simulating pipe breaks or fire response, creating customized element alerts. Creation of a digital twin will provide more reliable estimates of hydraulic quantities and can be used even for forecasting hydraulic emergencies in the future.

In the present work, the operation of the studied WDS was reproduced for a typical daily interval, calibrating the numerical model to represent the existing hydraulic parameters of the system. However, the analysis conducted did not consider operational costs (economical, environmental) of the system. Future work will focus on optimum pump scheduling to significantly decrease energy consumption costs and GHG emissions of the system's pumping stations. It is therefore a subject for future work to ensure a trade-off between energy usage, energy consumption costs, and pumping station efficiency, and to present a methodology for assessing optimum daily pumping schedules.

Author Contributions: Conceptualization, A.M. and P.G.; methodology, A.M. and P.G.; software, P.G.; formal analysis, P.G.; investigation, P.G., D.S., and A.S.; resources, A.M., P.G., and P.S.; data curation, P.G., D.S., and A.S.; writing—Original draft preparation, P.G.; writing—Review and editing, A.M., P.G., D.S., A.S., and P.S.; visualization, A.M., P.G., D.S., A.S., and P.S.; supervision, A.M. All authors have read and agree to the published version of the manuscript.

Funding: This research received no external funding.

Acknowledgments: The support of the Departments of Water Supply Installations and Water Supply Networks, and of the Division of Geoinformatics, Topography & Hydraulic Modeling of EYATH S.A. in acquiring all necessary data and information for all hydraulic simulations is highly acknowledged.

Conflicts of Interest: The authors declare no conflict of interest.

References

1. Rossman, L.A. *EPANET 2: Users' Manual*; National Risk Management Research Laboratory, Office of Research and Development, United Sates Environmental Protection Agency (EPA): Cincinnati, OH, USA, 2000.

2. Bentley Systems Incorporated. *Bentley WaterGEMS V8 XM Edition User's Guide*; Haestad Methods Solution Center: Watertown, CT, USA, 2007.

3. Yu-Kun, H.; Chun-Hui, Z.; Yu-Chung, H. A GIS-based water distribution model for Zhengzhou city, China. *Water Supply* **2011**, *11*, 497–503. [CrossRef]

4. Avesani, D.; Righetti, M.; Righetti, D.; Bertola, P. The extension of EPANET source code to simulate unsteady flow in water distribution networks with variable head tanks. *J. Hydroinform.* **2012**, *14*, 960–973. [CrossRef]

5. Jiang, B.; Zhang, F.; Gao, J.; Zhao, H. Building a water distribution network hydraulic model by using WaterGEMS. *ICPTT* **2012**, 453–461. [CrossRef]

6. Elsheikh, M.A.; Saleh, H.I.; Rashwan, I.M.; El-Samadoni, M. Hydraulic modelling of water supply distribution for improving its quantity and quality. *Sustain. Environ. Res.* **2013**, *23*, 403–411.

7. Okeya, I.; Kapelan, Z.; Hutton, C.; Naga, D. Online modelling of water distribution system using data assimilation. *Procedia Eng.* **2014**, *70*, 1261–1270. [CrossRef]

8. Alves, Z.; Muranho, J.; Albuquerque, T.; Ferreira, A.; Albuquerque, M. Water distribution network's modeling and calibration. A case study based on scarce inventory data. *Procedia Eng.* **2014**, *70*, 31–40. [CrossRef]

9. Sonaje, N.P.; Joshi, M.G. A review of modeling and application of water distribution networks (WDN) softwares. *Int. J. Tech. Res. Appl.* **2015**, *3*, 174–178.

10. Kara, S.; Karadirek, I.E.; Muhammetoglu, A.; Muhammetoglu, H. Hydraulic modeling of a water distribution network in a tourism area with highly varying characteristics. *Procedia Eng.* **2016**, *162*, 521–529. [CrossRef]

11. Świtnicka, K.; Suchorab, P.; Beata, K. The optimisation of a water distribution system using Bentley WaterGEMS software. In *ITM Web of Conferences*; EDP Sciences: Lublin, Poland, 2017; Volume 15, p. 3009.

12. Chatzivasili, S.; Papadimitriou, K.; Kanakoudis, V. Optimizing the formation of DMAs in a water distribution network through advanced modelling. *Water* **2019**, *11*, 278. [CrossRef]

13. Li, P.-H.; Kao, J.-J. Segment-based vulnerability analysis system for a water distribution network. *Civ. Eng. Environ. Syst.* **2008**, *25*, 41–58. [CrossRef]

14. Tabesh, M.; Yekta, A.H.A.; Burrows, R. An integrated model to evaluate losses in water distribution systems. *Water Resour. Manag.* **2008**, *23*, 477–492. [CrossRef]

15. Tabesh, M.; Jamasb, M.; Moeini, R. Calibration of water distribution hydraulic models: A comparison between pressure dependent and demand driven analyses. *Urban Water J.* **2011**, *8*, 93–102. [CrossRef]

16. Islam, M.S.; Sadiq, R.; Rodriguez, M.; Najjaran, H.; Hoorfar, M. Reliability assessment for water supply systems under uncertainties. *J. Water Resour. Plan. Manag.* **2014**, *140*, 468–479. [CrossRef]

17. Liserra, T.; Maglionico, M.; Ciriello, V.; Di Federico, V. Evaluation of reliability indicators for WDNs with demand-driven and pressure-driven models. *Water Resour. Manag.* **2014**, *28*, 1201–1217. [CrossRef]

18. Muranho, J.; Ferreira, A.; Sousa, J.; Gomes, A.J.; Marques, A.S. Pressure-dependent demand and leakage modelling with an EPANET extension—WaterNetGen. *Procedia Eng.* **2014**, *89*, 632–639. [CrossRef]

19. Gheisi, A.; Forsyth, M.; Naser, G. Water distribution systems reliability: A review of research literature. *J. Water Resour. Plan. Manag.* **2016**, *142*. [CrossRef]

20. Ingeduld, P.; Pradhan, A.; Svitak, Z.; Terrai, A. Modelling intermittent water supply systems with EPANET. In Proceedings of the Water Distribution Systems Analysis Symposium, Cincinnati, OH, USA, 27–30 August 2008.

21. Wu, Z.Y.; Walski, T.M. Pressure dependent hydraulic modelling for water distribution systems under abnormal conditions. In Proceedings of the IWA World Water Congress, Beijing, China, 10–14 September 2006.

22. Morosini, A.F.; Caruso, O.; Veltri, P.; Costanzo, F. Water distribution network management in emergency conditions. *Procedia Eng.* **2015**, *119*, 908–917. [CrossRef]

23. Wu, Z.Y.; Wang, R.H.; Walski, T.; Yang, S.Y.; Bowdler, D.; Baggett, C.C. Efficient pressure dependent demand model for large water distribution system analysis. In Proceedings of the 8th Annual Water Distribution Systems Analysis Symposium, Cincinnati, OH, USA, 27–30 August 2008.

24. Sayyed, M.A.H.A.; Gupta, R.; Tanyimboh, T.T. Noniterative application of EPANET for pressure dependent modelling of water distribution systems. *Water Resour. Manag.* **2015**, *29*, 3227–3242. [CrossRef]

25. Ang, W.K.; Jowitt, P.W. Solution for water distribution systems under pressure-deficient conditions. *J. Water Resour. Plan. Manag.* **2006**, *132*, 175–182. [CrossRef]

26. Babu, K.; Mohan, S. Extended period simulation for pressure-deficient water distribution network. *J. Comput. Civ. Eng.* **2012**, *26*, 498–505. [CrossRef]

27. Gorev, N.B.; Kodzhespirova, I.F. Noniterative implementation of pressure-dependent demands using the hydraulic analysis engine of EPANET 2. *Water Resour. Manag.* **2013**, *27*, 3623–3630. [CrossRef]

28. Lee, H.M.; Yoo, D.G.; Kim, J.H.; Kang, D. Hydraulic simulation techniques for water distribution networks to treat pressure deficient conditions. *J. Water Resour. Plan. Manag.* **2016**, *142*. [CrossRef]

29. Di Nardo, A.; Di Natale, M.; Giudicianni, C.; Greco, R.; Santonastaso, G.F. Complex network and fractal theory for the assessment of water distribution network resilience to pipe failures. *Water Sci. Technol. Water Supply* **2017**, *18*, 767–777. [CrossRef]

30. Batish, R. A new approach to the design of intermittent water supply networks. In Proceedings of the World Water and Environmental Resources Congress, Philadelphia, PA, USA, 23–26 June 2003.

31. Fontanazza, C.; Freni, G.; La Loggia, G. Analysis of intermittent supply systems in water scarcity conditions and evaluation of the resource distribution equity indices. *Des. Nat. III Comp. Des. Nat. Sci. Eng.* **2007**, *103*. [CrossRef]

32. Vairavamoorthy, K.; Gorantiwar, S.D.; Mohan, S. Intermittent water supply under water scarcity situations. *Water Int.* **2007**, *32*, 121–132. [CrossRef]

33. Soltanjalili, M.-J.; Haddad, O.B.; Mariño, M.A. Operating water distribution networks during water shortage conditions using hedging and intermittent water supply concepts. *J. Water Resour. Plan. Manag.* **2013**, *139*, 644–659. [CrossRef]

34. Agathokleous, A.; Christodoulou, S.E. Vulnerability of urban water distribution networks under intermittent water supply operations. *Water Resour. Manag.* **2016**, *30*, 4731–4750. [CrossRef]

35. EYATH S.A. *Master Plan for Water Management and Distribution in EYATH SA Area of Responsibility*; Corporate Report; EYATH S.A.: Thessaloniki, Greece, 2016. (In Greek)

36. EYATH S.A. *Analysis and Management of Thessaloniki Water Supply Networks and Leak Control Monitoring and Reduction*; Corporate Report; EYATH S.A.: Thessaloniki, Greece, 2002. (In Greek)

37. Tsakiris, G. *Hydraulic Works, Design and Management*; Symmetry Publications: Athens, Greece, 2006. (In Greek)

38. Balacco, G.; Carbonara, A.; Gioia, A.; Iacobellis, V.; Piccinni, A.F. Evaluation of peak water demand factors in Puglia (Southern Italy). *Water* **2017**, *9*, 96. [CrossRef]

39. Balacco, G.; Carbonara, A.; Gioia, A.; Iacobellis, V.; Piccinni, A.F. Investigation of peak water consumption variability at local scale in Puglia (Southern Italy). *Multidiscip. Digit. Publ. Inst. Proc.* **2018**, *2*, 674. [CrossRef]

40. Gato-Trinidad, S.; Gan, K. Characterizing maximum residential water demand. *Urban Water* **2012**, *1*, 15–24. [CrossRef]

Article

A Bi-Objective Approach for Optimizing the Installation of PATs in Systems of Transmission Mains

Enrico Creaco [1,*], Giacomo Galuppini [1], Alberto Campisano [2], Carlo Ciaponi [1] and Giuseppe Pezzinga [2,*]

[1] Dipartimento di Ingegneria Civile e Architettura, Università degli Studi di Pavia, Via Ferrata 3, 27100 Pavia, Italy; giacomo.galuppini01@universitadipavia.it (G.G.); ciaponi@unipv.it (C.C.)
[2] Dipartimento di Ingegneria Civile e Architettura, Università degli Studi di Catania, Via Santa Sofia 64, 95123 Catania, Italy; alberto.campisano@unict.it
* Correspondence: creaco@unipv.it (E.C.); giuseppe.pezzinga@unict.it (G.P.); Tel.: +39-(0)382-98-5317 (E.C.)

Received: 11 December 2019; Accepted: 21 January 2020; Published: 23 January 2020

Abstract: This paper proposes the bi-objective optimization for the installation of pumps operating as turbines (PATs) in systems of transmission mains, which typically operate at steady flow conditions to cater to tanks in the service of water distribution networks. The methodology aims to find optimal solutions in the trade-off between installation costs and generated hydropower, which are to be minimized and maximized, respectively. While the bi-objective optimization is carried out by means of a genetic algorithm, an inner optimization sub-algorithm provides for the regulation of PAT settings. The applications concerned a real Italian case study, made up of nine systems of transmission mains. The methodology proved able to thoroughly explore the trade-off between the two objective functions, offering solutions able to recover hydropower up to 83 KW. In each system considered, the optimal solutions obtained were postprocessed in terms of long-life net profit. Due to the large geodesic elevation variations available in the case study, this analysis showed that, in all systems, the optimal solution with the highest value of generated hydropower was the most profitable under usual economic scenarios, with payback periods always lower than 3 years.

Keywords: water distribution networks; transmission mains; pump as turbine; energy recovery; hydropower; multi-objective; optimization

---

1. Introduction

The contribution of the water industry is not negligible in the total demand of energy in the world [1]. Within this contribution, a major role is played by the extraction and treatment of source water [1], mainly due to the operation of pumping stations. To reduce this component of energy demand, water utilities are more and more often implementing practices to reduce wastes of water in distribution networks [2]. On the other hand, attempts have recently been made to recover energy from water networks and thus to compensate part of the intensive energy consumption associated with water supply. Energy recover can be accomplished by installing devices such as hydro-turbines [3–7] or pumps operating in reverse mode, also called pumps as turbines (PATs) [8–12], within pressurized pipes. Though featuring lower efficiencies than hydro-turbines, PATs feature much lower costs, thus offering themselves as a very cost-effective solution for energy recovery from water distribution networks.

The analysis of the scientific and grey literature concerning the use of hydro-turbines and PATs for energy recovery shows that research efforts have been made mainly in three different directions. The first direction is concerned with the analysis of the performance of the pump operating in reverse-mode, in terms of head-discharge and efficiency-discharge relationships. In fact, although acknowledging the possibility of using their devices in turbine mode, manufacturers of pumps do not usually provide the performance relationships. Therefore, several works in recent years were dedicated to analyzing, experimentally (e.g., [13,14]) or numerically through computational fluid dynamics (e.g., [15,16]), how the head-discharge and efficiency-discharge relationships of pumps relate to the relationships of the same devices operated in reverse mode, as well as to setting up procedures to derive the latter from the former (e.g., [17,18]).

The second research direction concerns the modeling, operation, and regulation of energy recovery devices in pressurized water networks in an attempt to both generate hydropower and reduce service pressure. In fact, the hydraulic regulation uses a bypass line and control valves to regulate flow, such that the PAT operates at or near the best efficiency point (BEP) [10]. Furthermore, the electric regulation of the PAT may be implemented by connecting the PAT to an inverter [11,12] to adjust the impeller rotational speed as a function of inflow and available head drop. Real time control algorithms can be applied to dynamically adjust the settings of PAT, control valves, and inverter under time varying operational conditions [19,20].

The third main research direction, to which the present paper belongs, concerns the optimization of sites and settings of energy recovery devices in pressurized pipes. In this context, energy production can also be optimized in combination with service pressure to help mitigating leakage in pressurized water networks. In this context, Giugni et al. [21] showed the application of a genetic algorithm to optimize location and settings. These authors considered the following two simplifying assumptions:

- the number of energy recovery devices is a surrogate for their cost;
- these devices have unitary efficiency (that is, their performance curves are neglected).

Additionally, the subsequent works by Corcoran et al. [22], Coelho and Andrade-Campos [23], and Fernández García and McNabola [24] use the number of devices as a surrogate for their total cost in the optimization phase, although removing the simplifying assumption of unitary efficiency made in [21]. Notably, Corcoran et al. [22] compared the effectiveness of three different kinds of algorithms, namely nonlinear programming, mixed integer nonlinear programming, and evolutionary optimization, in the maximization of produced hydropower. Coelho and Andrade-Campos [23] proposed a two-step optimization, which was carried out in a decoupled way. The first step concerned identification of site locations, which was performed using an optimization algorithm tailored to identifying the best locations of pressure reducing valves. A second manual step followed, aiming to identify the turbine models that could be installed at each site. Finally, a financial analysis was carried out to look at installation cost and at income associated with the sale of produced hydropower. Fernández García and Mc Nabola [24] proposed a nonlinear programming algorithm based on the sequential addition of energy recovery devices along the water paths present in a water distribution network. Device installation was iterated until there was no more significant power to be recovered from the network. A financial analysis was also carried out by the two authors to evaluate installation cost, annual income associated with PAT operation, and investment payback period. Unlike the works in [21–24], the works of Fecarotta and McNabola [25] and Tricarico et al. [26] proposed considering the installation cost in the context of the optimization, rather than the number of energy recovery devices. In fact, Fecarotta and McNabola [25] proposed an objective function combining installation costs and annual net income associated with device operation to be used inside a nonlinear mixed integer programming algorithm. The more recent work by Tricarico et al. [26] used a similar objective function into the context of a multi-objective optimization, leading to the identification of optimal sites for energy recovery devices. However, the device settings were not optimized in the methodology proposed in [26].

In this paper, a new multi-objective optimization algorithm is proposed for the installation of energy recovery devices in pressurized water networks, with the main focus on the installation of PATs in systems of transmission mains. As compared to work in [26], the novel algorithm enables optimization of both installation sites and settings. The algorithm is applied with two objective functions, namely the total installation cost of the devices and the generated hydropower, to be minimized and maximized, respectively. The Pareto fronts obtained with the algorithm lend themselves to being postprocessed in various economic scenarios to help decision makers in the choice of the ultimate solution.

In the following sections, first the methodology is described, followed by the applications to a real case study.

2. Methodology

2.1. PAT Installation

The methodology presented in this paper can be applied to systems of transmission mains that connect water sources to tanks feeding water distribution networks (Figure 1a). These systems are typically branched and operate under steady flow conditions over the whole day. While water distribution tanks are fed by transmission mains with a daily constant water discharge, their storage enables balancing the peaks of network consumption. In existing systems of transmission mains, the maximum water discharges depend only on the available heads at sources and tanks. In some cases, the values are larger than required or limited by source water availability. Therefore, the desired water discharges are enforced by valve regulation, which dissipates energy surpluses.

The redundant energy can be recovered through installation of rotary machines for electric power generation. The choice of number, sites, and models of machines represents an optimization problem aimed at minimizing installation costs and at maximizing the produced hydropower, under the hydraulic constraint of desired water discharges at mains, of fixed heads at source and tank nodes, and of desired heads at internal nodes. It is worth noting that the optimal solutions may not use the whole surplus of energy. In this case, the remaining surplus is dissipated through regulation valves.

In this work, the layout shown in Figure 1 was considered for PAT installation at the generic system main. The PAT was installed in-line with the system main inside a vault. Actually, the asynchronous PAT motor was connected to an inverter, which was in turn connected to the electrical grid (Figure 1b). The presence of the inverter enabled modification of the speed setting of the PAT and, therefore, the maximization of its production of hydropower for each daily value of water discharge in the main. Along each main of the system, a regulation valve was assumed to be present outside the vault to regulate flows as a function of fixed head values at system tanks and of desired head values at inner system nodes. Indeed, while the PAT was regulated to maximize the produced hydropower, some residual head surplus could be dissipated in the regulation valve. At a generic value of the PAT rotation speed, let Q (m^3/s) be the water discharge in the PAT. As a result of the PAT curves of performance, the PAT produces a head drop Y_{PAT} (m) and features a certain efficiency η (-), based on its performance curves. The hydropower P_{PAT} (KW) generated by the PAT is

$$P_{PAT} = \frac{\rho\, g\, Q\, Y_{PAT}\, \eta}{1000},\qquad(1)$$

where ρ = 1000 kg/m^3 is water density, and g = 9.81 m/s^2 is the gravity acceleration.

Finally, let Y_v (m) be the head drop across the regulation valve assumed present along the main line outside of the PAT vault, and let J (-) be the energy friction slope in the main, evaluated by means of the Strickler formula. The friction-related head loss in the main is equal to $J\,L$, where L (m) is the length of the main.

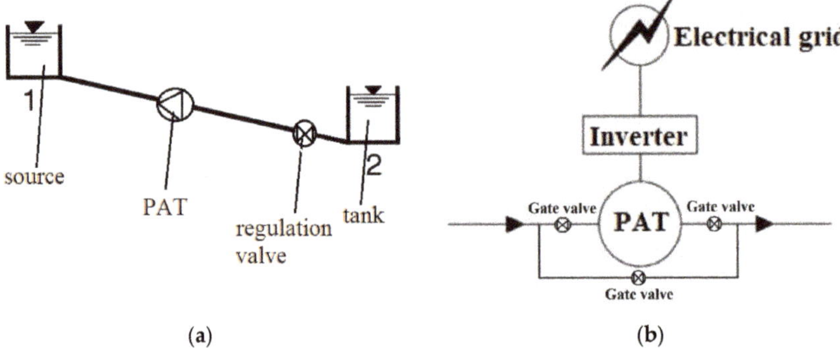

Figure 1. Example of transmission main (**a**). Layout of pump as turbine (PAT) installation, adapted from [11] (**b**).

2.2. Optimization Algorithm

The multi-objective genetic algorithm NSGAII [27] was used in this work to search for solutions in the trade-off between two objective functions, namely the total installation cost $C_{inst,tot}$ (€) of the PATs (to be minimized) and the total hydropower $P_{PAT,tot}$ (KW) generated (to be maximized) in each system of transmission mains. While the presence of the inverter enables PAT operation to adjust to yearly variable values of the daily constant water discharge, a yearly average value of the water discharge was considered in the optimizations of the present work.

The first objective function $C_{inst,tot}$ is calculated as the sum of the installation costs of the PAT at the various mains, i.e.,

$$C_{inst,tot} = \sum_{i=1}^{n_m} C_{inst,i} \tag{2}$$

where $C_{inst,i}$ (€) is the cost of the PAT installed in the i-th of the n_m mains. In the case of no PAT installation suggested by the optimizer at the generic main, $C_{inst,i} = 0$.

Indeed, the total installation cost should include other issues such as the costs associated with the approaching of the electrical grid to the PAT site in the case when the grid is not available at close sites.

The second objective function $P_{PAT,tot}$ is calculated as the sum of the hydropower contributions at each main, i.e.,

$$P_{PAT,tot} = \sum_{i=1}^{n_m} P_{PAT,i} \tag{3}$$

where $P_{PAT,i}$ is the hydropower generated in the i-th of the n_m mains. In the case of no PAT installation suggested by the optimizer at the generic main, $P_{PAT,i} = 0$.

Inside NSGAII, each individual is encoded in a set of n_m genes, i.e., decisional variables. Specifically, the i-th of the n_m genes represents the ID of the PAT installed in the i-th of the n_m system mains.

For each solution (set of PATs installed in the mains) proposed by NSGAII, an inner optimization is performed through an embedded single objective genetic algorithm (sub-algorithm) to search for the PAT revolution speeds that might maximize the total hydropower generation in the system. Starting from the curves $H(Q)$ and $P(Q)$ provided by the manufacturer, the corresponding curves $H(Q)$ and $P(Q)$ curves for the PAT were estimated using the methodology proposed in [13]. In this methodology, first the BEP of a PAT is estimated starting from the BEP of the pump operated in direct mode. Then, the performance of the PAT in terms of $H(Q)$ and $P(Q)$ away from the BEP is obtained through polynomial equations.

For each solution proposed by NSGAII and for each set of PAT speeds proposed by the sub-algorithm, a hydraulic simulator based on the global gradient algorithm [28] is used to search for feasible values of Y_v (i.e, $Y_v \geq 0$) for the regulation valves present at the system mains, while respecting the minimum desired heads at the inner nodes and the fixed heads at the system tanks.

The flowchart of instructions carried out for each individual in NSGAII is shown in Figure 2. First, the PATs are located in the system based on individual genes. Then, the total cost $C_{inst,tot}$ of the PATs installed in the system, which is the first objective function of the optimization, is calculated. Then, a sub-algorithm is applied to the PATs for estimating PAT speeds. Then, the total power $P_{PAT,tot}$ generated by the system is calculated, with a penalization applied to infeasible solutions (i.e., solutions with negative value(s) of Y_v).

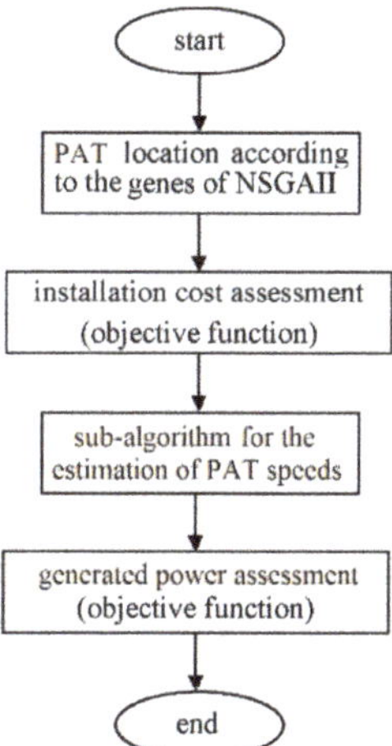

Figure 2. Flow chart of the instructions to carry out for each individual of NSGAII.

2.3. Economic Postprocessing

The optimization described in the previous subsection produces a Pareto front of optimal solutions, each of which features its own distribution of PATs in the system mains, in the trade-off between $C_{inst,tot}$ and $P_{PAT,tot}$. For the choice of the ultimate solution in the Pareto front, an economic criterion can be applied. In this context, the net profit C_p (€) of each solution can be calculated as follows:

$$C_P = C_{sale} - C_{inst,tot} - C_{add} \tag{4}$$

where C_{sale} (€) is the gross profit associated with the sale of hydropower, converted into value at the initial time. This variable can be evaluated as

$$C_{sale} = \sum_{i=1}^{T} \frac{8760\, c\, P_{PAT,tot}}{(1+r)^i} \tag{5}$$

where c (€/KW h), T (years), and r (-) are the unit selling price of the electrical energy produced, the useful life of the PAT, and the discount rate for price conversion, respectively. In Equation (5), the numerical value 8760 is the number of hours in a year. C_{add} (€) represents potential additional costs associated with the PAT (e.g., maintenance costs). Finally, the solution with the highest C_p value in the Pareto front can be selected as the ultimate solution.

Another meaningful postprocessing variable is the payback period, i.e., the amount of time T_b (years) it takes to recover the cost of the investment. While authors in [23,24] used a simplified calculation of T_b, which did not consider the decrease in the net present value as a function of the selling year, T_b was calculated here iteratively, by searching for the lowest value of T for which C_p is larger than 0 in Equation (4).

3. Applications

3.1. Case Study

The described methodology was applied to the system of networks operated by Azienda Consorziale Servizi Etnei (ACOSET), the major water distribution manager in Catania, Italy. The system was already analyzed in [3] and is characterized by transmission mains that cater to twenty urban centers situated at the foot of the volcano Etna. The main sources for these mains are the Ciapparazzo extraction tunnel, the Maniace source, and numerous ACOSET and private wells (Figure 3).

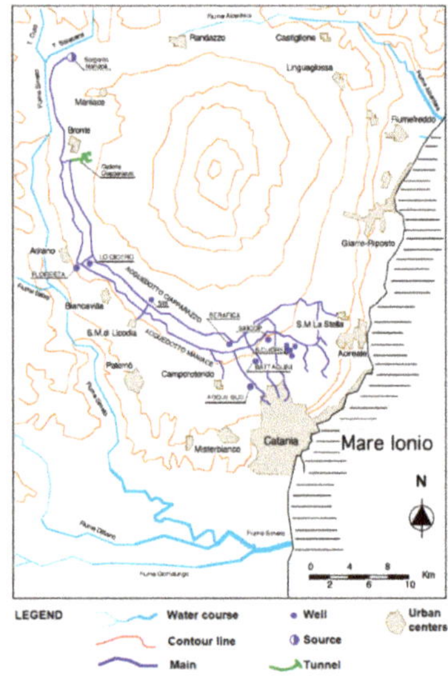

Figure 3. Planview of the water mains; adapted from Pezzinga and Tosto [3].

For the calculations in this paper, nine separate systems of water transmission mains (features reported in Table 1) were considered, in which no PATs are currently present, namely, the systems feeding Ciapparazzo, Adrano, Gravina, Mascalucia, S. Gregorio, S. M. La Stella, Maniace, Camporotondo, and S. Giovanni Galermo. As an example, the qualitative elevation layout of the Ciapparazzo system is shown in Figure 4.

Table 1. Features of the systems of transmission mains.

System	Main	Upstream Node	$i_{ts,un}$ (0/1)	Downstream Node	$i_{ts,dn}$ (0/1)	Q (m³/s)	H_{un} (m)	H_{dn} (m)	L (m)	D (m)	k (m$^{1/3}$/s)
	1	1	1	2	1	0.4	757	733	10,300	0.800	75
	2	2	1	3	0	0.4	733	719	4090	0.800	75
	3	3	0	4	0	0.325	719	716	1700	0.800	75
	4	4	0	5	0	0.37	716	715	590	0.800	75
	5	5	0	6	1	0.008	715	595	1000	0.200	75
	6	6	1	7	1	0.008	595	478	1600	0.200	75
	7	5	0	8	0	0.362	715	695	3730	0.700	75
Ciapparazzo	8	8	0	9	1	0.04	695	667	30	0.200	75
	9	8	0	10	0	0.322	695	695	4710	0.700	75
	10	10	0	11	1	0.092	695	666	170	0.400	75
	11	10	0	12	0	0.23	695	665	2300	0.700	75
	12	12	0	13	0	0.27	665	605	550	0.700	75
	13	13	0	14	1	0.02	605	493	2500	0.150	75
	14	13	0	15	0	0.25	605	670	1380	0.600	75
	15	15	0	16	1	0.171	670	683	3790	0.400	75

Table 1. *Cont.*

System	Main	Upstream Node	$i_{ts,un}$ (0/1)	Downstream Node	$i_{ts,dn}$ (0/1)	Q (m³/s)	H_{un} (m)	H_{dn} (m)	L (m)	D (m)	k (m$^{1/3}$/s)
Adrano	1	1	1	2	1	0.075	733	705	250	0.600	75
	2	2	1	3	1	0.061	705	619	1150	0.600	75
Gravina	1	1	1	2	0	0.024	435	410	900	0.150	75
	2	2	0	3	1	0.005	410	382	600	0.080	75
	3	2	0	4	1	0.019	410	322	1900	0.150	75
Mascalucia	1	1	1	2	0	0.049	665	583	1900	0.250	75
	2	2	0	3	1	0.007	583	577	350	0.150	75
	3	2	0	4	0	0.042	583	483	2100	0.250	75
	4	4	0	5	1	0.018	483	484	900	0.150	75
S Gregorio	1	1	1	2	0	0.038	476	398	1400	0.200	75
	2	2	0	3	0	0.019	398	303	1600	0.150	75
	3	3	0	4	1	0.01	303	389	600	0.125	75
	4	3	0	5	1	0.01	303	347	1600	0.125	75
S. M. La Stella	1	1	1	2	0	0.01	683	569	3600	0.125	75
	2	2	0	3	1	0.005	569	426	3100	0.125	75
Maniace	1	1	1	2	1	0.055	650	645	1678	0.450	90
	2	2	1	3	1	0.055	645	641	3690	0.450	90
	3	3	1	4	1	0.055	641	639	1737	0.450	90
	4	4	1	5	1	0.055	639	638	1710	0.450	90
	5	5	1	6	1	0.055	638	635	1650	0.450	90
	6	6	1	7	1	0.095	635	631	2796	0.450	90
	7	7	1	8	1	0.095	631	628	2836	0.450	90
	8	8	1	9	1	0.095	628	625	2043	0.450	90
	9	9	1	10	1	0.095	625	623	1928	0.450	90
	10	10	1	11	0	0.095	623	620	1456	0.450	90
	11	11	0	12	1	0.057	620	619	1456	0.400	90
	12	12	1	13	1	0.057	619	617	1945	0.400	90
	13	13	1	14	1	0.057	617	613	3116	0.400	90
	14	14	1	15	1	0.057	613	610	2365	0.400	90
	15	15	1	16	1	0.008	610	476	1900	0.250	90
	16	15	1	17	1	0.049	610	607	2567	0.400	90
	17	17	1	18	1	0.049	607	604	2873	0.400	90
	18	18	1	19	1	0.049	604	602	28.8	0.400	90
	19	19	1	20	0	0.049	602	600	1412	0.350	90
	20	20	0	21	1	0.02	600	599	50	0.200	90
	21	20	0	22	1	0.029	600	597	1412	0.350	90
	22	22	1	23	1	0.029	597	594	1987	0.300	90
	23	23	1	24	1	0.029	594	591	2696	0.300	90
Camporotondo	1	1	1	2	0	0.008	599	562	1750	0.150	75
	2	2	0	3	0	0.006	562	496	1650	0.125	75
	3	3	0	4	1	0.004	496	461	800	0.100	75
S. G. Galermo	1	1	1	2	0	0.036	420	351	1300	0.200	75
	2	2	0	3	1	0.018	351	249	220	0.150	75
	3	2	0	4	1	0.018	351	253	1900	0.150	75

Footnote: the columns represent the system name, the ID of the main, the upstream node of the main, the absence ($i_{ts,un} = 0$) or presence ($i_{ts,un} = 1$) of a tank/source at the upstream node, the downstream node, the absence ($i_{ts,dn} = 0$) or presence ($i_{ts,dn} = 1$) of a tank/source at the downstream node, the flowing water discharge Q, the head at the upstream and downstream nodes (H_{un} and H_{dn}) (prescribed for a tank/source and minimum desired value for an intermediate node), the main length L, diameter D, and Strickler roughness coefficient k.

Due to the high geodesic gradients, the considered systems of transmission mains always show a surplus of energy that is currently dissipated through already installed regulation valves (see Figure 4 as an example). As a virtuous alternative option to the full dissipation of the available surplus of energy, the installation of PATs in this system is proposed in this work.

For the pumps to be installed as turbines in the systems, the 84 pumps of the NM4 series in the catalogue of the manufacturer Calpeda were considered. In their normal operation as pumps, they operate at 1450 rpm in the 50-Hz electrical grid. In their operation as PATs, the device revolution speed was assumed to be variable within the range 1450 to 3000 rpm. Furthermore, the electrical grid was assumed to be available at all sites, thus incurring no extra installation costs.

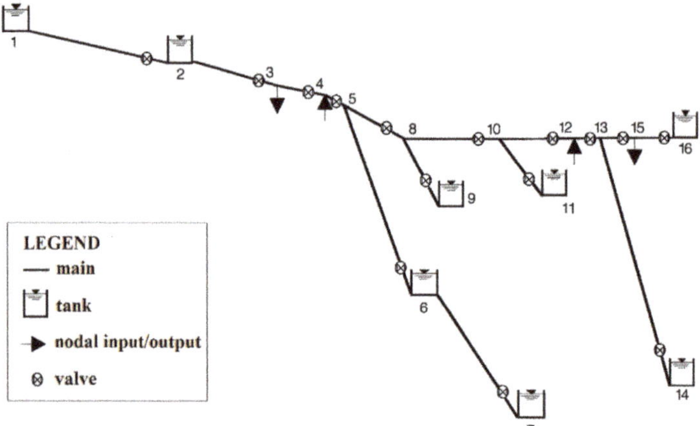

Figure 4. Qualitative elevation layout of the system of transmission mains of Ciapparazzo. The numbers indicate the nodes in Table 1, which also reports the geometrical data of the system.

The cost of the electro-mechanical equipment for the generic PAT was obtained starting from the Calpeda price list and by applying some steps of increase, including the replacement of the asynchronous motor, the installation of an inverter, and a standard amount of +10% for additional hydraulic and electric equipment (e.g., gates and by-pass pipe, and electric wiring and meter). As for the replacement of the asynchronous motor, this must be done because of the different performance curves of the PAT from the same device used in direct mode [29]. Since the cost of the electro-mechanical equipment typical adds up to about 40% of the total installation cost [30], the latter was obtained by multiplying the former by 100/40 = 2.5.

3.2. Results

3.2.1. Optimization

The methodology described above yielded the Pareto fronts shown in the graphs in Figure 5. As expected, each front reports increasing values of the produced hydropower $P_{PAT,tot}$ as $C_{inst,tot}$ grows. Overall, the maximum values of $P_{PAT,tot}$, at the right end of the Pareto fronts, ranged from about 6 KW to 83 KW in the systems of Camporotondo and Ciapparazzo, respectively. As expected, the larger maximum values of $P_{PAT,tot}$ were associated with larger geodesic height variations and larger water discharges available in the system.

The analysis of the results revealed irregularities in the growth of $P_{PAT,tot}$ as a function of $C_{inst,tot}$, with changes in slope in the Pareto front for all systems. To shed light on this aspect, the Pareto front in the system of Camporotondo, made up of nine solutions (see Table 2), was taken as an example. The first solution was made up of no PAT in the system. As $C_{inst,tot}$ grew, the optimizer first suggested installing only one PAT in main three, which was the last main of the system with the lowest water discharge. For a larger value of $C_{inst,tot}$ = 6468 €, the installation of the PAT at main two was proposed. When the cost further increased, combinations of two PATs, involving first mains one and three and then mains two and three, were proposed. Finally, the two most expensive solutions featured PATs installed at all sites. Globally, the only PAT model suggested for main one was NM4 40/16C, which was operated at different speeds in the various solutions. The same consideration could be made for main two, in which the only PAT model installed was the NM4 32/20A. Various PAT models, operated at different speeds, were proposed, instead, for main three. For growing values of $C_{inst,tot}$, the change in the PAT model and PAT speed, as well as the introduction of the PAT in new sites, was responsible for slope changes in the Pareto front $P_{PAT,tot}(C_{inst,tot})$.

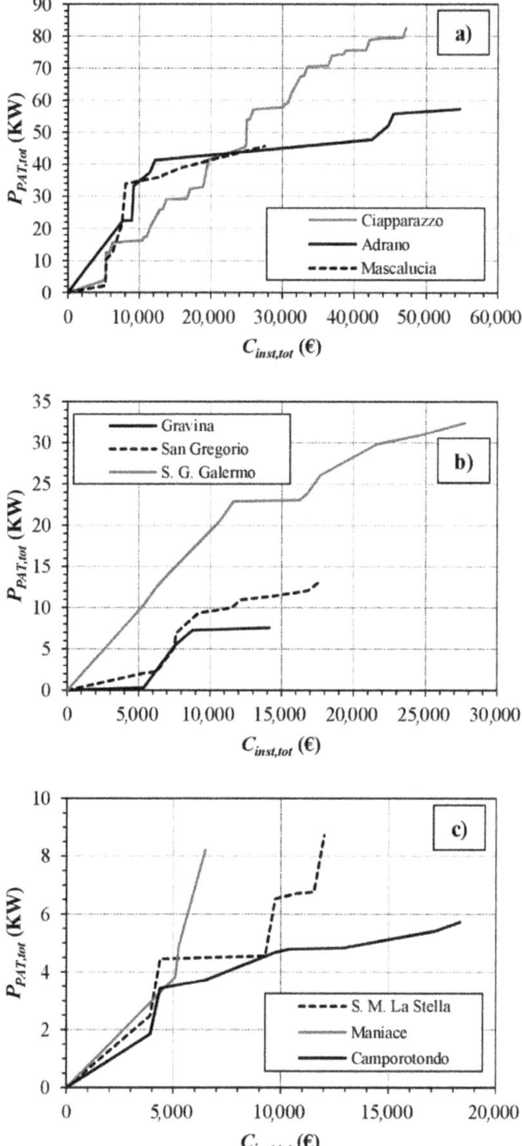

Figure 5. Pareto front of optimal solutions in the trade-off between $C_{inst,tot}$ and $P_{PAT,tot}$ in the systems of (**a**) Ciapparazzo, Adrano, Mascalucia, (**b**) Gravina, S. Gregorio, S. G. Galermo, (**c**) S. M. La Stella, Manciace, and Camporotondo.

Table 2. Solutions in the Pareto front for the system of Camporotondo. Features in terms of PATs installed in the system, PAT speeds, installation cost $C_{inst,tot}$, and hydropower $P_{PAT,tot}$.

Solution	PAT in Main 1	PAT in Main 2	PAT in Main 3	PAT Speed in Main 1	PAT Speed in Main 2	PAT Speed in Main 3	$C_{inst,tot}$ (€)	$P_{PAT,tot}$ (KW)
1							0	0
2			NM4 25/12A				3000	3920	1.85
3			NM4 25/160B				3000	4391	3.43
4		NM4 32/20A			3000		6468	3.72
5	NM4 40/16C		NM4 25/160B	2577		3000	9748	4.68
6		NM4 32/20A	NM4 25/12A		2499	1988	10,388	4.79
7		NM4 32/20A	NM4 32/20A		3000	2083	12,936	4.84
8	NM4 40/16C	NM4 32/20A	NM4 32/16A	2537	2510	2132	17,094	5.40
9	NM4 40/16C	NM4 32/20A	NM4 32/20A	2316	2510	1671	18,293	5.73

The solutions obtained in the various systems for the highest values of $C_{inst,tot}$ (and therefore of $P_{PAT,tot}$) are presented in Table 3, which reports the features in terms of installed PAT models and adopted speed settings, hydraulic variables, and produced hydropower. As an example of these solutions, the system of S. M. La Stella is shown in Figure 6, also including the qualitative pattern of the head grade line along the system mains. This pattern points out that the head decreased due to friction losses in the mains as well as to local head dissipations at the two PATs and at the two regulation valves.

Table 3. Solutions obtained in the various systems for the highest values of $C_{inst,tot}$.

System	Main	PAT	$C_{inst,i}$ (€)	$H_{act,un}$ (m)	$H_{act,dn}$ (m)	Y_v (m)	JL (m)	Y_{PAT} (m)	Speed (rpm)	P_{PAT} (kW)
	1			756.50	733.00	13.58	9.92			
	2			733.00	729.06	0.00	3.94			
	3			729.06	727.98	0.00	1.08			
	4			727.98	727.49	0.00	0.49			
	5	NM4 32/16B	5107	727.49	595.00	72.21	0.63	59.65	3000	3.83
	6	NM4 32/16A	5269	595.00	477.50	30.26	1.00	86.24	3000	4.90
	7			727.49	721.50	0.00	6.00			
Ciapparazzo	8	NM4 65/16S	10,908	721.50	667.30	0.00	0.47	53.72	2932	16.71
	9			721.50	715.50	0.00	5.99			
	10	NM4 125/25E	19,697	715.50	665.50	2.56	0.35	47.09	3000	41.58
	11			715.50	714.01	0.00	1.49			
	12			714.01	713.52	0.00	0.49			
	13	NM4 40/16A	6262.5	713.52	493.40	15.77	45.38	158.96	3000	15.62
	14			713.52	711.11	0.00	2.41			
	15			711.11	683.00	1.37	26.74			
Total			47,243							82.65
Adrano	1	NM4 125/25A	42,428	733.00	705.00	0.00	0.04	27.96	1511	16.05
	2	NM4 80/20C	12,206	705.00	619.00	4.89	0.12	80.99	3000	41.24
Total			54,634							57.29

Table 3. *Cont.*

System	Main	PAT	$C_{inst,i}$ (€)	$H_{act,un}$ (m)	$H_{act,dn}$ (m)	Y_v (m)	$J\,L$ (m)	Y_{PAT} (m)	Speed (rpm)	P_{PAT} (kW)
Gravina	1			434.00	410.47	0.00	23.53			
	2	NM4 40/16C	5357	410.47	382.00	0.85	19.45	8.18	1618	0.30
	3	NM4 50/20C	8784	410.47	322.00	0.93	31.13	56.42	2976	7.27
Total			14,141							7.58
Mascalucia	1	NM4 80/20B	13,675	665.00	588.85	0.00	13.30	62.85	3000	24.76
	2			588.85	577.00	11.18	0.67			
	3	NM4 65/20A	13,858	588.85	508.32	0.00	11.03	69.50	3000	21.01
	4			508.32	484.00	11.08	13.23			
Total			27,533							45.77
S Gregorio	1	NM4 80/20C	12,206	476.00	421.71	0.00	19.78	34.51	2672	10.97
	2			421.71	395.50	0.00	26.21			
	3			395.50	389.00	0.00	6.50			
	4	NM4 40/16C	5357	395.50	347.00	2.09	17.33	29.08	3000	2.07
Total			17,563							13.04
S. M. La Stella	1	NM4 40/20B	7630	683.00	569.30	14.20	38.99	60.51	2820	4.29
	2	NM4 25/160B	4391	569.30	425.70	12.61	7.53	123.45	3000	4.45
Total			12,021							8.74
Maniace	1			650.00	645.40	4.14	0.46			
	2			645.40	641.40	3.00	1.00			
	3			641.40	638.80	2.13	0.47			
	4			638.80	637.70	0.63	0.47			
	5			637.70	634.85	2.40	0.45			
	6			634.85	631.00	1.58	2.27			
	7			631.00	627.75	0.95	2.30			
	8			627.75	625.30	0.79	1.66			
	9			625.30	622.85	0.89	1.56			
	10			622.85	620.00	1.67	1.18			
	11			620.00	619.10	0.10	0.80			
	12			619.10	616.90	1.13	1.07			
	13			616.90	613.00	2.19	1.71			
	14			613.00	610.20	1.50	1.30			
	15	NM4 32/20A	6468	610.20	475.50	5.77	0.25	128.68	3000	8.21
	16			610.20	607.00	2.16	1.04			
	17			607.00	603.50	2.34	1.16			
	18			603.50	602.30	0.04	1.16			
	19			602.30	601.14	0.00	1.16			
	20			601.14	599.00	2.00	0.14			
	21			601.14	596.85	3.88	0.41			
	22			596.85	593.85	1.69	1.31			
	23			593.85	591.00	1.08	1.77			
Total			6468							8.21
Camporotondo	1	NM4 40/16C	5357	599.00	574.43	0.00	5.08	19.49	2316	1.20
	2	NM4 32/20A	6468	574.43	497.19	0.00	7.13	70.12	2510	3.49
	3	NM4 32/20A	6468	497.19	461.00	0.00	5.05	31.14	1671	1.03
Total			18,293							5.73
S. G. Galermo	1	NM4 65/16S	10,908	420.10	362.76	0.00	16.49	40.85	2410	11.79
	2	NM4 50/25B	11,437	362.76	248.55	0.00	3.23	110.98	2616	12.91
	3	NM4 40/16C	5357	362.76	253.00	0.00	27.94	81.83	2424	7.74
Total			27,702							32.44

Footnote: the columns represent the system name, the main ID, the PAT model installed, the installation cost $C_{inst,i}$, the calculated upstream ($H_{act,un}$) and downstream ($H_{act,dn}$) head, the head drop Y_v in the valve, the head loss $J\,L$ in the main, the head drop Y_{PAT} in the PAT, the PAT speed setting, and produced hydropower P_{PAT}.

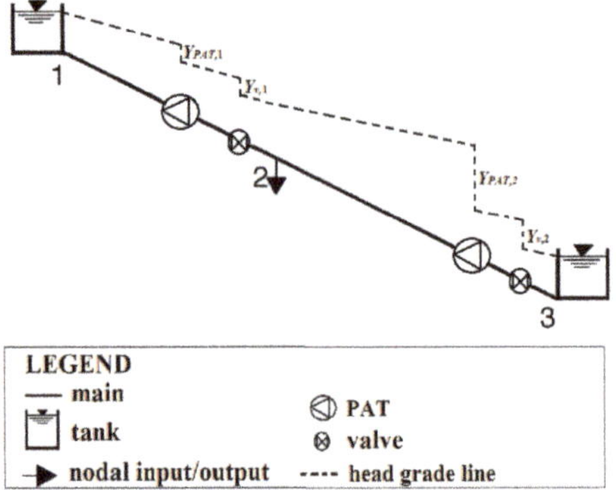

Figure 6. Layout of the system of transmission mains of S. M. La Stella. Qualitative pattern for the head grade line for the solution with $C_{inst,tot}$ = 12,021 € and $P_{PAT,tot}$ = 8.74 KW. See Table 3 for numerical values.

3.2.2. Economic Postprocessing

The Pareto fronts of optimal solutions shown in Figure 5 were postprocessed to obtain curves of net profit C_p as a function of the total installation cost $C_{inst,tot}$ (see Figure 7). In these calculations, a useful life T = 20 years was considered for the PAT. C_{add} in Equation (4) was assumed to include the annual maintenance costs, which were set at 3% of $C_{inst,tot}$ and converted into present value at year 0. As an example, the Pareto front for the system of Camporotondo (reported in Figure 5c) was postprocessed by considering four alternative scenarios.

The main features of the four scenarios are:

- scenario 1, with c = 0.1561 €/KW h, r = 0.04;
- scenario 2, with c = 0.1561 €/KW h, r = 0.02;
- scenario 3, with c = 0.1561 €/KW h, r = 0.06;
- scenario 4, with c = 0.08 €/KW h, r = 0.04.

In scenario 1, the values of c and r represent currently typical conditions in Italy for the sale of hydropower and for the discount rate for cost conversion, respectively. Scenario 2 differs from scenario 1 in the lower value of r, thus resulting in higher present worth values for sold hydropower and for maintenance costs. Compared to scenario 1, scenario 3 has a larger value of r, resulting in lower present worth values for sold hydropower and for maintenance costs. Finally, scenario 4 differs from scenario 1 because of its lower value of c.

Overall, Figure 7 shows that, in scenario 1, C_p tends to grow, over almost the whole range of $C_{inst,tot}$ values, up to a value of about 81,000 €. Therefore, the most profitable solution in terms of C_p is that featuring the highest value of $C_{inst,tot}$ = 18,293 € (solution 9 in Table 2). In comparison with scenario 1, due to the different values of r, the pattern of $C_p(C_{inst,tot})$ in scenarios 2 and 3 is slightly higher and lower, respectively. In both scenarios, the superiority of solution 9 persists because of its large generated hydropower. In fact, for all the considered values of r, the sale of hydropower pays back the large investment and maintenance costs associated with solution 9. Under scenario 4, C_p rapidly grows up to a quite stable value around 28,000 €, and the maximum of C_p is at 30,964 €, obtained for $C_{inst,tot}$ = 10,388 € (solution 6 in Table 2). In this case, due to the low value of c, the large investment and maintenance costs of solution 9 are less paid back in the long run, in comparison with other solutions.

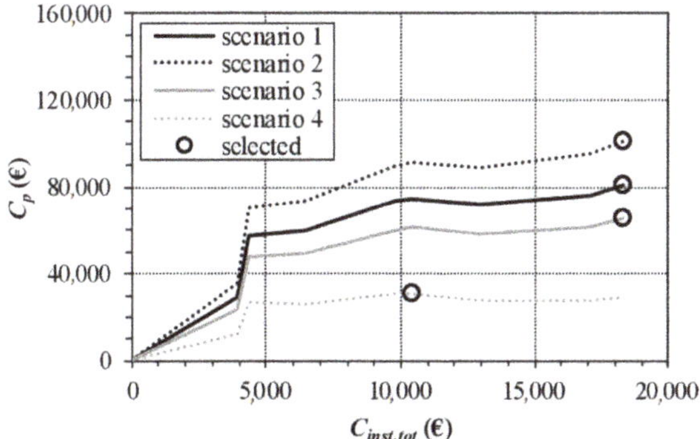

Figure 7. Curves of total net profit C_p as a function of $C_{inst,tot}$, obtained in scenarios 1–4, by postprocessing the Pareto front in Figure 5c for the system of Camporotondo.

Overall, the economic postprocessing in the system of Camporotondo shows that the most expensive solution in terms of $C_{inst,tot}$ yielded by the optimization tends to be the most profitable solution unless the economic scenario used for the postprocessing differs largely from the typical ones in terms of c.

Scenario 1 for the economic postprocessing was applied for all the considered systems. In all systems, the most profitable solution was the right end solution in the Pareto front in Figure 5, thus confirming that the largest optimal investment is well paid back by the sale of electrical energy. Overall, the net profit values C_p ranged from about 8.1×10^4 € to about 1.5×10^6 €, for the systems of Camporotondo and Ciapparazzo, respectively. Higher values of net profit were associated with larger values of the produced hydropower $P_{PAT,tot}$. The payback period T_b was lower than 2 years in all cases except for the system of Camporotondo, for which $T_b = 3$ years. Overall, these values of T_b were even lower than those observed in [24], due to the more favorable conditions of geodesic elevation variations in the present case study.

4. Discussion and Conclusions

In this paper, the bi-objective genetic optimization was proposed for tackling the problem of optimal installation of PATs in existing systems of transmission mains. The proposed methodology is able to search for optimal locations and speed settings of PATs in each system in the trade-off between the total installation cost and the produced hydropower, to be minimized and maximized, respectively. The applications concerned various topologically simple systems of transmission mains managed by ACOSET in Italy, while including in the analysis the estimated installation costs and the expected reverse-mode performance of the pumps produced by a manufacturer.

The application of the bi-objective optimization enabled Pareto fronts of optimal PAT installation solutions in the trade-off between installation costs and hydropower produced to be obtained. Even if the optimization of PAT locations and settings was performed in this case on yearly average water discharges at the mains, the presence of the inverter in the installation enables adjustment of the operation of each PAT to variable values of the daily water discharge in the year. The analysis of these Pareto fronts revealed that, as the cost grows, the power produced grows irregularly. In fact, as the installation cost varies in the generic system, optimal solutions are found with variable numbers, models, and speeds of PATs. Based on this result, it is expected that algorithms based on the sequential addition of PATs and keeping PAT speeds unchanged during the iterations, such as that proposed in [24], may lead to sub-optimal solutions. However, it must be remarked that the application of the bi-objective optimization took computational time with order of magnitude of hours/days for the simple case systems considered in this work. Therefore, computational times would become prohibitive for more topologically complex systems.

Starting from the trade-off between installation costs and generated hydropower, which is explored in a single optimization on each system, a quick economic postprocessing enables analysis of how optimal solutions perform in terms of long-cycle net profit and payback periods in various economic scenarios, thus supporting decision makers in the choice of the ultimate design solution. This marks a significative improvement compared to the single objective optimization, in which one optimization must be carried out for each considered economic scenario.

The following conclusions can be drawn from this work:

- the installation of PATs confirms itself as a valuable solution for energy recovery from pressurized water networks;
- in the case-study considered, made up of transmission mains featuring high geodesic elevation variations, the payback period for the installation investment ranges from 1 to 3 years, lower than those previously remarked in the scientific literature;
- in the case-study considered, the most profitable solutions at all systems in typical economic scenarios are those associated with the maximum hydropower recovered, which ranges from about 6 KW to 83 KW.

Future work will be dedicated to making the methodology feasibly applicable to more topologically complex systems under time varying operating conditions.

Author Contributions: Conceptualization, E.C. and G.P.; methodology, E.C. and G.G.; writing—original draft preparation, E.C. and G.G.; writing—review and editing, A.C., C.C., and G.P. All authors have read and agreed to the published version of the manuscript.

Funding: This research received no external funding.

Conflicts of Interest: The authors declare no conflict of interest.

References

1. IEA (International Energy Agency). Water Energy Nexus: Excerpt from the World Energy Outlook. 2016. Available online: https://www.iea.org/publications/freepublications/publication/WorldEnergyOutlook2016 ExcerptWaterEnergyNexus.pdf (accessed on 22 May 2019).
2. Xu, Q.; Chen, Q.; Ma, J.; Blanckaert, K.; Wan, Z. Water saving and energy reduction through pressure management in urban water distribution networks. *Water Resour. Manag.* **2014**, *28*, 3715–3726. [CrossRef]
3. Pezzinga, G.; Tosto, G. Adeguamento energetico di reti di adduzione idrica in pressione. *L'Acqua* **2001**, *1*, 27–34. (In Italian)
4. Sammartano, V.; Sinagra, M.; Filianoti, P.; Tucciarelli, T. A Banki-Michell turbine for in-line water supply systems. *J. Hydraul. Res.* **2017**, *55*, 686–694. [CrossRef]
5. Samora, I.; Manso, P.; Franca, M.; Schleiss, A.; Ramos, H. Energy recovery using micro-hydropower technology in water supply systems: The case study of the city of Fribourg. *Water* **2016**, *8*, 344. [CrossRef]
6. Sinagra, M.; Sammartano, V.; Morreale, G.; Tucciarelli, T. A new device for pressure control and energy recovery in water distribution networks. *Water* **2017**, *9*, 309. [CrossRef]
7. Bonthuys, G.J.; van Dijk, M.; Cavazzini, G. Water distribution system energy recovery and leakage reduction optimisation through hydro turbines. *J. Water Resour. Plan. Manag.* **2016**, *142*, 04015045.
8. Ramos, H.; Borga, A. Pumps as turbines: An unconventional solution to energy production. *Urban Water* **1999**, *1*, 261–263. [CrossRef]
9. Agarwal, T. Review of pump as turbine (PAT) for microhydropower. *Int. J. Emerg. Technol. Adv. Eng.* **2012**, *2*, 163–169.
10. Carravetta, A.; Del Giudice, G.; Fecarotta, O.; Ramos, H. Energy production in water distribution networks: A PAT design strategy. *Water Resour. Manag.* **2012**, *26*, 3947–3959. [CrossRef]
11. Carravetta, A.; Del Giudice, G.; Fecarotta, O.; Ramos, H. PAT design strategy for energy recovery in water distribution networks by electrical regulation. *Energies* **2013**, *6*, 411–424. [CrossRef]
12. Alberizzi, J.C.; Renzi, M.; Nigro, A.; Rossi, M. Study of a Pump-as-Turbine (PaT) speed control for a Water Distribution Network (WDN) in South-Tyrol subjected to high variable water flow rates. *Energy Procedia* **2018**, *148*, 226–233. [CrossRef]

13. Derakhshan, S.; Nourbakhsh, A. Experimental study of characteristic curves of centrifugal pumps working as turbines in different specific speeds. *Exp. Therm. Fluid Sci.* **2008**, *32*, 800–807. [CrossRef]

14. Pugliese, F.; De Paola, F.; Fontana, N.; Giugni, M.; Marini, G. Performance of vertical-axis pumps as turbines. *J. Hydraul. Res.* **2018**, *56*, 482–493. [CrossRef]

15. Nautiyal, H.; Kumar, V.; Thakur, A. CFD Analysis on Pumps working as Turbines. *Hydro Nepal* **2010**, *6*, 35–37. [CrossRef]

16. Fecarotta, O.; Carravetta, A.; Ramos, H. CFD and comparisons for a pump as turbine: Mesh reliability and performance concerns. *Int. J. Energy Environ.* **2011**, *2*, 39–48.

17. Marchiori, I.N.; Lima, G.M.; Brentan, B.M.; Junior, E.L. Effectiveness of methods for selecting pumps as turbines to operate in water distribution networks. *Water Supply* **2018**, *19*, 417–423. [CrossRef]

18. Novara, D.; McNabola, A. A model for the extrapolation of the characteristic curves of pumps as turbines from a datum best efficiency point. *Energy Convers. Manag.* **2018**, *174*, 1–7. [CrossRef]

19. Fontana, N.; Giugni, M.; Glielmo, L.; Marini, G. Real time control of a prototype for pressure regulation and energy production in water distribution networks. *J. Water Resour. Plan. Manag.* **2016**, *142*, 04016015. [CrossRef]

20. Fontana, N.; Giugni, M.; Glielmo, L.; Marini, G.; Zollo, R. Hydraulic and Electric Regulation of a Prototype for Real-Time Control of Pressure and Hydropower Generation in a Water Distribution Network. *J. Water Resour. Plan. Manag.* **2018**, *144*, 04018072. [CrossRef]

21. Giugni, M.; Fontana, N.; Ranucci, A. Optimal location of PRVs and turbines in water distribution systems. *J. Water Resour. Plan. Manag.* **2014**, *140*, 06014004. [CrossRef]

22. Corcoran, L.; McNabola, A.; Coughlan, P. Optimization of water distribution networks for combined hydropower energy recovery and leakage reduction. *J. Water Resour. Plan. Manag.* **2015**, *142*, 04015045. [CrossRef]

23. Coelho, B.; Andrade-Campos, A. Energy recovery in water networks: Numerical decision support tool for optimal site and selection of micro turbines. *J. Water Resour. Plan. Manag.* **2018**, *144*, 04018004. [CrossRef]

24. Fernández García, I.; Mc Nabola, A. Maximizing Hydropower Generation in Gravity Water Distribution Networks: Determining the Optimal Location and Number of Pumps as Turbines. *J. Water Resour. Plan. Manag.* **2020**, *146*, 04019066. [CrossRef]

25. Fecarotta, O.; McNabola, A. Optimal location of pump as turbines (PATs) in water distribution networks to recover energy and reduce leakage. *Water Resour. Manag.* **2017**, *31*, 5043–5059. [CrossRef]

26. Tricarico, C.; Morley, M.S.; Gargano, R.; Kapelan, Z.; Savic, D.; Santopietro, S.; Granata, F.; De Marinis, G. Optimal energy recovery by means of pumps as turbines (PATs) for improved WDS management. *Water Sci. Technol. Water Supply* **2018**, *18*, 1365–1374. [CrossRef]

27. Deb, K.; Pratap, A.; Agarwal, S.; Meyarivan, T. A Fast and Elitist Multiobjective Genetic Algorithm: NSGA-II. *IEEE Trans. Evolut. Comput.* **2002**, *6*, 182–197. [CrossRef]

28. Todini, E.; Pilati, S. A gradient algorithm for the analysis of pipe networks. Computer Application in Water Supply. In *System Analysis and Simulation*; Coulbeck, B., Choun-Hou, O., Eds.; John Wiley & Sons: London, UK, 1988; Volume I, pp. 1–20.

29. Novara, D.; Carravetta, A.; McNabola, A.; Ramos, H. Cost Model for Pumps as Turbines in Run-of-River and In-Pipe Microhydropower Applications. *J. Water Resour. Plan. Manag.* **2019**, *145*, 04019012. [CrossRef]

30. Chapallaz, J.M.; Eichenberger, P.; Fischer, G. *Manual on Pumps Used as Turbines*; VIEWEG, 1992. Available online: http://skat.ch/book/manual-on-pumps-used-as-turbines-volume-11/ (accessed on 22 January 2020).

Article

Geospatial Information System-Based Modeling Approach for Leakage Management in Urban Water Distribution Networks

Parima Mirshafiei [1,†], Abolghasem Sadeghi-Niaraki [2,3,*,†], Maryam Shakeri [2] and Soo-Mi Choi [3]

[1]   Computer Engineering, Shariaty University, Tehran 18918-16851, Iran
[2]   Geoformation Technology Center of Excellence, Faculty of Geodesy & Geomatics Engineering,
      K.N. Toosi University of Technology, Tehran 19697, Iran
[3]   Department of Computer Science and Engineering, Sejong University, Seoul 143-747, Korea
*   Correspondence: a.sadeghi.ni@gmail.com
†   These authors contributed equally to this work.

Received: 22 June 2019; Accepted: 13 August 2019; Published: 20 August 2019

Abstract: The purpose of this paper is to model one of the urban network problems, the issue of water leakage. In order to manage water leakage, the specific area should be partially isolated from the rest of the network. As Geospatial Information System (GIS) is a powerful technology in spatial modeling, analysis and visualization of the water network management, a web GIS system for finding optimal valves to close in the event of an incident was developed. The system consists of a new GIS based algorithm for identifying the ideal valves to isolate the desired pipeline. The algorithm is able to identify optimum valves in a water distribution network in the shortest time by using the traceability in GIS web services. The system uses the functions of storing and managing the spatial data by expert users based on web 2.0 technology. The system was implemented and evaluated for Tehran's district 5 water distribution network using Silverlight, C# and ArcGIS SDK (Software Development Kit). The evaluations demonstrated the accuracy of the algorithm and the operational viability of the system developed.

Keywords: GIS modeling; leakage management; urban water network management; valve closing algorithm; web 2.0

---

1. Introduction

The rapid technological and economic developments of the recent century have led to the emergence and expansion of a large number of human settlements across the world. However, for a human settlement to be called a proper city, it should contain the infrastructures necessary for comfortable living. One of the basic and yet most complex infrastructures needed in every city is a water distribution system, which can be described as a multi-source and non-directional network [1,2].

In the last few years, urban water network modeling has been boosted to face the new challenges of modern society. The challenges include, among others, leakage management of water distribution network and urban drainage systems. Despite immense progress in human technological ability, tens of thousands of one common leakage problem that regularly causes interruption in water supply is pipes bursting. Following a pipe burst, network maintenance personnel should be able to address the issue as soon as possible while keeping the number of affected citizens to a minimum. Although strong management of repair and maintenance efforts is of paramount importance, the method of choosing the fastest, and most effective way to solve the issue can also play an important role in this regard. This approach is quite time-consuming and error-prone.

A possible solution to improve the response time to such problems is the utilization of a Geospatial Information System (GIS) as an enhanced technology in the field of water distribution management.

GIS technology can be used for spatial modeling of the urban water network and as an interactive user environment for daily water management tasks [3,4]. Given the spatial nature of the data used and generated in GIS, it can be used to effectively enhance water resource modeling. Another merit of GIS is its ability to determine the water network condition and capacity information from existing databases, transform them into a geo-analysis environment, and produce reports and graphic information accordingly [5].

With the rapid growth of Internet and online devices, there have been many successful efforts that take advantage of these developments to solve operational problems in different fields [6,7]. A web-based online system provides ubiquitous access to software data and utilities without needing to install any specific application, and can execute the operations requested by a large number of clients simultaneously [8]. At present, the majority of software applications related to water distribution networks are desktop-based and their data depend on the software version [9]. Considering the aforementioned issues, one may ask whether it is possible to develop a web GIS software with a flexible platform, multi-user capability, and proper upgradability.

This study provides a GIS based leakage management model to establish a partially isolated area from the rest of the water distribution network. In order to achieve the aforementioned benefit, we developed a web GIS application based on the Silverlight platform. This web application is founded upon a relational database, an intermediate layer for the management of geospatial information, two layers of web interface for customizing services, and other software components. This application consists of web services for managing, visualizing and storing geospatial data, support of the water spatial data model, and a GIS based algorithm for optimal burst pipe isolation. This web application was successfully tested on the database of Tehran's district 5 water distribution network. This test demonstrated the operational viability of the application and its potential as a platform for future web-based solutions.

The rest of this paper is organized as follows. Section 2 reviews the previous works on burst pipe isolation and analysis, GIS solutions and related issues. Section 3 describes the steps of the present work, the system architecture, and its constituting services. Section 4 provides the details of system implementation and the related diagrams. Section 5 outlines the test details and provides the test results; finally, Section 6 presents the conclusions and some suggestions for future research.

2. Related Works

The internet is very much alive and kicking with regards to related works. The first generation of internet sites primarily gave information. Most people have become familiar with web 2.0, blogging, tagging, social networking and social bookmarking have paved a way to the next step in the development of the web. The next step was to the intelligent and omnipresent web 3.0. The Ubiquitous Computing System is a concept in software engineering and computer science, in which computations are performed at any location and time. Contrary to the computing of desktop computers, computational calculations can be done on any device, anywhere, and in any format. In this technology, the user interacts with the computer in a variety of shapes such as lobes, tablets, and everyday objects from a refrigerator to a pair of glasses. Ubiquitous Geographic Information Systems (UbiGIS) are geographic information systems that provide users or systems at any location, at any time, through GIS services [10]. With the advent and expansion of Web 2.0, the user no longer needs to passively accept the author's limited information, and can also work with customized online maps [11]. For example, a user can utilize the Open Geospatial Consortium (OGC), a Web Map Service (WMS), and Google API (application program interface) interface to integrate geospatial data provided in different sources and generate a web 2.0 online map in the form of mashup [12] or mapping hack [13].

At present, Web Mapping 2.0 has been expressed as citizen science and social media geotags. Additionally, the web contents produced by Twitter or Flickr communities is more than just Big Data [14] and has generated a new paradigm for socio-spatial research [15]. This paradigm, which Lansley and Langley call the Twitter geography, has been successfully used in recent geography studies.

Initially, Web 2.0 maps were used only in graphical representations of the results at the end of the research. However, as data representations became more dynamic and online maps became more detailed, Web 2.0 maps became a part of the research workflow [16–18].

Web GIS is the product of the combination of GIS with Internet and can be considered a major development in this field. In its essence, Web GIS is a type of Dynamic Web Application [8]. One of many ways to implement Web GIS is the use of Silverlight. Silverlight, as a cross-browser and cross-platform plugin, has served as a platform for a new generation of rich interactive applications based on NET-media experiences [19]. In the traditional web development paradigm, the only task of the browser was to display information, and the extraction of the data requested by the user from the database was a task of the server, but this arrangement does not meet the requirements of Web GIS. The use of Silverlight in web development enables the user to experience and work with new multimedia, images and graphics [7]. Vaitis et al. designed a spatial database and a web-GIS application based on Silverlight in order to visualize and analyse the Greece climate data [20]. In another research, WANG and LI utilized the GIS data processing capability to implement a web-GIS tourism system [21]. The study carried out by Vitianingsih et al. provided a web-GIS application based on Silverlight to indicate the mapping of roads with heavy traffic [22].

One of the areas where the features of Web GIS can be used to address an urban management problem is the water distribution network. Several investigations on Iranians abilities in maintaining and managing the water resource throughout the history have been done [23,24]. According to these surveys, renewable water per capita in Tehran is now reaching a critical threshold level. Moreover, due to old water pipe systems, a significant fraction of the treated water is leaked. In order to predict a burst pipe in urban water distribution in Tehran, Alizadeh et al. evaluated a data driven model. Based on the result, Gaussian Process Regression performed better than Grasshopper Optimization Algorithm-based Support Vector Regression and Artificial Neural Network [25]. In addition, Motiee and Ghasemnejad applied four statistical models to evaluate pipe vulnerability variables with the goal of discovering equations to predict likely future pipe accidents [26]. Moreover, Tchorzewska-Cieslak et al. proposed a possibilistic model for analyzing the failure risk in water supply network. The technique described was based on the fact hypothesis (Shafer) which was based on the notion of the incorrect probability, which incorporated the distribution capacities of its own Fuzzy Set function. This model is useful when the risk factor data are uncertain or incomplete [27]. A typical water distribution network consists of many components including water pipes, valves, etc. The major problems of a water distribution network are pipe bursts and leakages, and the consequent disruption in citizen access to water. Therefore, following such events, immediate steps should be taken to isolate the failed pipeline by closing the linked valves in order to prevent water loss and replace the damaged pipe as soon as possible. Naturally, it is unwise to close the valve of the main sources because that will affect all consumers served by that source, including the critical buildings. Therefore, closing the valves in an optimal way to minimize the number of affected buildings is of great importance. The burst pipe analysis and emergency repair have been the subjects of much research.

A review of literature shows that a number of studies have proposed different algorithms for optimal valve closing at the site of pipe burst. These include the study of [28], where the genetic algorithm was used to find the shortest path, and thereby the valves that should be closed following a pipe burst incident. By simplifying the scale of the problem and dismembering the complex pipeline into a simple network structure, they increase search effectiveness. Ouyang et al. [29] also developed a valve-closing optimization algorithm and implemented it using the ArcEngine. The study carried out by [30] provided another such algorithm that operated based on the status of check valves. Jian-chuan et al. and Jiankang, W.J. [1,31] developed other such algorithms by optimizing and modifying the BFS (Breadth-first search) algorithm. Despite the variety of works carried out on this subject, the present study was the first to implement its proposed algorithm based on a web platform to allow simultaneous use by a large number of users.

3. Architecture and Development Stages

The web application development stages are illustrated in Figure 1. The development stages of Web and GIS sections started at the same time and joined together at the Web GIS Integration stage, where they proceeded as a single integrated implementation process. In the end, the implementation was evaluated by a number of tests, the test results were reviewed, and the overall conclusion regarding the web application were presented.

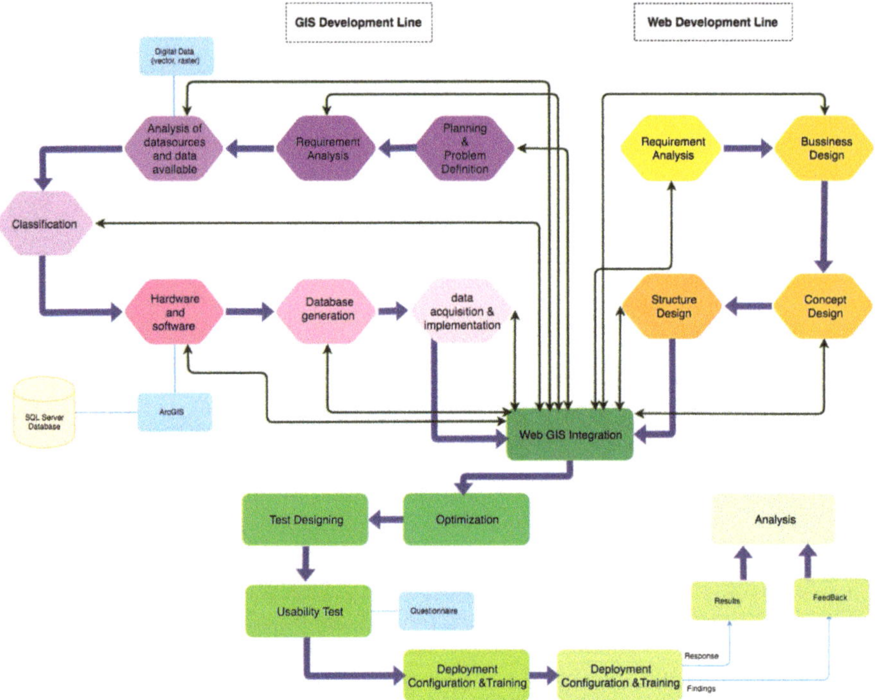

Figure 1. Development stages.

The authors decided to support all system components with three web services. In this system, data transfer was asynchronous. The web services included in the system are described in the following:

3.1. Web Service for Managing, Visualizing and Storing Geospatial Data

This web service operates based on two components: (1) a relational database called Water Spatial Database Engine (WRSDE), which was created in Microsoft SQL Server; (2) ArcGIS Server with 9 geospatial vector layers. The WRSDE database contains data elements with labels pump, valve, pipe, and flow source as well as their Attribute Table. The water network data model is shown in Figure 2. According to the data mode, each layer contains a code, description, and a geometry field to save the necessary information. The pipeline layer, contains some specific fields like diameter, usage, cover type, and material. Fields like pump type, power, head, phase, and voltage are specified for the pump layer. The source layer has capacity, height, floater, floater type, and model as well. At last, the valve layer has diameter, material, serial number, and gate material. The ArcGIS Server developed by Esri (Redlands, CA, USA) is a software application for sharing the data of geographic information system [32]. Here, ArcGIS Server used the WRSDE database to create and manage GIS web services, applications, and data. The main task of ArcGIS Server was to serve as an intermediate layer for linking the WRSDE relational database to the developed web services.

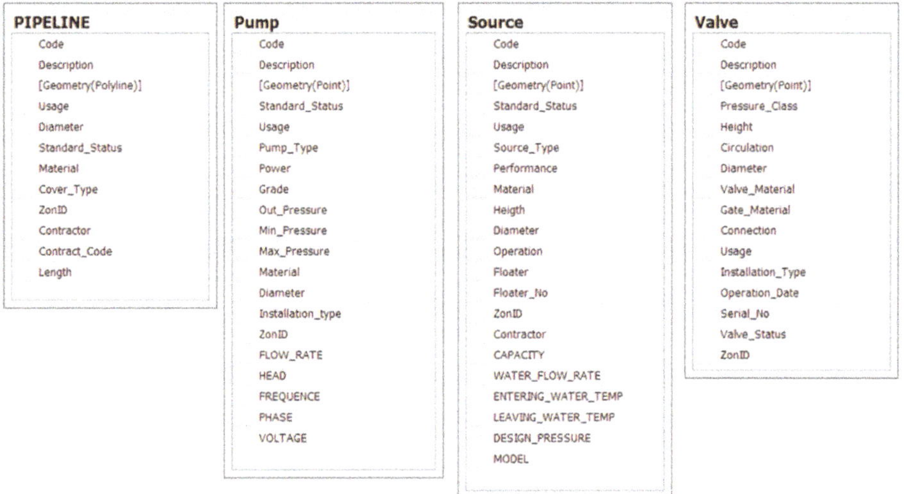

Figure 2. Water network data model.

3.2. Web Application for Supporting Water Resource Modeling

The web application user interface must provide user access to web tools and services. Here, the user interface provides the tools for using nine geospatial vector layers that display the components and structure of the water network and sources. This user interface has a tool for viewing the points that are critical for the isolation of any given section of the pipe system, a number of tools for modifying features and markers on the map, and a section that shows a list of active layers and allows the user to activate/deactivate the desired layers. The elements provided for the visualization in the web application consists of multiple types of geospatial data. Geospatial objects are represented by their geographic information and their corresponding attributes. Water pipes are of polyline type and are represented by certain attributes such as code, label, name, diameter, material, ID, etc.

This web application has a 3-layer architecture. The Presentation Layer is the visible and interactive component of the application and is in direct contact with the user. This layer was designed with Silverlight to take advantage of its technological merits including high compatibility, cross-browser and cross-platform features, variety of media experiences, support of rich interactive applications, flexible programming, low cost, fast speed, and other technical specifications [33,34]. The business layer is responsible for running and organizing all processing routines of the software application. Based on the software requirements, this layer creates an interface called I[BizName]Service and implements a class of this interface, where the processing routine will be called. For the internal modules such as feature editing operation, the module will be created in the form of a class library and will be called in the desired business class. The Data Access Layer provides a mapping of database structure in the software application. Here, the data model was implemented by the use of Entity Framework technology. In the application, this layer is utilized by implementing the desired table model in the class structure and creating a DbSet from the table model.

3.3. Web Service for Valve-Closing

The web application provides a tool for identifying the valves that should be closed to isolate a certain pipe in an optimal way. The web service responsible for this operation utilizes an algorithm for this purpose. Details and method of implementation of this service and its algorithm are described in the next section.

The selected architecture covers the key requirements in terms of size, performance, and runtime. Experience has shown that the selected architecture allows approximately 50 users to simultaneously use the application. It should however be noted that this estimation does not account for hardware and network constraints and is based on the assumption that an ideal platform is in place. The proposed system can also edit, remove, and create features and markers, and identify optimal valve-closing (pipe isolation) solutions in a reasonable time. These features are irrespective of hardware and network constraints. Because of the web-based nature of the software application and the use of Silverlight, this application can be executed by any online device containing a standard web browser with Silverlight plugin.

The quality of the developed architecture can be examined in terms of extensibility and security. The main patterns employed in this architecture contribute to the extensibility of the software application. Furthermore, because of component-based nature of this architecture, components can be extended without much dependency or change in other layers. In general, security segments are provided through .NET authentication mechanisms, and the Web map service is provided by the use of tokens.

These technical specifications have provided a strong technical support for the extension of burst pipe analysis Web GIS service. In this web application, the ability to visualize different map layers and perform desired operations on the desired layers are provided by the use of web map services (WMS). The interface of the application contains several segments, most notably a layer list, where the user can activate/deactivate the desired layers, and a Pipe-Isolation tool, which allows the user to identify the valves that should be closed to isolate a certain point or section.

4. Implementation

This section describes the method of implementation of the developed valve-closing service for Tehran's District 5 water distribution network. The algorithm was implemented using ArcGIS for Silverlight SDK, Silverlight, ArcGIS Server, and Visual Studio 2015. As outlined in Section 3, the system was based on a three-layer architecture. Figure 3 illustrates an overview of the system architecture and its interactions with other components.

When a user logs in, the system sends a map display request to the main server. The user can then use the editing tools to edit, delete, or add a feature or marker on the map. Upon the use of this tool, the system sends a request through WCF to the main server and then connects to the ArcGIS Server by a map Service address in order to perform the requested operation. The user will be notified of the result via the same path. When the user clicks on the Pipe-Isolation tool and selects the pipeline that should be isolated, the system sends the information of the selected pipeline through the webService to the main server. At the server, the system uses the database to collect a list of associated valves and sources, and then uses the SOE (Server object extensions) to connect to the ArcGIS Server and analyzes the network on the map. Server object extensions (SOE) allow users to create service operations to expand the base functionality of map or image services. SOEs are suitable if customers have to conduct some well-defined business logic that is not readily achieved using ArcGIS client APIs. Most SOEs do this by utilizing custom code to work with geospatial data and maps [32]. With the list of valves and sources, the system then runs the valve-closing algorithm on the main server side while maintaining a connection through the SOE to the ArcGIS Server as long as needed. This operation continues until the algorithm ends. Finally, the list of valves in the algorithm solution will be sent through WCF to the browser in order to provide a visualization for the user.

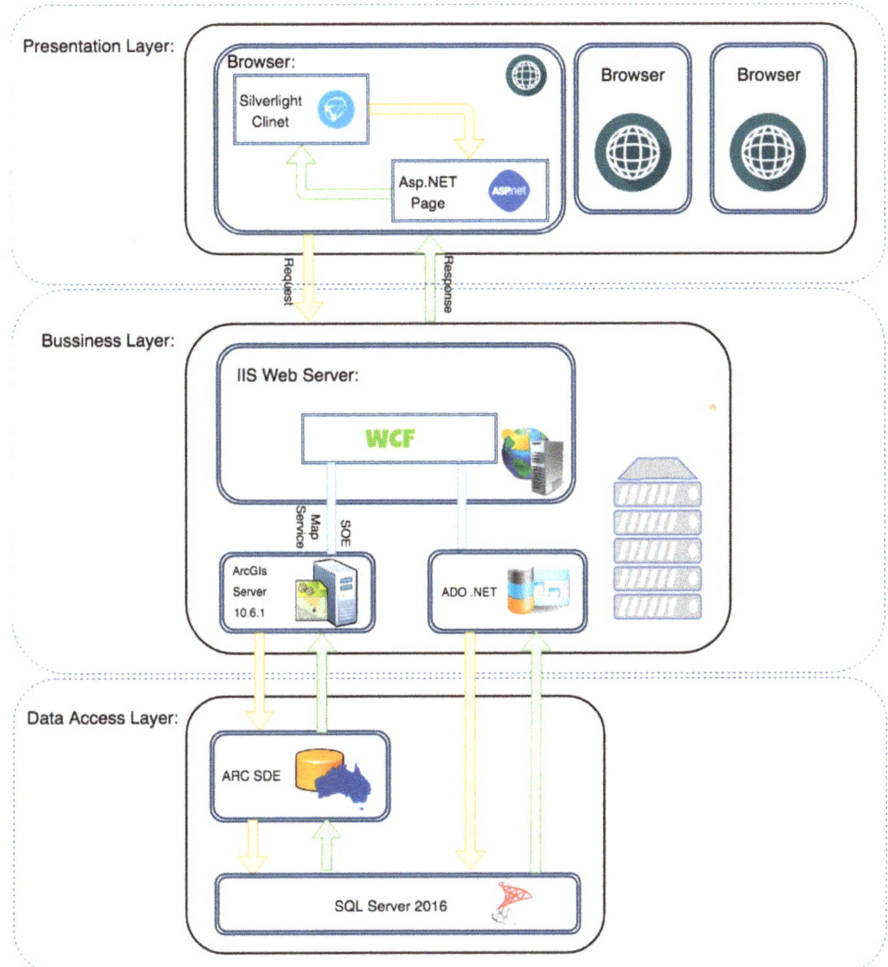

Figure 3. System architecture.

Figure 4 shows the map of Tehran's district 5 water distribution network and an overview of the tools included in the interface. As mentioned, the interface included a Layer-List for activating/deactivating any combination of layers as needed, some tools for editing, removing and adding features and markers, and the Pipe-Isolation tool. In addition, Figure 4 shows a view of the area in the vicinity of a damaged pipe. The burst pipe is located in the section marked with *. The system selects this section as the point of interest and marks it as the burst point.

To illustrate the algorithm more accurately, a schematic design of the water network is shown in Figure 5. In this schema, the green circle is the source point, the gray icons are the valves, and the black lines are pipes. The algorithm input parameter is the burst point. The system starts tracing upstream from the burst point to the related source point and finds the valves in this path. Once the Trace operation is completed, the system will find (several) potential valve(s). The candidate valves in Figure 5 are 1, V2, V3, V4, and V9. Since the trace output is out of sorts, the system will select one of the valves randomly and start the trace from the valve to the burst point. If the trace operation finds any valves in this trace, the current valve will be removed from the final list. In this scenario, the system selected the valve V2 and started tracing. The valve V1 was found in the path, so the V2 would be removed from the final list. Afterwards, the valve V1 would be selected and the system starts tracing from V1 to the burst point. Since no valves were found in this operation, V1 would be added to the final list. In the next step, the system would use V4 for tracing. Because there are no valves in this path, V4 will be added to the final list too. This operation will be done for V3 and if valve V4 is found in the trace result, thus V3 will be deleted from the final list. V9 will be processed like other valves and will be added to the final list. As no more candidate valves exists, the system traced downstream from the burst point and found the candidate valves: V5, V6, V7, and V8. The aforementioned operations will be done for the current candidate valves and V5 will be appended to the final list. When there won't be any candidate valves from the downstream operation the algorithm is finished. Finally, the valves in the final list will be exported as the list of valves that should be closed to isolate the damaged pipe.

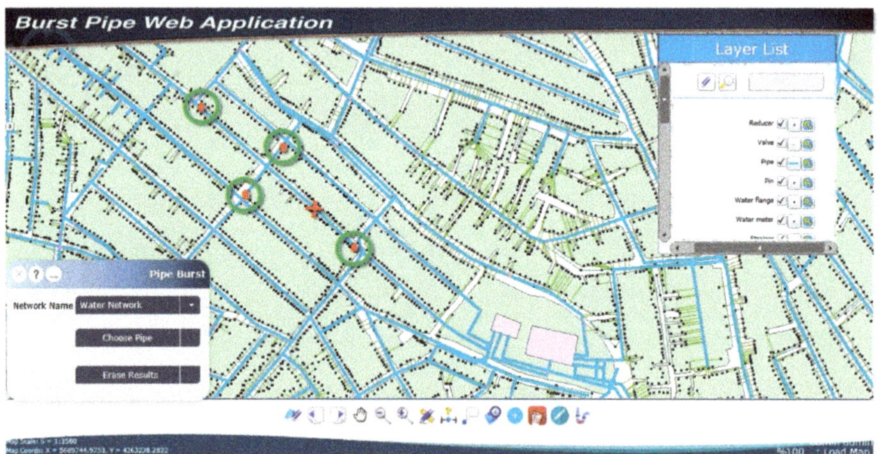

Figure 4. Interface, view of the area around damaged pipe and valves linked to damaged pipeline.

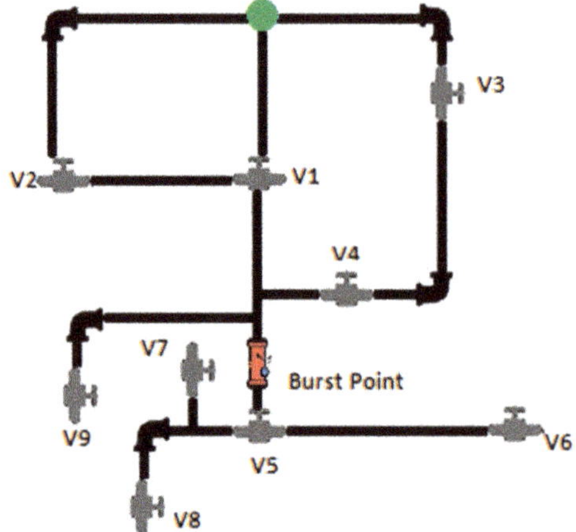

Figure 5. Sample water distribution network.

5. Test and Results

The developed software was deployed on a server with 16 GB of RAM (random access memory), 200 GB of hard disk, Windows Server 2019 operating system, and was successfully tested for Tehran's water network data. The main objective of the test was to evaluate the accuracy of the output of the pipe-isolation tool and the system's ability to support real-time multi-user activity. To embed Tehran's water distribution network in the application, it was modeled as a web service. In addition to the current application, any other software application with the ability to use WFS or MapService could connect to the map server and manage, modify and update the geospatial data. This feature demonstrates the flexibility of the system components and how conveniently they can be connected and integrated with other applications.

To evaluate system accuracy and its multiuser service capability, 32 experts on water networks were asked to use the system simultaneously for 10 h. The objective of this test was to check the stability of the web application, its capacity to serve multiple users simultaneously, and the accuracy of the pipe-isolation operation. All participants had the same access to the application and were asked to undertake the following activities:

1. Use the services to search and download any desired variety of maps and manage, view, and store geospatial data
2. Use the services to draw, edit and delete geospatial objects and their attributes
3. Use the provided pipe-isolation tool to find the optimal valve-closing solutions for any desired section of the water network

Before the test, participants received training on how to use the application and increase the workload when the system was in simultaneous use by multiple users. The test results demonstrated the system's ability to respond to simultaneous requests of multiple users to manage, visualize and store geospatial data. As expected, the services exhibited a satisfactory performance in performing the search, visualization and download operations requested by multiple users. Additionally, the performance of the pipe-isolation tool was tested simultaneously by multiple users. The tests were performed first for the simple sections of the network and then for the more complex sections, and the solutions were checked and approved by the experts. Despite system use by multiple users, there was no interruption in searching, downloading, managing, and viewing maps and storing geospatial data. Figure 6 shows the result of this test based on Average Response Time and Average Processing Time.

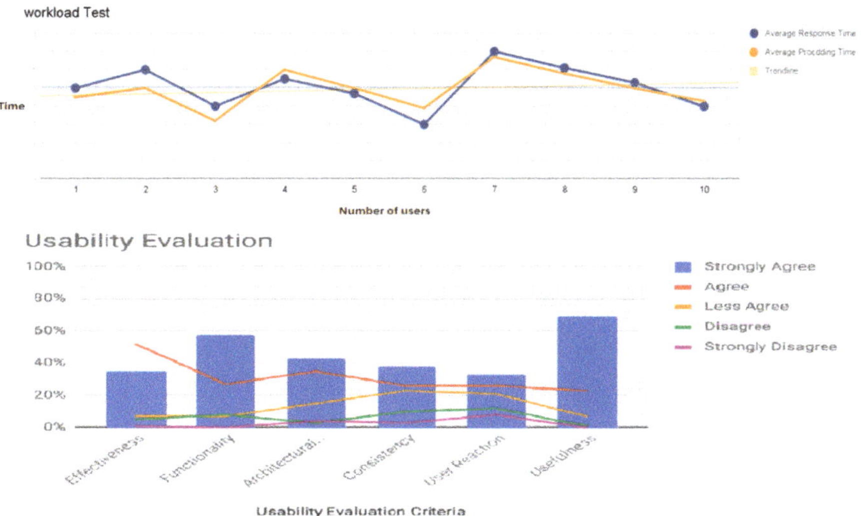

Figure 6. Results of the usability evaluation questionnaire and system performance test.

As can be seen, the test results were satisfactory and the low slope of the graph indicated the system's ability to provide simultaneous support to multiple users. For further evaluation of the system, test participants were asked to fill out the usability evaluation questionnaire developed by [35]. As shown in Figure 6, 35%, 58%, 43%, 38%, 33% and 69% of participants expressed a strong positive opinion with regard to Effectiveness, Functionality, Architectural and Visual Clarity, Consistency, User Reaction, and Usefulness of the system, respectively.

The performance of the valve-closing algorithm was further evaluated by the use of sample data provided by [29], who introduced an algorithm for the same purpose. After executing the present algorithm with the aforementioned data, the results were found to be identical to the results reported in that article. From these tests, it can be concluded that the application is stable, operationally viable and can support multiple users simultaneously under a heavy workload. A number of other system functions including system performance and other features were not evaluated.

6. Conclusions

This paper modeled leakage management in an urban water distribution network using the GIS based algorithm to detect the valves that should be closed to optimally isolate a burst pipe. The GIS based model can improve repair and maintenance operations. By using this GIS algorithm, a web GIS application was developed with access to multilayered maps and the ability to edit, create, and delete features and markers. The provided algorithm is also applicable to other networks with similar characteristics, such as gas distribution networks. The results obtained from the accuracy evaluation test, workload test, and usability questionnaire showed good accuracy of the proposed algorithm and the desirable qualities of the developed web application. To improve and accelerate pipe repair and the maintenance operation, the presented system can be linked to hardware sensors mounted on pipe valves so that they could be closed automatically. The system can also be implemented using the ArcGIS JavaScript API. For further development of the system, it can be linked to water pressure sensors on the pipes so that leakage detection and valve closing operations could be done swiftly and automatically. Naturally, such automatic and real-time management of pipe bursts can result in significant time and cost saving in repair and maintenance operations.

Author Contributions: These authors contributed equally to this work: P.M. and A.S.-N. (co-first authors). Conceptualization, P.M., A.S.-N. and M.S.; methodology, P.M. and A.S.-N.; software, P.M.; formal analysis, P.M. and M.S.; investigation, A.S.-N. and M.S.; visualization, P.M. and A.S.-N.; resources, A.S.-N.; data curation, P.M. and M.S.; writing–original draft preparation, P.M.; writing–review and editing, A.S.-N., M.S. and S.-M.C.; supervision, A.S.-N.; project administration, S.-M.C.; funding acquisition, S.-M.C.

Funding: This research was supported by the MSIT (Ministry of Science and ICT), Korea, under the ITRC (Information Technology Research Center) support program (IITP-2019-2016-0-00312) supervised by the IITP (Institute for Information & communications Technology Planning & Evaluation).

Conflicts of Interest: The authors declare no conflicts of interest.

References

1. Jian-chuan, L.; Yong-shu, L.; Guo-ling, C. Algorithmic optimization and implementation of pipe burst analysis based on ArcGIS. *Sci. Surv. Mapp.* **2008**, *1*, 70.
2. Ugarelli, R.; Venkatesh, G.; Brattebø, H.; Di Federico, V.; Sægrov, S. Historical analysis of blockages in wastewater pipelines in Oslo and diagnosis of causative pipeline characteristics. *Urban Water J.* **2010**, *7*, 335–343. [CrossRef]
3. AbdelBaki, C.; Touaibia, B.; Ammari, A.; Mahmoudi, H.; Goosen, M. Contribution of GIS and Hydraulic Modeling to the Management of Water Distribution Network. In *Geospatial Challenges in the 21st Century*; Springer: Cham, Switzerland, 2019; pp. 125–150.
4. Arnold, U.; Datta, B.; Haenscheid, P. *Intelligent Geographic Information Systems (IGIS) and Surface Water Modeling*; IAHS: Wallingford, UK, 1989; p. 81.
5. Neji, H.B.B.; Turki, S.Y. GIS based multicriteria decision analysis for the delimitation of an agricultural perimeter irrigated with treated wastewater. *Agric. Water Manag.* **2015**, *162*, 78–86. [CrossRef]

6.	Jia, Y.; Zhao, H.; Niu, C.; Jiang, Y.; Gan, H.; Xing, Z.; Zhao, X.; Zhao, Z. A WebGIS-based system for rainfall-runoff prediction and real-time water resources assessment for Beijing. *Comput. Geosci.* **2009**, *35*, 1517–1528. [CrossRef]

7.	Shi, Q.; Wang, H.; Wang, H.; Chen, J. Design of WebGIS Rendering Engine Based on Silverlight-based RIA. In Proceedings of the 2011 International Conference on Intelligent Computation Technology and Automation (ICICTA), Shenzhen, China, 28–29 March 2011; pp. 1050–1053.

8.	Li, P.; Ma, L.; Cai, C.; Zhu, L. Research on Application of Ajax and Silverlight Technology in WebGIS. In Proceedings of the 2009 1st International Conference on Information Science and Engineering (ICISE), Nanjing, China, 26–28 December 2009; pp. 2149–2152.

9.	Delipetrev, B.; Jonoski, A.; Solomatine, D.P. Development of a web application for water resources based on open source software. *Comput. Geosci.* **2014**, *62*, 35–42. [CrossRef]

10.	Rad, T.G.; Sadeghi-Niaraki, A.; Abbasi, A.; Choi, S.M. A methodological framework for assessment of ubiquitous cities using ANP and DEMATEL methods. *Sustain. Cities Soc.* **2018**, *37*, 608–618.

11.	Atabekova, A.; Belousov, A.; Shoustikova, T. Web 3.0-Based Non-formal Learning to Meet the Third Millennium Education Requirements: University Students' Perceptions. *Procedia Soc. Behav. Sci.* **2015**, *214*, 511–519. [CrossRef]

12.	Purvis, M.; Sambells, J.; Turner, C. *Beginning Google Maps Applications with PHP and Ajax*; Apress: New York, NY, USA, 2006.

13.	Erle, S.; Gibson, R.; Walsh, J. *Mapping Hacks: Tips & Tools for Electronic Cartography*; O'Reilly Media Inc.: Sebastopol, CA, USA, 2005.

14.	Laney, D. 3D data management: Controlling data volume, velocity and variety. *META Group Res. Note* **2001**, *6*, 70.

15.	Jiang, B.; Thill, J.C. Volunteered Geographic Information: Towards the establishment of a new paradigm. *Comput. Environ. Urban Syst.* **2015**, *53*, 1–3. [CrossRef]

16.	Fox, P.; Hendler, J. Changing the equation on scientific data visualization. *Science* **2011**, *331*, 705–708. [CrossRef]

17.	Tsou, M.H. Revisiting Web Cartography in the United States: The Rise of User-Centered Design. *Cartogr. Geogr. Inf. Sci.* **2011**, *38*, 250–257. [CrossRef]

18.	Zastrow, M. Data visualization: Science on the map. *Nature* **2015**, *519*, 119. [CrossRef] [PubMed]

19.	Liu, Y.; Liu, X.F.; Mao, J.H. Research on the Integration of Silverlight and WebGIS Based on REST. In Proceedings of the 2010 International Conference on Multimedia Technology (ICMT), Ningbo, China, 29–31 October 2010; pp. 1–4.

20.	Vaitis, M.; Feidas, H.; Symeonidis, P.; Kopsachilis, V.; Dalaperas, D.; Koukourouvli, N.; Taskaris, S. Development of a spatial database and web-GIS for the climate of Greece. *Earth Sci. Inform.* **2019**, *12*, 97–115. [CrossRef]

21.	Wang, X.J.; Li, C.H. Design and Implementation of Tourism System Based on WebGIS. *Comput. Technol. Dev.* **2018**, *8*, 31.

22.	Vitianingsih, A.V.; Cahyono, D.; Choiron, A. Web-GIS Application Using Multi-Attribute Utility Theory Method as an Alternative Classification of New Highway Development with Heavy Traffic. *Adv. Sci. Lett.* **2018**, *24*, 9186–9192. [CrossRef]

23.	Ardalan, A.; Rad, M.K.; Hadi, M. Urban Water Issues in the Megacity of Tehran. In *Urban Drought*; Springer: Singapore, 2019; pp. 263–288.

24.	Saatsaz, M. A historical investigation on water resources management in Iran. *Environ. Dev. Sustain.* **2019**, 1–37. [CrossRef]

25.	Alizadeh, Z.; Yazdi, J.; Mohammadiun, S.; Hewage, K.; Sadiq, R. Evaluation of data driven models for pipe burst prediction in urban water distribution systems. *Urban Water J.* **2019**, *16*, 136–145. [CrossRef]

26.	Motiee, H.; Ghasemnejad, S. Prediction of pipe failure rate in Tehran water distribution networks by applying regression models. *Water Supply* **2019**, *19*, 695–702. [CrossRef]

27.	Tchórzewska-Cieślak, B.; Boryczko, K.; Piegdoń, I. Possibilistic risk analysis of failure in water supply network. In *Safety and Reliability: Methodology and Applications*; CRC Press: Boca Raton, FL, USA, 2015; pp. 1473–1480.

28. Ye, D.; Liu, Q. Simulation for Emergency Treatment of Squib Based on Genetic Algorithm. In Proceedings of the 2010 Second International Workshop on Education Technology and Computer Science (ETCS), Wuhan, China, 6–7 March 2010; pp. 296–299.

29. Ouyang, P.; Jian, J.; Zhang, Z. Pipe Burst Analysis Based On GIS. In *ICPTT 2013: Trenchless Technology*; American Society Of Civil Engineers: Reston, WV, USA, 2013; pp. 10–18.

30. Chen, R.; Shen, X. Analysis and Realization of Burst Pipe's Closing Valve in Gas Pipe System Based on ArcGIS Engine. *J. Kunming Metall. Coll.* **2011**, *1*, 5.

31. Jiankang, W.J. Valve Closing Analysis of Water Distribution Network in Case of Multi-accidents. *Ship Electron. Eng.* **2009**, *7*, 46.

32. ArcGIS Server. Available online: https://enterprise.arcgis.com/en/server/latest/develop/windows/about-extending-services.htm (accessed on 22 July 2019).

33. Cheng, G.X.; Hu, S.Q. System architecture and pattern research of RIA based on Silverlight. *Comput. Eng. Des.* **2010**, *8*, 1706–1709.

34. Gao, S.; Qin, F.; Liu, J.; Qin, L.; Li, J. Design of the tourism service WebGIS based on Silverlight. In Proceedings of the 2013 21st International Conference on Geoinformatics, Kaifeng, China, 20–22 June 2013; pp. 1–5.

35. Khan, Z.A.; Adnan, M. Usability evaluation of web-based GIS Applications. In Proceedings of the 11th International Conference on Information Integration and Web-based Applications & Services-iiWAS'09, Paris, France, 8–10 November 2010.

Article

Pattern Recognition and Clustering of Transient Pressure Signals for Burst Location

Daniel Manzi [1], Bruno Brentan [2], Gustavo Meirelles [2], Joaquín Izquierdo [3,*] and Edevar Luvizotto Jr. [1]

[1]   School of Civil Engineering, Architecture and Urban Planning, University of Campinas, 951 Albert Einstein Av., 13.083-189 Campinas, SP, Brazil; dmanzi@gmail.com (D.M.); edevar@fec.unicamp.br (E.L.J.)

[2]   Hydraulic and Water Resources Department, Federal University of Minas Gerais, 6627 Antônio Carlos Av., 31270-901 Belo Horizonte, MG, Brazil; brunocivil08@gmail.com (B.B.); gustavo.meirelles@ehr.ufmg.br (G.M.)

[3]   Fluing-Institute for Multidisciplinary Mathematics, Universitat Politècnica de València, Camino de Vera s/n, 46022 Valencia, Spain

*   Correspondence: jizquier@upv.es; Tel.: +34-628-028-804

Received: 19 September 2019; Accepted: 25 October 2019; Published: 30 October 2019

Abstract: A large volume of the water produced for public supply is lost in the systems between sources and consumers. An important—in many cases the greatest—fraction of these losses are physical losses, mainly related to leaks and bursts in pipes and in consumer connections. Fast detection and location of bursts plays an important role in the design of operation strategies for water loss control, since this helps reduce the volume lost from the instant the event occurs until its effective repair (run time). The transient pressure signals caused by bursts contain important information about their location and magnitude, and stamp on any of these events a specific "hydraulic signature". The present work proposes and evaluates three methods to disaggregate transient signals, which are used afterwards to train artificial neural networks (ANNs) to identify burst locations and calculate the leaked flow. In addition, a clustering process is also used to group similar signals, and then train specific ANNs for each group, thus improving both the computational efficiency and the location accuracy. The proposed methods are applied to two real distribution networks, and the results show good accuracy in burst location and characterization.

Keywords: water distribution systems; pipe bursts; hydraulic transients; real-time control; machine learning

---

1. Introduction

A significant part of the water produced for urban consumption is lost across supply systems between sources and final consumers. These losses range from less than 20% in developed countries up to 50% in developing nations [1]. Non-revenue water losses naturally increase through the normal operation of systems because of their gradual deterioration, which gives rise to physical losses. In this paper we are interested in breaks occurring in water networks that lead those systems to operate under conditions sometimes far away from the design conditions. Specifically, we are concerned about fast restoration of system efficiency, in other words, to make the time elapsed between the report of a new break and its effective location and repair—defined as run time—be as short as possible.

Due to the importance of this problem, many studies have sought for solutions to reduce water loses derived from leakage and pipe bursts; see, for example, [2–7]. Moreover, during the last decades, the improvements in the information and communication technologies applied to water distribution systems (WDSs) have enabled the production of substantial amounts of data, most of it related to the hydraulic state of the network. A number of smart solutions have been developed in the wake.

In urban hydraulics, for example, pressure and flow data have been used to identify and locate leakage and bursts, as observed in many works in the literature [4–7]. Special attention has been given to transient pressure signals [8,9]. Machine learning techniques and statistical inference of the inflows into a system have also been applied, especially in real-time control and monitoring [10–13].

A transient flow model to detect and locate pipe bursts is proposed by [14]. The method is integrated by two modules. By filtering demand fluctuations, the first module is responsible for monitoring and evaluating the inflows into the network. The second module locates the water loss. Flows identified as losses are added to the consumption demands, and an objective function, defined as the square of the difference between measured and observed pressures, is evaluated. The lowest value of this function points to the node in which the burst occurred. All the nodes are considered potential loss points, although more than one burst occurring simultaneously is not studied.

Calibration processes of water networks have also been used to detect and locate pipe bursts and leaks. Calibration is a fundamental process for many applications of water distribution system analysis. With calibration methods, open parameters, such as roughness, leakage, and pressure wave speed propagation are adjusted. Online calibration of nodal demand [5,15] typically uses least squares and geographically allocated demand. The increase of the final nodal demand results in impaired identification.

The inverse transient method was used in [5] for roughness and leakage calibration and used a genetic algorithm (GA) in the optimization process. The authors proposed a hybrid optimization model based on a GA and Levenberg–Marquardt theory, resulting in a more accurate solution, when compared with a single optimization algorithm.

During the last decade, the use of machine learning approaches for leakage detection and location has witnessed a notable increase. For example, [16] used simulated data of steady state flow to train a multi-class support vector machine aiming to identify leakage areas. The authors highlighted the efficiency of their approach using nonlinear pattern recognition tools to locate water losses.

Flow and pressure data were processed by both static and dynamic artificial neural networks (ANNs) to detect pipe bursts in water systems in [17]. Bursts were simulated by hydrant maneuvers, with data generated for every minute. The accuracy to detect pipe bursts was closely related to the capacity of the ANNs to process the non-linearities associated to the hydraulic parameters.

Frequency analyses of transient pressure signals can also be used for leakage location, as shown in [18]. However, the authors highlighted that these analyses heavily rely on an extremely precise transient model for a good evaluation of the integrity of the system—something which is challenging due to the existence of leaks in several locations, and to the nature of various magnitudes, thus limiting the method's application to real systems.

The standing wave difference method was applied in [19] to locate leaks in pipelines. The authors used pressure signal and spectral analysis of maximum pressure amplitude to identify the leakage locations.

A wavelet-based analysis was performed in [20] to process transient pressure data to detect and locate pipe bursts in WDSs. The location of a burst was accomplished by a graph-based algorithm, which used the arrival time of the pressure wave to locate the various measurement points.

A study of bursts in WDSs with various loading conditions was presented in [12]. The transient pressure signals were evaluated at various points, and the results revealed, for the same burst, different behaviors of the observed pressures at those points. This technique enabled a correlation to be identified between the amplitude of the pressure signal and its proximity to the lossy point.

Considering the intrinsic behavior of a transient pressure signal, the present work proposes water loss location based on the pressure signal information resulting from arbitrary pipe bursts. The representation of the pressure signals is considered under three arrangements: (i) the entire pressure series, including data of the initial and final steady-state conditions; (ii) a symbolic representation of this time series using the SAX technique, with a discrete series based on 20 pressure levels; and (iii) the peak information of the pressure signal at the sensors, represented by just the maximum pressure drop, and the time interval between the initial steady-state condition and that peak. Using simulated data for a range of bursts in all nodes of the network, an ANN is trained using as input the representation of the pressure signal of a limited number of sensors placed in the network. The output is the node where the burst occurred together with the leaked volume. To improve the location accuracy, a clustering process is performed using a hybridization of a self-organizing map (SOM) and a k-means methodology, whereby similar pressure signals are grouped. Finally, specific ANNs are trained for each pressure cluster, which improves the computational efficiency and the location accuracy. This tool, as a real-time control mechanism using pressure signals from the distributed sensors, will improve management in water distribution by rapidly identifying location and magnitude of bursts.

The paper is organized as follows. Section 2 provides a succinct description of materials and methods. Then the two real-world networks used for testing are described in Section 3, and the obtained results are also reported and discussed. Finally, conclusions are given in Section 4, before closing the paper with the Section of references.

2. Materials and Methods

2.1. Pipe Bursts and Transient Flow Modeling

The literature presents several methods for modeling leaks in a water network, using both steady and unsteady state modeling, and a suitable orifice area approximation for leakage flows [21–23]. In this paper, to create a relevant and reliable database of transient pressure signals, simulated results from pipe bursts were used. Also, a realistic leakage modeling, which considers pressure-dependent flow through the break, following the model proposed by [23], was used (Equation 1). To accommodate leaked flow within an expected range of 5–10% of the total inflow, the break area, A_0, was adjusted for each simulation.

$$Q_l = C_d \cdot \sqrt{2 \cdot g} \cdot \left(A_0 \cdot h^{0.5} + m \cdot h^{1.5}\right) \tag{1}$$

In Equation (1), Q_l is the leaked flow, C_d is the discharge coefficient, g is the gravity acceleration, A_0 is the area of the break, h is the available head, and m the pressure–area coefficient.

A pipe burst causes an instantaneous pressure drop at the break point, which is transmitted to the entire network. The event creates a pressure signal in each node with a unique amplitude and time delay according to its distance from the burst location, the size of the break and the various pipes characteristics (diameter, material and thickness). The signal propagation is modeled from the Eulerian viewpoint by means of the system of partial differential equations obtained when the momentum and mass conservation laws are applied to a pipe flow. The method of characteristics (MOC) [24] is used to solve that system of equations. The MOC transforms the system into an ordinary differential problem, and integration can be performed using some discretization in the space-time plane. This discretization should meet the Courant stability criterion [14], namely $\Delta x = a \cdot \Delta t$, where Δx and Δt are the discretization step sizes for space and time, respectively, and a is the wave speed.

Differently from single pipelines, in water distribution networks the upstream and downstream sections of a pipe can be connected to various other pipes and control elements, such as pumps and valves. Pipe internal calculation points were defined by the pipe discretization, and pressure and flow values at those points were calculated using information propagated by the positive and negative characteristic lines (Equations (2)–(4)) [25].

$$H_P = (H_A + B \cdot Q_A) - (B + R \cdot |Q_A|) \cdot Q_P = CA - BA \cdot Q_P \tag{2}$$

$$H_P = (H_B - B \cdot Q_B) + (B + R \cdot |Q_B|) \cdot Q_P = CB + BB \cdot Q_P \tag{3}$$

$$Q_P = \frac{CA - CB}{BA + BB} \tag{4}$$

In these equations, H_P and Q_P are the hydraulic head at an internal point and the flow through an internal point, respectively, at time $t + 1$; H_A and H_B are the hydraulic head of the upstream and downstream internal points, respectively, at time t; Q_A and Q_B are the flow of the upstream and downstream internal points, respectively, at time t; B is a pipe constant; R is the pipe resistance coefficient, linked to the hydraulic head loss equation; CA and BA are the coefficients of the positive characteristic line; and CB and BB are the coefficients of the negative characteristic line.

For the end points, the procedure presented in [18] was used, where the continuity law is applied for a generic node, as presented in Figure 1. The flow of a convergent pipe has a positive signal and can be calculated using a positive characteristic line, while a divergent pipe has a negative signal and the negative characteristic line is used. Accordingly, Equation (5) is obtained, which is applied to any node.

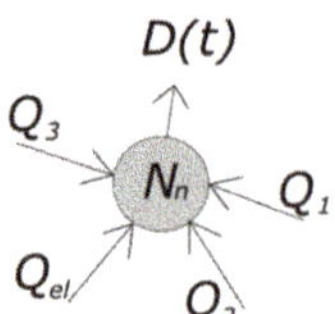

Figure 1. Generic node of a water distribution system (WDS).

$$\left[\sum_{i=1}^{CP}\frac{CA(i)}{BA(i)} + \sum_{j=1}^{DP}\frac{CB(j)}{BB(j)} - D(t)\right] - \left[\sum_{i=1}^{CP}\frac{1}{BA(i)} + \sum_{j=1}^{DP}\frac{1}{BB(j)}\right]\cdot HN - Q_{el} = EN - BN\cdot HN - Q_{el} = 0 \quad (5)$$

Here, CP and DP are the numbers of converging and diverging pipes, respectively, connected to the node; (i) and (j) indicate that coefficients correspond to pipes i and j, respectively; HN is the hydraulic head on the node; Q_{el} is the flow through a non-pipe element; EN and BN are the expressions in brackets in Equation (5); and $D(t)$ is the nodal demand at time t.

2.2. Database Creation and Proposed Models

To create the database to train the various ANNs, each node of the studied networks was considered as a potential bursting point. Ten values for the leaked flow were simulated, considering that their intensity may range from 5% to 10% of the total inflow of the network. The signal includes information of the initial and final steady-state conditions. A discretization time step of 0.01 s was used. For all the presented methods, the time series used as input corresponded to 60 s and contained information about the entire relevant transient.

For the first model, named LOCP (standing for location method using pressures), the original series of the pressure signals was used as input, resulting in 6000 to 60,000 data points. This method contains very detailed information about the signal and can result in redundancies for the ANN.

The second method, named LOCSAX (standing for location method using SAX), which represents time series though strings, tries to fix this drawback by using symbolic aggregate approximation (SAX) [26] to reduce the amount of information by using a small number of symbols. The representation of time series using symbols has attracted interest in several areas of knowledge, such as computer science, astronomy and medicine, to deal with problems of clustering, classification, indexing and detection of anomalies. Several techniques of symbolic representation of time series for data mining have been proposed in the last decades. Some of them exhibit difficulties derived from the dimensionality of the problems, as they maintain the same dimension of the original data, thus also maintaining the same scale of the problem [26]. SAX is applied in [27] to forecast water demand in WDSs. This technique transforms a time series into a string with reduced data dimensionality, while still providing great ability to compare similarities between time series and to discriminate between them. The pressure time series is divided into a number of n segments, and the average value of each segment is classified into one of the m letters of an alphabet, as exemplified in Figure 2.

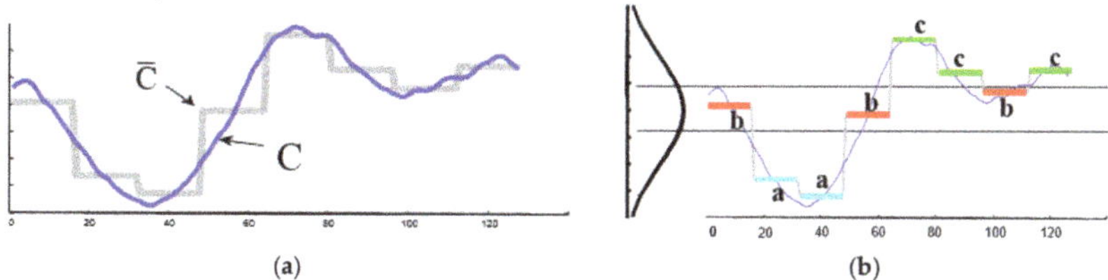

Figure 2. Converting a time series (blue line) to symbolic aggregate approximation (SAX) code using an alphabet of three letters: (**a**) original time series (C) and discretization ($\overline{C}$) for $n = 8$; (**b**) string (baabccbc) conversion of $\overline{C}$, using an alphabet of $m = 3$ symbols (source: adapted from [27]).

In this work, the number of characters representing the time series varied from 5 to 20, with $n = 10$ and $m = 20$ being the best combination. Thus, the input for the ANN is a vector containing just $n = 10$ characters.

Finally, the third method, named LOCPEAKS (standing for location method using peak pressure information), uses only the most salient information of the transient pressure signal, namely the maximum pressure drop observed (p) and the time delay for the pressure sensor to record this value (o), as illustrated in Figure 3. Although this method contains the essential information about the pressure transient, its limited information can make it difficult for the ANN to distinguish bursts occurring in nodes with similar conditions of flow, elevation and distance to the monitoring points.

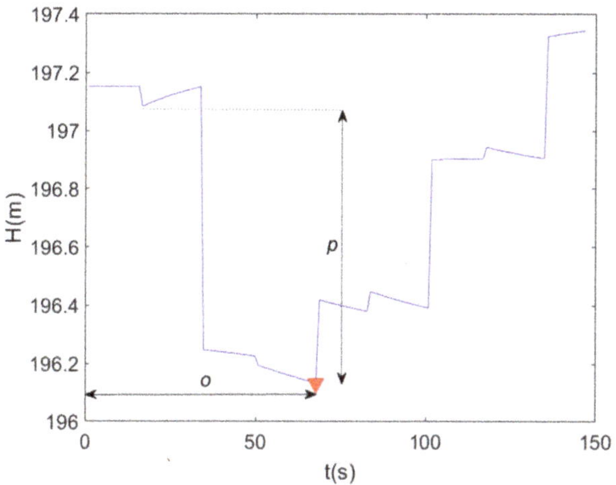

Figure 3. Pressure drop amplitude (p) and its time delay during a pipe burst (o) at certain sensor (red triangle). Axes are in seconds and water column meters for total head.

In the three methods, the output response of the ANN is the index of the node where the burst occurred and its leaked flow.

2.3. ANN Training for Burst Location

ANNs have been successfully applied in non-linear multivariable function approximations, and as classifiers. The ANN's ability to predict patterns has found success in practical applications such as WDS calibration and demand forecasting [28,29]. The ANN training process followed the flowchart presented in Figure 4. The database containing the pressure signal together with the location and intensity of the lost flows was loaded and then the representation method was selected. After the preprocessing of the pressure signal according to the selected method, 50% of the database was used for training, 35%

for validation, and the remaining 15% was used for testing. The ANN was built with three hidden layers, and with a (20, 40, 20) layer-architecture of neurons. The Levenberg–Marquardt technique was used to obtain the synaptic weights of the neurons, and the stop criteria was set to 500 epochs or a relative change of less than 10^{-8}.

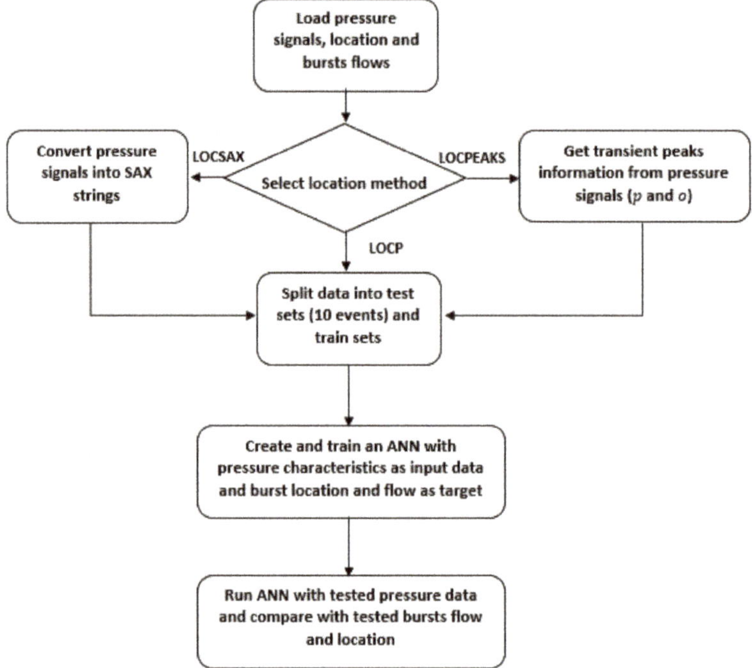

Figure 4. Flowchart of artificial neural network (ANN) training for burst location.

2.4. Clustering Pressure Signal to Improve ANN Efficiency

Each pipe burst creates a unique pressure signal input in each sensor placed in the WDS. Yet, when a pipe burst occurs with different intensity, the shape of the pressure signal is very similar, with just a slight difference in the peaks. In addition, when bursts in two pipes have the same intensity, the respective pressure signals are very similar as well, with some transmission lag. As a result, it is possible to cluster the observed signals, and train specific ANNs for each group, thus reducing the error in the burst location.

To define the clusters, the hybrid methodology proposed by the authors in [30], using SOMs coupled with the k-means algorithm, was used. The neurons obtained using SOMs cluster the input data by their similarity. Calculating the distance between each neuron and the input data, a dissimilarity matrix can be created:

$$D == \begin{bmatrix} \|w_1 - x_1\|^2 & \cdots & \|w_1 - x_n\|^2 \\ \vdots & \ddots & \vdots \\ \|w_l - x_1\|^2 & \cdots & \|w_l - x_n\|^2 \end{bmatrix}. \tag{6}$$

This matrix can be used as input for the k-means algorithm, where various numbers of clusters are considered. Here $w = (w_i)$ is the neuron l-vector of synaptic weights; and $x = (x_j)$ is the input n-vector. For this study, the architecture of the SOM was defined following the discussions presented in [30], based on a tradeoff between the quantization error and the training time. The topology followed a hexagonal distribution. The SOM with the best tradeoff had 16 neurons.

Finally, the *CH* index [31], shown in Equation (7), was calculated to determine the optimal number of clusters.

$$CH = \frac{\left[\frac{\sum_{k=1}^{K} n_{k.}\|c_k - c\|^2}{K-1}\right]}{\left[\sum_{k=1}^{K} \sum_{i=1}^{n_k} \frac{\|x_i - c_k\|^2}{n-K}\right]} \tag{7}$$

In Equation (7), n_k is the number of elements of cluster k; $c = (c_k)$ is an element in cluster k; c is the centroid of all input data; K is the number of clusters; and n is the number of input data.

3. Results

3.1. Jardim Laudissi Network

The Jardim Laudissi network is a Brazilian district metered area (DMA) in Piracicaba, São Paulo, having 222 user connections, 2.7 km total pipe length, and three points for pressure monitoring. One monitoring point is at the entrance of the network, and the two others are at critical points in the network (highest and lowest elevations), representing 12% of the 25 nodes of the hydraulic model. Pressure sensors are located to monitor both minimum and maximum pressures inside the DMA and to drive a pressure reduction valve (PRV) installed at the DMA entrance, so as to maintain a suitable pressure control. The network topology and pressure sensor locations (noted as green circles —"Obs nodes (3)") are presented in Figure 5b. The results of the methods' application were evaluated both in terms of the Euclidean distance between estimated and real bursting node and the error between estimated and real leaked flow.

The application of LOCP, LOCSAX and LOCPEAKS methods to the Jardim Laudissi network indicated good accuracy in locating events and in assessing their respective flows, as shown in Table 1. The LOCSAX presented the smallest average error for the leaked flow, while the shortest average distance for burst localization was performed by LOCP. Figure 5 shows the results for the LOCP method applied for 10 test scenarios, showing the true burst position and the estimated one.

Table 1. Jardim Laudissi network results for distance and leaked flow.

Method	LOCP	LOCSAX	LOCPEAKS
Average error in burst flow forecasting (%)	25	1	4
Average error in burst location (m)	55	80	83

3.2. Campos do Conde Network

The Campos do Conde network is also a Brazilian DMA with 854 user connections, and has a total network length of approximately 12 km, with diameters varying from 50 to 200 mm. As the Campos do Conde network is a bigger and more complex network, 16 pressure monitoring nodes were considered. This represents 13.5% of the 118 nodes of the hydraulic model of the network, thus following a similar proportion as that used in the previous case. The topology and pressure sensor locations are presented in Figure 6.

The three proposed methods, LCOP, LOCSAX and LOCPEAKs, were applied for this network, and the results are shown in Table 2. In contrast to the Jardim Laudissi network results, here the LOCP method showed the greatest accuracy both for location and forecasting of the leaked flow. Figure 6 shows the results for the LOCP method applied for 10 test scenarios, showing the true burst position and the estimated one.

Table 2. Campos do Conde results for leaked flow and distance.

Method	LOCP	LOCSAX	LOCPEAKS
Average error in burst flow forecasting (%)	2	16	9
Average error in burst location (m)	82	109	280

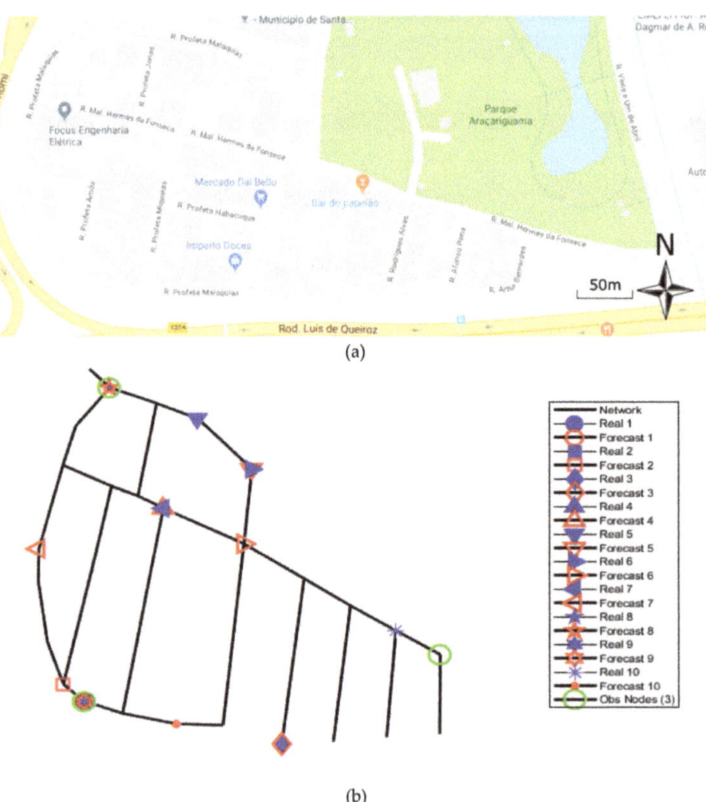

(a)

(b)

Figure 5. (**a**) Scale map of the district metered area (DMA) Laudissi: adapted from Google Maps (**b**) Results for test scenarios using LOCP method in Jardim Laudissi network.

As the database is much larger for Campos do Conde network than it is for the Laudissi network, the above-mentioned clustering procedure using SOMs and k-means was used. As a result, ten groups of transient pressure signals collected by the sensors were created, as shown in Figure 7. In an attempt to simplify the methodology for real networks, the use of a reduced number of sensors was also evaluated. Thus, in addition to the case with sixteen sensors, a midway scenario with eight sensors representing 6.8% of the network nodes and a low-level monitoring scenario with only four sensors representing 3.4% of the network nodes were also studied. The results obtained for the Campos do Conde network are summarized in Table 3. The improvement is clear for all the three methods when the clustering is made, but the LOCP still had the best performance. It is interesting to observe that when using this approach, reducing the number of sensors did not harm the performance when compared with the case with no clustering. In fact, the best performance was observed for the midway (eight sensors) condition, probably due to the reduction of redundant information together with the maintenance of essential sensors for the ANN. Figure 7 presents an instance of the final clustered signals using measurements from eight sensors. It is possible to highlight visible pattern differences between clusters which identify different network regions.

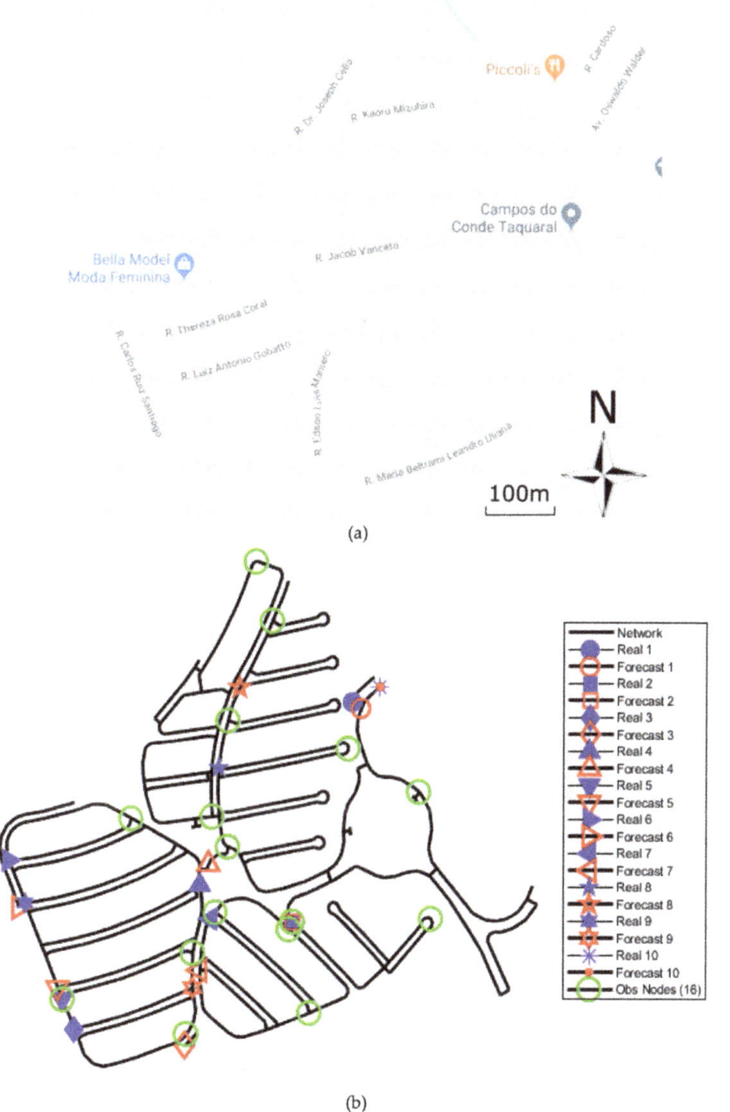

Figure 6. (**a**) Scaled map of the DMA Campos do Conde: adapted from Google Maps (**b**) Results for test scenarios using LOCP method in Campos do Conde network.

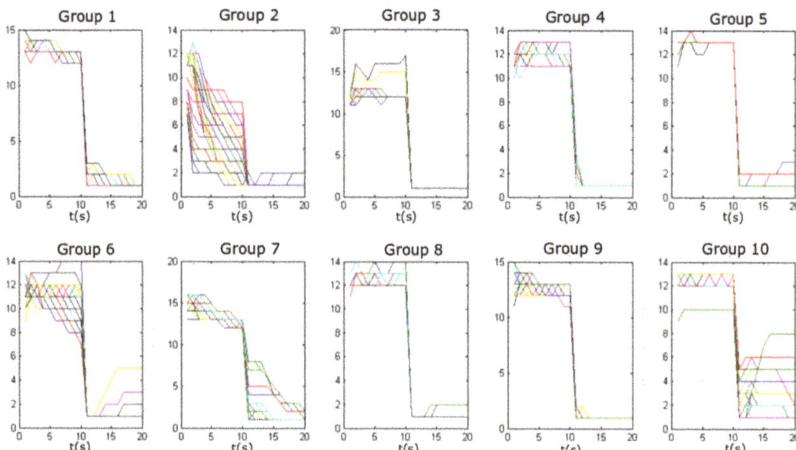

Figure 7. Clustering for the LOCP method using eight sensors in the Campos do Conde network.

Table 3. Campos do Conde results with clustering.

Number of Sensors	Method	Average Error in Burst Location (m)
16	LOCP	67
	LOCSAX	89
	LOCPEAKS	156
8	LOCP	52
	LOCSAX	156
	LOCPEAKS	171
4	LOCP	73
	LOCSAX	133
	LOCPEAKS	172

4. Conclusions

The resilience of a WDS strongly depends on good conditions of design, installation and maintenance. However, in operational terms, the reduction of run times of potential bursts allows the reduction of volumes lost in break events which also helps maintain good resilience standards. The pressure signals caused by bursts have an important correlation with their location and magnitude, thus stamping a "hydraulic signature" of these events. For this signature to provide reliable information, the network must be suitably calibrated. Otherwise, the associated uncertainty in pipe characteristics, such as material, age and effective diameter, with well-known great influence on the propagation speed of transient signals, would blur the obtained results, thus turning them useless. In our study we have performed simulations on two networks with perfectly defined characteristics.

In this paper, the use of ANNs as pattern recognizers has succeeded in the identification of hydraulic patterns of various tested bursts. The proposed methods allow the detection and location of bursts in WDSs, after analyzing the characteristics of the transient signals using data acquisition rates compatible with commercial equipment (of the order of 1 kHz), and without the need to measure the inlet flow to the system, which is more difficult to acquire at higher frequencies. The results showed that the LOCP method is the most accurate method to locate bursts. Although the data used to train the ANN are significantly larger for this method, the complete signal appears to contain relevant information that is not well described by the other two methods. Both LOCSAX and LOCPEAKS are able to estimate the burst location with less effort, however these methods are less accurate than LOCP. Even though the use of the complete transient data increases the effort to train the ANNs, this procedure has to be done only once, with the network re-trained sporadically upon the arrival of new data or after modifications in the network topology. Furthermore, the use of clusters to improve the accuracy for burst location estimation reduces the number of data to train the ANNs since thanks to data similarities, the training process is easier. Finally, the clustering of transient signals increased the performance of all three methods and showed that, even in a reduced monitoring scenario, burst location was satisfactory.

In practical terms, this situation contributes to technical and economic feasibility of networks for fast location of breaks, by means of commercially available pressure measurement equipment with very affordable costs compared to flowmeters with the same accuracy. The methods have been tested on two small and medium size networks, which could be seen as district metered areas of larger networks. As a result, we recognize that the application of the proposed methods may be limited by the network size. However, when using transient pressure signals, the effect of a water loss probably will not be detected by a sensor far from the DMA where the burst occurs. This is because of the attenuation process of the pressure wave due to the friction effects. In addition, a larger network is likely to have more pressure sensors available, thus there is more information to process in the ANN training, allowing a broader coverage of the network. As future works, it is recommended to evaluate the methods considering entire large networks using different sampling frequencies for sensors, which will enrich the methodology presented with analyses of the uncertainty derived from pressure sensors in high frequency data acquisition.

Author Contributions: Conceptualization: D.M., B.B., G.M. and E.L.J.; software: D.M. and B.B.; investigation: D.M., B.B., G.M., J.I. and E.L.J.; supervision: B.B., J.I., E.L.J.; writing, review and editing: B.B. and J.I.

Funding: This research received no external funding.

Conflicts of Interest: The authors declare no conflict of interest.

Abbreviations

The following abbreviations are used in this manuscript:

Abbreviation	Explanation
ANN	artificial neural network
CH	Caliński-Harabasz (index for cluster quality)
DMA	district metered area
GA	genetic algorithm
LOCP	location method using pressures
LOCPEAKS	location method using peak pressure information
LOCSAX	location method using SAX
MOC	method of characteristics
PRV	pressure reduction valve
SAX	symbolic aggregate approximation
SOM	self-organizing map
WDS	water distribution system

References

1. Farley, M.; Trow, S. *Losses in Water Distribution Networks*; IWA Publishing: London, UK, 2003.
2. Creaco, E.; Walski, T. Economic Analysis of Pressure Control for Leakage and Pipe Burst Reduction. *J. Water Resour. Plan. Manag.* **2017**, *143*, 4017074. [CrossRef]

3. Campisano, A.; Creaco, E.; Modica, C. RTC of valves for leakage reduction in water supply networks. *J. Water Resour. Plan. Manag.* **2009**, *136*, 138–141. [CrossRef]

4. Campisano, A.; Modica, C.; Reitano, S.; Ugarelli, R.; Bagherian, S. Field-Oriented Methodology for Real-Time Pressure Control to Reduce Leakage in Water Distribution Networks. *J. Water Resour. Plan. Manag.* **2016**, *142*, 04016057. [CrossRef]

5. Vítkovský, J.P.; Simpson, A.R.; Lambert, M.F. Leak Detection and Calibration Using Transients and Genetic Algorithms. *J. Water Resour. Plan. Manag.* **2000**, *126*, 262–265. [CrossRef]

6. Pérez, R.; Puig, V.; Pascual, J.; Quevedo, J.; Landeros, E.; Peralta, A. Methodology for leakage isolation using pressure sensitivity analysis in water distribution networks. *Control Eng. Pract.* **2011**, *19*, 1157–1167. [CrossRef]

7. Jung, D.; Kim, J.H. Robust Meter Network for Water Distribution Pipe Burst Detection. *Water* **2017**, *9*, 820. [CrossRef]

8. Covas, D.; Ramos, H. Hydraulic transients used for leakage detection in water distribution systems. In Proceedings of the 4th International Conference in Water Pipeline Systems, York, UK, 28–30 March 2001.

9. Colombo, A.F.; Lee, P.; Karney, B.W. A selective literature review of transient-based leak detection methods. *HydroResearch* **2009**, *2*, 212–227. [CrossRef]

10. Choi, D.Y.; Kim, S.-W.; Choi, M.-A.; Geem, Z.W. Adaptive Kalman Filter Based on Adjustable Sampling Interval in Burst Detection for Water Distribution System. *Water* **2016**, *8*, 142. [CrossRef]

11. Christodoulou, S.E.; Kourti, E.; Agathokleous, A. Waterloss detection in water distribution networks using wavelet change-point detection. *Water Resour. Manag.* **2017**, *31*, 979–994. [CrossRef]

12. Misiunas, D. Failure Monitoring and Asset Condition Assessment in Water Supply Systems. Ph.D. Thesis, Vilniaus Gedimino Technikos Universitetas, Vilnius, Lithuania, 2008.

13. Guo, X.; Yang, K.; Guo, Y. Leak detection in pipelines by exclusively frequency domain method. *Sci. China Ser. E Technol. Sci.* **2012**, *55*, 743–752. [CrossRef]

14. Holloway, M.B.; Chaudhry, M.H. Stability and accuracy of waterhammer analysis. *Adv. Water Resour.* **1985**, *8*, 121–128. [CrossRef]

15. Sanz, G.; Pérez, R.; Kapelan, Z.; Savic, D. Leak detection and localization through demand components calibration. *J. Water Resour. Plan. Manag.* **2015**, *142*, 04015057. [CrossRef]

16. Zhang, Q.; Wu, Z.Y.; Zhao, M.; Qi, J.; Huang, Y.; Zhao, H. Leakage Zone Identification in Large-Scale Water Distribution Systems Using Multiclass Support Vector Machines. *J. Water Resour. Plan. Manag.* **2016**, *142*, 04016042. [CrossRef]

17. Mounce, S.R.; Machell, J. Burst detection using hydraulic data from water distribution systems with artificial neural networks. *Urban Water J.* **2006**, *3*, 21–31. [CrossRef]

18. Almeida, A.B.; Koelle, E. *Fluid Transients in Pipe Networks*; Computational Mechanics Publications: Southampton, UK, 1992.

19. Covas, D.I.C.; Ramos, H.M.; De Almeida, A.B. Standing Wave Difference Method for Leak Detection in Pipeline Systems. *J. Hydraul. Eng.* **2005**, *131*, 1106–1116. [CrossRef]

20. Srirangarajan, S.; Allen, M.; Preis, A.; Iqbal, M.; Lim, H.B.; Whittle, A.J. Wavelet-based burst event detection and localization in water distribution systems. *J. Signal Process. Syst.* **2013**, *72*, 1–16. [CrossRef]

21. Liggett, J.A.; Chen, L. Inverse Transient Analysis in Pipe Networks. *J. Hydraul. Eng.* **1994**, *120*, 934–955. [CrossRef]

22. Caputo, A.C.; Pelagagge, P.M. An inverse approach for piping networks monitoring. *J. Loss Prev. Process. Ind.* **2002**, *15*, 497–505. [CrossRef]

23. Van Zyl, J.E. Theoretical modeling of pressure and leakage in water distribution systems. *Procedia Eng.* **2014**, *89*, 273–277. [CrossRef]

24. Wylie, E.B.; Streeter, V.L.; Suo, L. *Fluid Transients in Systems*; Prentice Hall: Englewood Cliffs, NJ, USA, 1993; Volume 1, p. 464.

25. Izquierdo, J.; Iglesias, P. Mathematical modelling of hydraulic transients in complex systems. *Math. Comput. Model.* **2004**, *39*, 529–540. [CrossRef]

26. Lin, J.; Keogh, E.; Wei, L.; Lonardi, S. Experiencing SAX: A novel symbolic representation of time series. *Data Min. Knowl. Discov.* **2007**, *15*, 107–144. [CrossRef]

27. Navarrete-López, C.; Herrera, M.; Brentan, B.M.; Luvizotto, E.; Izquierdo, J. Enhanced Water Demand Analysis via Symbolic Approximation within an Epidemiology-Based Forecasting Framework. *Water* **2019**, *11*, 246. [CrossRef]

28. Meirelles, G.; Manzi, D.; Brentan, B.; Goulart, T.; Luvizotto, E. Calibration Model for Water Distribution Network Using Pressures Estimated by Artificial Neural Networks. *Water Resour. Manag.* **2017**, *31*, 4339–4351. [CrossRef]

29. Adamowski, J.; Chan, H.F. A wavelet neural network conjunction model for groundwater level forecasting. *J. Hydrol.* **2011**, *407*, 28–40. [CrossRef]

30. Brentan, B.; Meirelles, G.; Luvizotto, E., Jr.; Izquierdo, J. Hybrid SOM+ k-Means clustering to improve planning, operation and management in water distribution systems. *Environ. Model. Softw.* **2018**, *106*, 77–88. [CrossRef]

31. Caliński, T.; Harabasz, J. A dendrite method for cluster analysis. *Commun. Stat. Theory Methods* **1974**, *3*, 1–2. [CrossRef]

 *water*

Case Report

Application of Rehabilitation and Active Pressure Control Strategies for Leakage Reduction in a Case-Study Network

Camillo Bosco *, Alberto Campisano, Carlo Modica and Giuseppe Pezzinga

Department of Civil Engineering and Architecture, University of Catania, Via Santa Sofia 64, 95123 Catania, Italy; alberto.campisano@unict.it (A.C.); carlo.modica@unict.it (C.M.); giuseppe.pezzinga@unict.it (G.P.)
* Correspondence: camillo.bosco@phd.unict.it

Received: 8 June 2020; Accepted: 3 August 2020; Published: 6 August 2020

Abstract: The paper discusses the results of a simulation analysis to evaluate the potential of rehabilitation measures and active pressure control strategies for leakage reduction in a water distribution network (WDN) in southern Italy. The analysis was carried out by using a simulation model developed under the EPANET-MATLAB environment. The model was preliminarily calibrated based on pressure and flow measurements acquired during a field monitoring campaign in two districts of the WDN. Three different scenarios of leakage reduction including (i) pipe rehabilitation (scenario S1), (ii) implementation of pressure local control (S2), and (iii) introduction of remote real-time pressure control (RTC) (S3) were simulated and compared with the current scenario of network operation (S0). Results of the simulations revealed that a combination of the used strategies can improve network performance by a significant reduction of water leakage. Specifically, 16.7%, 35.0%, and 37.5% leakage reductions (as compared to S0) can be obtained under scenarios S1, S2, and S3, respectively.

Keywords: water distribution systems; rehabilitation; pressure control; real-time control; leakage reduction strategies

---

1. Introduction

The level of service in water distribution networks (WDNs) is related to critical issues such as the frequency of breaks and the amount of background leakage in the network. Very often such issues are the cause of pressure drops in the network pipes, which are eventually detrimental to users' water demand satisfaction [1]. Modeling of the network coupled with information gained through measurement of flows and pressures can significantly contribute to the identification of anomalies in the WDN and the proper estimation of leakage levels, as well as to the selection of actions for network rehabilitation and technological renewal [2,3].

For instance, strategies include methods for identifying pipes with high rehabilitation priority [4], as well as methods to reduce pressure (and, thus, leakage) in the WDN [5,6]. Pipe rehabilitation/replacement programs require acquiring knowledge about the characteristics of the pipes (typically, material, diameter, age, current status, and historical frequency of breaks) [7]. Most pressure control methods are based on dividing the WDN into district metered areas (DMAs) and on installing pressure control valves (PCVs) at their entrance for limiting pressure values in the downstream conduits [5]. Normally, PCVs allow local pressure control (they allow controlling the pressure value locally at the valve outlet) with the aim to control pressure levels in the whole downstream district. Therefore, local control relies on the use of hydraulic models of the network to predict pressure levels in the downstream nodes of the district [8,9]. Hydraulic models are also adopted to determine the optimal placement of control valves for leakage reduction [10,11]. However, use of models does not assure the proper control of pressure in all circumstances. Indeed, because of the uncertainty in the estimation of the spatiotemporal distribution of nodal water demands and energy losses in the WDN, large network portions might exhibit pressure excess/deficit during the day as compared to setpoint values.

In recent years, the adoption of remote real-time control (RTC) has been shown to outperform methods based on architectures of local control with increased potential for pressure control and leakage reduction in WDNs [12–15]. RTC has been used in various modalities in the field of water engineering for the control of both urban drainage systems [16] and water distribution systems.

Differently from systems based on local control, remote RTC systems use (distributed) remote information about the present situation of the WDN in order to improve the effectiveness of pressure (and, thus, leakage) control.

Typically, pressure sensors are installed in the network (in nodes that are far from the control valve site) and acquire pressure measurements in a continuous way. Then, pressure measurements are transmitted in real time (normally using GSM) to controllers, specific devices that are programmed to dynamically adjust the PCVs [17,18]. Adjustments are carried out based on appropriate strategies and algorithms, in order to drive the pressure at the remote monitoring node/s to the desired setpoint. Many of the available studies on adoption of techniques of remote RTC of pressure in WDNs (e.g., [17–21]) have mainly invoked the use of simulation approaches. However, in most of the cases, models have been applied to WDNs, without being preliminarily calibrated on the basis of measurements in the network. In line of principle, such an approach cannot always assure accurate evaluation of the level of water leakage in the network due to the predictive error of the used model.

In this paper, a simulation analysis was carried out to evaluate the potential of rehabilitation measures and active pressure control strategies for leakage reduction in a WDN in southern Italy. The reliability of the obtained results is corroborated by the preliminary calibration of the model used for the simulations based on measurements acquired in the WDN during a specific campaign of monitoring. The experimental campaign was carried out separately for two network districts and included monitoring of pressure and flows at different nodes of the network.

The paper is structured in sections. First, the methods are presented, including the description of the water distribution network, the details of the monitoring campaign, the used simulation model, as well as the considered scenarios of simulation for leakage reduction. Secondly, results of the model calibration are presented, with emphasis on the obtained level of accuracy of the model. Third, simulation results are discussed for the purpose of comparing scenarios of application of strategies of rehabilitation and active pressure control for leakage reduction in the WDN.

2. Materials and Methods

2.1. Case-Study Network

A portion of the municipal WDN of the town of San Giovanni la Punta (Italy) was selected as case-study to evaluate benefits of some rehabilitation and active pressure control strategies.

The WDN (see Figure 1) consists of about 39 km of pipes and supplies about 6100 users (about 2400 households). Two main DMAs were identified in a conjunct work carried out with the water company that manages the water distribution in the network. The two districts supply the northwest area (namely district DMA1, pipes reported in red in Figure 1) and the southeast area (namely district DMA2, pipes reported in blue in Figure 1) of the town.

The whole system is supplied by the reservoir "Alto", which is located in the northern part of the town at 422 m a.s.l. (maximum capacity approximately equal to 100 m^3). The reservoir is supplied by well pumps operated through use of an inverter-based system that adjusts the inlet to maintain a constant water level of 2 m in the reservoir during a period of 24 h. A cast iron conduit conveys flow from the reservoir to the main branches and loops of the WDN including a 2 km long steel pipeline with the function of north–south main distribution line (reported in orange in Figure 1). Most of the pipes larger than DN200 in the WDN are made of steel and cast iron, and date back to the 1980s and 2000s, respectively. Conversely, the small conduits are almost entirely in HDPE and were installed in the early 2000s. Various surveys carried out during the last decade by the water company have revealed the occurrence of leakages in several branches of the WDN. In this regard, a point of weakness of the network concerns the described north–south main distribution line. Indeed, this line is affected by high leakage levels that are the cause of pressure problems (and inadequate water supply) for several households belonging to DMA1.

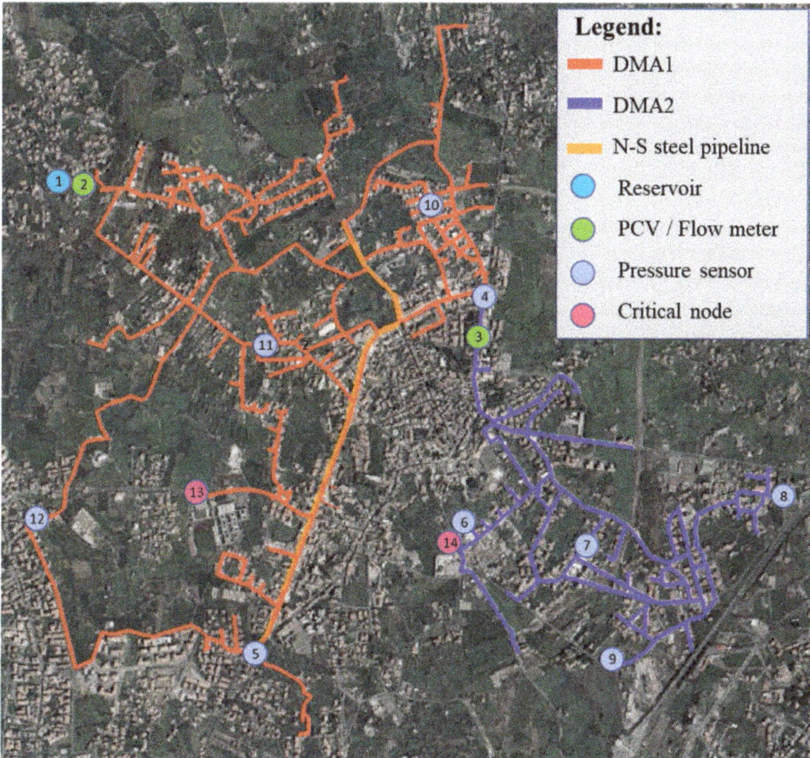

Figure 1. Sketch of the water distribution network (WDN) with identification of significant nodes (with critical nodes in magenta), districts district metered area 1 (DMA1) (in red) and DMA2 (in blue), and north–south main distribution line (in orange).

2.2. Monitoring Campaign

A campaign of monitoring was carried out during the research work to explore the behavior of the network in terms of water consumption and pressure in the two districts.

The Specific objective of the experimental monitoring was to identify critical issues in the current operation of the WDN (e.g., identify areas of leakage, hydraulic malfunctioning, etc.). A further objective was to set up a dataset of measurements to be used for the calibration of the simulation model adopted for the analysis of the WDN.

The experimental campaign was carried out separately for the two districts (in the months of October 2018 for DMA2 and April 2019 for DMA1). Installation of pressure sensors and devices required preliminary works aiming at isolation of network sectors and assessment of the condition and settings of existing valves in the network. A flow meter (at the inlet of the district) and five pressure sensors were installed in nodes of each DMA (see Figure 1) for two consecutive weeks. During such periods, flows and pressures were monitored at intervals of 1 and 10 min, respectively. The flow meters used allowed flows to be measured with an accuracy of 0.5%, while pressure sensors provided pressure measurements with 0.1% accuracy. All the data were stored locally and downloaded at the end of the monitoring period for the successive analyses. Recorded measurements enabled determining the daily pattern of the pressure in the two districts, thus, identifying areas at low or high pressure (thus, more or less prone to leakage) in the network during the 24 h. Furthermore, the analysis of the data allowed determining the daily pattern of flows supplied to the two DMAs, thus enabling a measure of the total sum of household consumption and of water leakages in the two districts.

Moreover, the comparison of the data of flow with quarterly data of grouped billed consumption (provided by the water company for each household of the WDN) allowed to unbundle leakage contribution by the total measured flow.

2.3. Model of the Network

A hydraulic model of the network was set up to assess benefits determined by implementation of scenarios of rehabilitation and active pressure control in order to reduce leakage levels in the WDN.

The extended period simulation (EPS) of the WDN was performed, i.e., the simulation was run assuming successive conditions of steady state for 24 h of the day.

Simulations were carried out under the MATLAB environment through the EPANET-MATLAB Toolkit [22]. The toolkit allows exploiting the tools of EPANET software, developed by the EPA (U.S. Environmental Protection Agency) for the simulation of water distribution networks [23]. The adopted skeletonization of the network was taken from the GIS owned by the water utility, which includes not only the primary level of loops but also the secondary one (i.e., the level of pipe branches before the building/household level). Overall, the network was skeletonized using 921 links and 869 nodes. The adopted skeletonization allowed considering a level-wise description of the network (primary and secondary branches and loops) appropriate for the correct allocation of nodal demands consistently with both acquired measurements and available billing data.

Preliminarily, pipe roughness was estimated based on the available information from surveys concerning pipe material and level of pipe corrosion. Notably, the inspection of pipes in different manholes of the network revealed a (high) level of corrosion of the north–south distribution pipeline typical of pipes in operation for several years. Differently, the plastic pipes and those in cast-iron showed better conditions, consistent with their lesser age. The inspection was carried out in about twenty manholes distributed in a rather uniform way in the two DMAs, thus providing an idea of the global conditions of the whole WDN. Roughness values for the three types of pipe materials were properly selected from the literature [24] based on the observed conditions from the field survey.

On this background, values of Hazen–Williams roughness coefficients were set equal to 140, 95, and 75 for pipes in HDPE, cast iron, and old steel, respectively.

Network leakage was evaluated using a pressure-driven approach [25], based on the following equation:

$$Q_k = \beta P_k^\alpha \sum_{J=1}^{n_{j,k}} \frac{L_{j,k}}{2} \tag{1}$$

where Q_k (L/s) is the leak at the k-node and P_k (m) is the corresponding pressure; $n_{j,k}$ is the total number of j-pipes converging to node k; $L_{j,k}$ (m) is the length of pipe j converging to node k; and α and β are leakage coefficients to be calibrated.

The analysis of the scientific literature shows efforts to characterize the behavior of leakage in WDNs in terms of values of the leakage exponent α. The value 1.18 has been proposed based on field data [26]. Values ranging between 0.5 and 2.5 have been found depending on pipe material, on the background soil hydraulic characteristics, and on the prominent type of leak [27–30]. Instead, β has been shown to depend on the level of leakage and on the description of the WDN. Very often, given their large variability, the values of these two coefficients have been chosen in the literature based on the use of specific calibration procedures [31].

A pressure control module was developed in MATLAB. The module was coupled to the hydraulic model to allow the simulation of potential benefits deriving from adoption of active control of pressure in the WDN based on remote RTC [12]. The control module allows implementing architectures of RTC systems and to use various types of PCVs (e.g., screw-based valves; plunger valves, etc.) for pressure control in the WDN. A control strategy (i.e., the sequence of control actions to drive the pressure to the setpoint) derived from the literature [17] was implemented in the module, which assumes that pressure at each of the two DMAs is controlled on the basis of the pressure value in one node (critical node) of the district. Specifically, in the pressure control module, it is assumed that pressure measurements acquired at the critical node are remotely transmitted in real time to the controller that operates the adjustment of the PCV. The control algorithm implemented in the pressure control module provides (at each control time step) the valve shutter displacement Δa (-) based on the deviation e_t (m) at time t between the current pressure value and the related setpoint value at the critical node:

$$\Delta a = a_{t+\Delta t_c} - a_t = -K\, e_t,$$ (2)

where a_t and $a_{t+\Delta t_c}$ are valve opening degrees at time t and $t + \Delta t_c$, respectively; Δt_c is the control time step, and K (m^{-1}) is the controller gain.

Equation (2) shows that Δa is proportional to e_t through K. Therefore, with respect to the shutter position a, the adopted algorithm shows the characteristics of an integral-type controller. Moreover, the negative sign in Equation (2) allows considering the negative proportionality between Δa and e_t, if gain K is assumed to be intrinsically positive.

Finally, valve regulation is constrained by the limits 0 and 1 (saturation), corresponding to valve fully closed (0) and fully open (1). The control module allows also including limits (of the valve manufacturer) on the mechanical velocity of the shutter of the valve, to prevent risks of unwanted transients in the network [12].

2.4. Model Calibration

The availability of measurements in the network allowed the calibration of the simulation model, thus, increasing the reliability of the simulation results concerning the potential benefits due to implementation of leakage control strategies in the WDN.

Calibration was carried out using the Genetic Algorithm Toolbox available in MATLAB environment. The adopted procedure is based on a recent application [31] and consists in calibrating simultaneously (through a genetic algorithm) the optimal values of the hourly multipliers of the daily curve of consumption and the optimal values of parameters α and β of Equation (1) for leakage evaluation. The values of the leakage parameters were assumed the same for all the network pipes, except for the coefficient β related to the steel north–south steel pipeline that was evaluated separately, to improve model results.

In order to correctly describe the demand dynamics in the network, hourly demand multipliers of each DMA were considered, i.e., coefficients that multiply the average daily consumption and provide the hourly demand in the nodes of the WDN.

Average hourly measured values of flow and pressure, recorded during the monitoring campaign at each DMA, were provided as model input with the aim of identifying the optimal leakage parameters and hourly demand multipliers by minimizing the following objective function:

$$OF = \sum_{h=1}^{n_h} \sum_{i=1}^{n_i} \left(\frac{P_{C_{i,h}} - P_{M_{i,h}}}{P_{M_{i,h}}} \right)^2,$$ (3)

where P_C (m) and P_M (m) are computed and measured pressure values, respectively; n_i is the total number of installed pressure sensors; and n_h is the number of hourly values considered.

Use of Equation (3) was subject to the constraint that total inflow to each DMA is always equal to the sum of household consumption and leakage in the district, at any time step of the simulation.

2.5. Scenarios of Simulation

Three main scenarios were considered for the simulations that include options of rehabilitation/active pressure control of the WDN. The first scenario (S1) concerns the rehabilitation of the described north–south steel pipeline in order to eliminate leakages identified during the field surveys. Such scenario includes improvement of water supply to users of DMA1 as determined by the re-arrangement of circulation of flows in the conduits.

The second scenario (S2) adds up to S1 the installation of two PCVs to allow local control of the pressure in the network. The two PCVs are assumed to be conventional screw-based valves and to be installed at the inlet of each DMA (PCV1 at node 2 of the DMA1 and PCV2 at node 3 of the DMA2, as shown in Figure 1). In this scenario, each valve is set with the objective of reducing as much as possible pressure levels in the respective DMA, while ensuring the full demand satisfaction to all the users of the district. This is obtained by preliminarily identifying the critical node of each DMA, that is the node with the lowest value of pressure in the district during 24 h (node 13 for DMA1 and node 14 for DMA2). Then, the scenario considers that pressure at the critical nodes can never drop down below the minimum value of 30 m. Accordingly, the local pressure setpoint at PCV1 and PCV2 outlet was set to 7.5 and 33 m for DMA1 and DMA2, respectively.

The third scenario (S3) considers the adoption of a remote RTC for pressure control in the WDN. Two plunger valves (DN300 for DMA1 and DN80 for DMA2) are assumed to be installed at the same sites as in scenario S2. Recent laboratory experiments have shown this type of valve to provide potential for accurate RTC in WDNs [32]. The scenario assumes the same pressure setpoint (30 m) at the two critical nodes as for scenario S2 (nodes 13 and 14). However, in comparison to scenario S2, the valves are directly controlled on the basis of the remote pressure measurements acquired in real time at the critical nodes. Simulation of scenario S3 included preliminary identification of a suitable value of the controller gain K to perform effective pressure control in the two districts without the occurrence of permanent pressure oscillations [33]. In agreement with previous literature results (e.g., [20,34], the simulations under RTC were carried out using control time $\Delta t_c = 5$ min.

Scenarios S1, S2, and S3 were compared to scenario zero (S0), the current reference scenario in which no actions are taken to reduce leakage levels in the network.

3. Results

3.1. Results of the Monitoring Campaign and Leakage Level Estimation

On the one hand, results of the analysis of the data of the monitoring campaign with specific reference to flow measurements at node 2 show that average daily flow to the whole WDN is equal to 56.5 L/s. Moreover, flow measurements at the inlet of DMA2 reveal that average daily inflow to this district is 11.8 L/s. By difference, nodes of DMA1 are supplied with total average daily flow of 44.7 L/s.

On the other hand, the analysis of the billed consumption (from the database of the households) points out consumption for the whole WDN corresponding to an average value of the flow of 14.9 L/s. The same analysis at the scale of DMA, provides 9.9 L/s for DMA1 and 5.0 L/s for DMA2. This last result is consistent with the spatial distribution of population, since about 2/3 of the total population served by the WDN belongs to DMA1.

Based on the above, a comparison of the results of the monitoring campaign and of the billed consumption would confirm a huge level of leakage in the WDN (about 73.6% of the total inflow, i.e., about 92.2 m^3/day/km). The same comparison at the scale of DMA shows that leakage levels for DMA1 and DMA2 are equal to 78% (107.4 m^3/day/km) and 58% (53.4 m^3/day/km), respectively.

3.2. Results of Model Calibration

Figure 2 reports optimal values of the hourly demand multipliers, M_h, for consumption at both DMA1 and DMA2 as a result of the model calibration in the scenario of simulation S0.

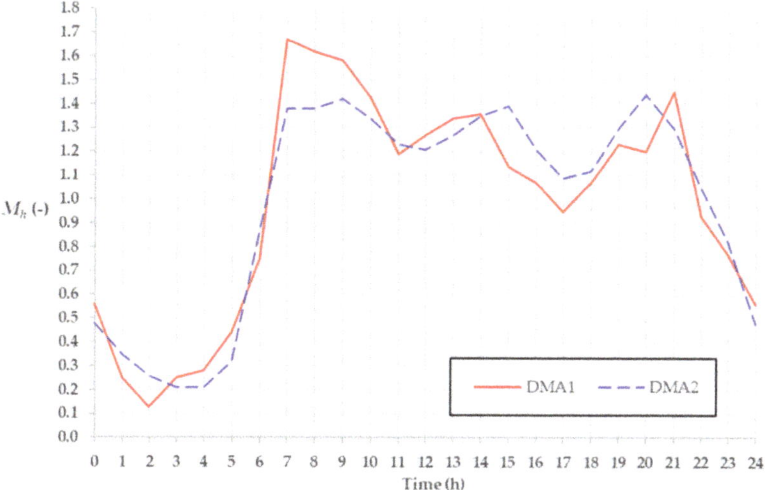

Figure 2. Results of model calibration. Hourly demand multipliers for the two DMAs.

The figure shows that the two patterns are rather similar with the occurrence of three main peaks of consumption (the first at early morning, the second at lunch time, and the third at dinner time) as well as with the minimum flow value occurring during the night. Slight differences are shown between the two DMAs with the first peak being larger and in advance for DMA1 as compared to DMA2 due to the widespread presence of commercial activities in DMA1.

Additionally, the calibration returned the optimal values of leakage coefficients of Equation (1). Specifically, the calibration procedure returned $\alpha = 0.87$ and $\beta = 2.85 \times 10^{-5}$. Such values were attributed to both types of pipes, i.e., stainless steel and HDPE. It was not considered the opportunity to differentiate the coefficients for the two types of pipe material given the very modest incidence of pipes in stainless steel as compared to that of pipes in HDPE. As expected, according to the results of the optimization, a larger value of the coefficient was obtained ($\beta = 4.77 \times 10^{-4}$) for the north–south steel pipeline.

Results are consistent with previous literature results from similar applications [31–37], showing values of demand multipliers to range between 0.1 and 1.7, α values close to 1 and β values between 10^{-5} and 10^{-4}.

Values of the absolute mean percent error (MAPE) and of the root mean squared percent error (RMSPE) as obtained by the optimization process for flow and pressure values at the various nodes of installation are shown in Table 1 for DMA1 and DMA2. Remarkably, the table shows maximum value of MAPE for pressure values to be 4% for DMA1 and 2% for DMA2, while the largest value for the flow is 1.7% as obtained for DMA1. Consistently, maximum RMSPE values obtained for the two DMA are 5.1% and 2.5% at nodes 11 and 8, respectively.

Table 1. Results of model calibration. Values of absolute mean percent error (MAPE) and root mean squared percent error (RMSPE) for flow and pressure.

	DMA1						DMA2				
	Flow			Pressure			Flow		Pressure		
Node	2	10	11	12	4	5	3	6	7	8	9
MAPE	0.017	0.037	0.040	0.032	0.031	0.022	0.006	0.016	0.014	0.020	0.014
RMSPE	0.020	0.045	0.051	0.038	0.039	0.026	0.008	0.020	0.017	0.025	0.018

3.3. Results of Simulation Scenarios

Scenarios of intervention described previously in Section 2.5 were simulated in order to evaluate the potential for leakage reduction. The simulations were run with reference to water demands of a "average" day as obtained based on the measured flow discharges during the monitoring campaign. The results of the simulations are summarized in Figure 3.

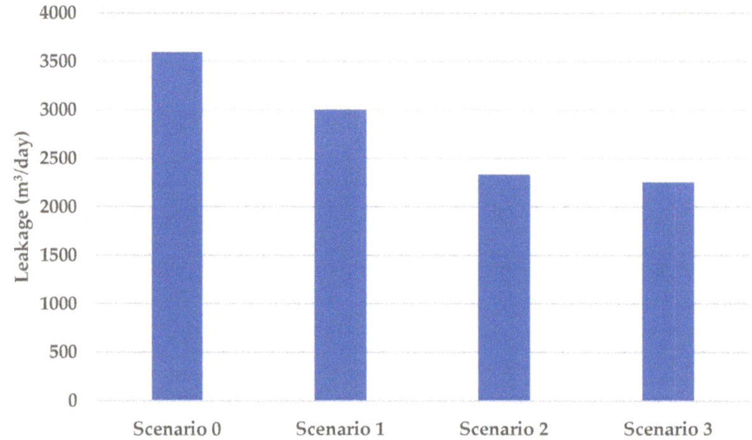

Figure 3. Potential leakage reduction under implementation of the different scenarios.

The figure shows that scenario S1 of rehabilitation of the leaking segments of the north–south steel pipeline allows to reduce total daily leakage volume from about 3600 m^3/day to about 3000 m^3/day (about 2082 m^3/day for DMA1 and 918 m^3/day for DMA2). Globally, this would mean recovering about 16.7% (about 6.95 L/s) of the current network leakage. In addition, the rehabilitation of the pipeline leads to improved pressure levels in the WDN, that is reflected into an improved satisfaction of the nodal water demands with specific reference to nodes of DMA1 that were characterized by supply deficit.

As scenario S2 is considered, Figure 3 shows that the total leakage volume in the WDN drops down to 2330 m^3/day (1726 m^3/day in DMA1 and 604 m^3/day in DMA2), that is a 35% leakage reduction with respect to the current scenario S0 (about 14.7 L/s recovered). The figure shows also that replacement of the two PCVs with valves controlled by remote RTC (scenario S3) has further potential to improve the pressure control action with respect to scenario S2. The total leakage volume in the WDN drops down to about 2252 m^3/day, that is a 37.5% leakage reduction as compared to S0 (about 15.6 L/s recovered).

Leakage reduction obtained in both scenarios S2 and S3 is strictly depending on the decreasing of the pressure level in the network during the 24 h of the day. The results of local control of the pressure as obtained by using the two PCVs are shown in Figure 4. The figure shows pressure to decrease at the two critical nodes 13 and 14 for DMA1 and DMA2, respectively. However, the desired setpoint of 30 m is achieved only in some hours, while the pressure remains higher than the setpoint for the rest of the day (on average 31.5 and 31.3 m for the two critical nodes). Figure 4 also reports the results of the simulation of scenario S3 with the adoption of remote RTC. The simulation was performed using $K = 0.005$ m^{-1} (as obtained by preliminary runs aimed at the tuning of the controller). The figure shows that the used value of the gain allows driving the pressure (at both critical nodes) to the setpoint without incurring in the generation of oscillations of the pressure during the day. Remarkably, the RTC system is able to maintain each pressure setpoint in an accurate way during the 24 h. Specifically, the figure highlights that the maximum pressure deviation from the setpoint is smaller than 1 m at both DMAs (against 3–4 m in the case of local control of S2). Figure 4 also reports the curve of the valve opening settings *aDMA1* and *aDMA2* in the two districts. As shown, openings/closures of the valve shutter are consistent with the daily water demand pattern at the two DMAs. In fact, the two plunger valves are adjusted by the control algorithm in order to close majorly during night hours (when pressure levels are normally higher) and to open during the peaks of the water demand.

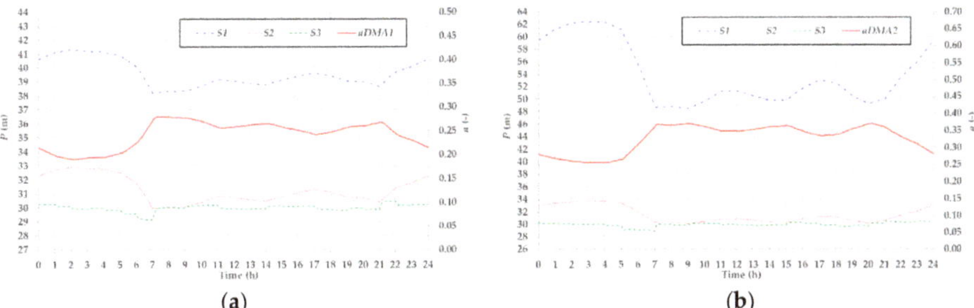

Figure 4. Pressure levels at the critical nodes of (**a**) DMA1, and (**b**) DMA2 under scenarios S1, S2, S3.

As already discussed, the simulations under EPS were performed considering a day with average characteristics among the days of the monitoring campaign. However, it should be stressed that benefits of remote RTC as compared to local control of pressure are normally emphasized under conditions of high spatiotemporal variability of water demands in the WDN. In fact, while local control remains affected by uncertainty in water demands, in principle, remote RTC systems are able to assure appropriate pressure control in real time in all the circumstances.

An important aspect that deserves discussion concerns the concurrent use of the two PCVs installed at the two DMAs. In fact, PCV2 significantly reduces pressure levels at DMA2 while sustaining pressure in the upstream DMA1, thus reducing beneficial effects of pressure reduction as determined by PCV1. In this regard, the results of a further simulation (not presented here) of the WDN that includes implementation of PCV1 only, allowed quantifying that the pressure value at node 4 (upstream of PCV2) would be about 4 m less than the pressure value determined at the same node in scenario S2.

The accuracy (thus the transferability into practice) of the simulation results obtained deserves specific discussion. The availability of measurements from the monitoring campaign allowed appropriate calibration of the simulation model, with the average MAPE (Table 1) corresponding to absolute error in pressure in the order of 0.5 m (for a 30 m setpoint). Based on the results of the simulations, such an error is generally smaller than pressure differences as obtained by the comparison of the different scenarios, thus, confirming the reliability of the results of leakage reduction.

Finally, it should be stressed that, although strategies of active pressure control such as those discussed in this paper may help reducing leakage levels in WDNs, they can also be a cause of reduction of the network resilience during emergency conditions (e.g., firefighting scenarios). However, strategies of remote RTC may allow performing a real-time adjustment of the pressure control valves in the network, according to the requirements, preventing water hammer conditions and restoring appropriate levels of pressure in the WDN to cope with the emergency.

4. Conclusions

A simulation analysis was carried out to evaluate the potential of rehabilitation measures and active pressure control strategies for leakage reduction in a WDN in southern Italy.

Three main scenarios were considered for the simulations. The first scenario (S1) concerned the rehabilitation of a steel pipeline of the WDN in order to eliminate leakages that were identified during the survey of the network. The second scenario (S2) added to S1 the installation of two PCVs to allow local control of the pressure in the network. The third scenario (S3) considered the adoption of a remote RTC for pressure control in the WDN by means of two plunger valves in place of the PCVs adopted in S2. The three scenarios were compared to scenario zero (S0), i.e., the current reference scenario in which no actions are taken to reduce leakage levels in the network.

The analysis has shown that a combination of the used strategies can significantly improve the network performance. Specifically, rehabilitation of the steel pipeline (scenario S1) would provide a significant reduction of the current leakage level of the WDN. Addition of local pressure control (scenario S2) would add further benefits. The adoption of remote RTC (scenario S3) in comparison with local pressure control increases leakage reduction. Additional benefits of remote RTC as compared to local control of PCVs would be observed under conditions of high spatiotemporal variability of water demands in the WDN.

Two main aspects played an important role in corroborating the reliability of the results of the analysis. Firstly, simulations were based on the results of the accurate calibration of model parameters relying on a program of measurements carried out during the monitoring campaign. Secondly, the modeling of the pressure control processes by using local control of PCVs and remote RTC was carried out based on the literature results of consolidated experimental investigations.

Evidently, the results of this analysis must be considered as preliminary and open to further research work. Transferability of the obtained results requires testing the methodology on other WDNs with different hydraulic characteristics and topology. In addition, in real cases, the choice of measures of intervention to adopt would require the evaluation of the benefit in terms of leakage reduction, but cannot prescind also a comprehensive analysis of the costs (including economic, environmental, and social costs) of each scenario of implementation.

Author Contributions: Methodology, C.B., A.C., C.M. and G.P.; development of the models, C.B.; analysis of results, C.B., A.C., C.M. and G.P.; writing and editing, C.B., A.C., C.M. and G.P. All authors have read and agreed to the published version of the manuscript.

Funding: This research was partially supported by the Department of Civil Engineering and Architecture of the University of Catania, under the Research Program "Piano Triennale della Ricerca 2016/2018".

Conflicts of Interest: The authors declare no conflict of interest.

References

1. Berardi, L.; Laucelli, D.B.; Simone, A.; Mazzolani, G.; Giustolisi, O. Active Leakage Control with WDNetXL. *Procedia Eng.* **2016**, *154*, 62–70. [CrossRef]
2. Wu, Z.Y.; Sage, P.; Turtle, D. Pressure-Dependent Leak Detection Model and Its Application to a District Water System. *J. Water Resour. Plan. Manag.* **2010**, *136*, 116–128. [CrossRef]
3. Romano, M.; Kapelan, Z.; Savic, D. Geostatistical techniques for approximate location of pipe burst events in water distribution systems. *J. Hydroinform.* **2013**, *15*, 634–651. [CrossRef]
4. Berardi, L.; Giustolisi, O.; Kapelan, Z.; Savic, D. Development of pipe deterioration models for water distribution systems using EPR. *J. Hydroinform.* **2008**, *10*, 113–126. [CrossRef]
5. Thornton, J. *Water Loss Control Manual*; McGraw Hill Professional: New York, NY, USA, 2002.
6. Ciaponi, C.; Franchioli, L.; Murari, E.; Papiri, S. Procedure for Defining a Pressure-Outflow Relationship Regarding Indoor Demands in Pressure-Driven Analysis of Water Distribution Networks. *Water Resour. Manag.* **2015**, *29*, 817–832. [CrossRef]
7. Le Gat, Y.; Eisenbeis, P. Using maintenance records to forecast failures in water networks. *Urban Water* **2000**, *2*, 173–181. [CrossRef]

8. Prescott, S.L.; Ulanicki, B. Improved Control of Pressure Reducing Valves in Water Distribution Networks. *J. Hydraul. Eng.* **2008**, *134*, 56–65. [CrossRef]

9. Nicolini, M.; Zovatto, L. Optimal Location and Control of Pressure Reducing Valves in Water Networks. *J. Water Resour. Plan. Manag.* **2009**, *135*, 178–187. [CrossRef]

10. Pezzinga, G.; Gueli, R. Discussion of "Optimal Location of Control Valves in Pipe Networks by Genetic Algorithm" by LFR Reis, RM Porto, and FH Chaudhry. *J. Water Resour. Plan. Manag.* **1999**, *125*, 65–67. [CrossRef]

11. Creaco, E.; Pezzinga, G. Comparison of Algorithms for the Optimal Location of Control Valves for Leakage Reduction in WDNs. *Water* **2018**, *10*, 466. [CrossRef]

12. Campisano, A.; Modica, C.; Reitano, S.; Ugarelli, R.; Bagherian, S. Field-Oriented Methodology for Real-Time Pressure Control to Reduce Leakage in Water Distribution Networks. *J. Water Resour. Plan. Manag.* **2016**, *142*, 04016057. [CrossRef]

13. Berardi, L.; Simone, A.; Laucelli, D.B.; Ugarelli, R.M.; Giustolisi, O. Relevance of hydraulic modelling in planning and operating real-time pressure control: Case of Oppegård municipality. *J. Hydroinform.* **2018**, *20*, 535–550. [CrossRef]

14. Page, P.R.; Abu-Mahfouz, A.M.; Mothetha, M.L. Pressure Management of Water Distribution Systems via the Remote Real-Time Control of Variable Speed Pumps. *J. Water Resour. Plan. Manag.* **2017**, *143*, 04017045. [CrossRef]

15. Creaco, E.; Walski, T. Operation and Cost-Effectiveness of Local and Remote RTC. *J. Water Resour. Plan. Manag.* **2018**, *144*, 04018068. [CrossRef]

16. Campisano, A.; Schilling, W.; Modica, C. Regulators' setup with application to the Roma–Cecchignola combined sewer system. *Urban Water* **2000**, *2*, 235–242. [CrossRef]

17. Campisano, A.; Creaco, E.; Modica, C. RTC of Valves for Leakage Reduction in Water Supply Networks. *J. Water Resour. Plan. Manag.* **2010**, *136*, 138–141. [CrossRef]

18. Giustolisi, O.; Ugarelli, R.M.; Berardi, L.; Laucelli, D.B.; Simone, A. Strategies for the electric regulation of pressure control valves. *J. Hydroinform.* **2017**, *19*, 621–639. [CrossRef]

19. Bakker, M.; Rajewicz, T.; Kien, H.; Vreeburg, J.H.G.; Rietveld, L. Advanced control of a water supply system: A case study. *Water Pr. Technol.* **2014**, *9*, 264–276. [CrossRef]

20. Page, P.R.; Abu-Mahfouz, A.M.; Yoyo, S. Parameter-Less Remote Real-Time Control for the Adjustment of Pressure in Water Distribution Systems. *J. Water Resour. Plan. Manag.* **2017**, *143*, 04017050. [CrossRef]

21. Creaco, E.; Campisano, A.; Modica, C. Testing behavior and effects of PRVs and RTC valves during hydrant activation scenarios. *Urban Water J.* **2018**, *15*, 218–226. [CrossRef]

22. Eliades, D.G.; Kyriakou, M.; Vrachimis, S.; Polycarpou, M.M. EPANET-MATLAB toolkit: An open-source software for interfacing EPANET with MATLAB. In Proceedings of the 14th International Conference on Computing and Control for the Water Industry (CCWI 2016), Amsterdam, The Netherland, 1–4 November 2016.

23. Rossman, L.A. *EPANET 2: Users Manual*; National Risk Management Research Laboratory: Cincinnati, OH, USA, 2000.

24. Williams, G.S.; Hazen, A. *Hydraulic Tables: The Elements of Gagings and the Friction of Water Flowing in Pipes, Aqueducts, Sewers, etc. as Determined by the Hazen and Williams Formula and the Flow of Water Over Sharp-Edged and Irregular Weirs, and the Quantity Discharged, as Determined by Bazin's Formula and Experimental Investigations Upon Large Models*; J. Wiley & Sons: Hoboken, NJ, USA, 1908.

25. Germanopoulos, G. A technical note on the inclusion of pressure dependent demand and leakage terms in water supply network models. *Civ. Eng. Syst.* **1985**, *2*, 171–179. [CrossRef]

26. Goodwin, S.J. *The Results of the Experimental Programme on Leakage and Leakage Control*; WRC Environmental Protection: Swindon, UK, 1980.

27. Greyvenstein, B.; Van Zyl, J.E. An experimental investigation into the pressure—leakage relationship of some failed water pipes. *J. Water Supply Res. Technol.* **2007**, *56*, 117–124. [CrossRef]

28. Farley, M.; Trow, S. Losses in Water Distribution Networks: A Practitioners' Guide to Assessment, Monitoring and Control. *Water Intell. Online* **2015**, *4*. [CrossRef]

29. Al-Ghamdi, A.S. Leakage–pressure relationship and leakage detection in intermittent water distribution systems. *J. Water Supply Res. Technol.* **2011**, *60*, 178–183. [CrossRef]

30. Ferrante, M. Experimental Investigation of the Effects of Pipe Material on the Leak Head-Discharge Relationship. *J. Hydraul. Eng.* **2012**, *138*, 736–743. [CrossRef]

31. Di Nardo, A.; Di Natale, M.; Gisonni, C.; Iervolino, M. A genetic algorithm for demand pattern and leakage estimation in a water distribution network. *J. Water Supply Res. Technol.* **2015**, *64*, 35–46. [CrossRef]

32. Campisano, A.; Modica, C.; Musmeci, F.; Bosco, C.; Gullotta, A. Laboratory experiments and simulation analysis to evaluate the application potential of pressure remote RTC in water distributions networks. *Water Res.* **2020**, *183*, 116072. [CrossRef] [PubMed]

33. Ziegler, J.G.; Nichols, N.B. Optimum Settings for Automatic Controllers. *J. Dyn. Syst. Meas. Control.* **1942**, *115*, 220–222. [CrossRef]

34. Creaco, E.; Franchini, M. A new algorithm for real-time pressure control in water distribution networks. *Water Supply* **2013**, *13*, 875–882. [CrossRef]

35. Letting, L.; Hamam, Y.; Abu-Mahfouz, A.M. Estimation of Water Demand in Water Distribution Systems Using Particle Swarm Optimization. *Water* **2017**, *9*, 593. [CrossRef]

36. Tucciarelli, T.; Criminisi, A.; Termini, D. Leak Analysis in Pipeline Systems by Means of Optimal Valve Regulation. *J. Hydraul. Eng.* **1999**, *125*, 277–285. [CrossRef]

37. Maskit, M.; Ostfeld, A. Leakage Calibration of Water Distribution Networks. *Procedia Eng.* **2014**, *89*, 664–671. [CrossRef]

 *water*

Article

An Improved Genetic Algorithm for Optimal Layout of Flow Meters and Valves in Water Network Partitioning

Yu Shao [1], Huaqi Yao [1], Tuqiao Zhang [1], Shipeng Chu [1] and Xiaowei Liu [1,2,*

[1] Institute of Municipal Engineering, College of Civil Engineering and Architecture, Zhejiang University, Hangzhou 310058, China; shaoyu1979@zju.edu.cn (Y.S.); yaohuaqi@zju.edu.cn (H.Y.); ztq@zju.edu.cn (T.Z.); chushipeng@zju.edu.cn (S.C.)
[2] Institute of Water Resources & Ocean Engineering, Ocean College, Zhejiang University, Hangzhou 310058, China
* Correspondence: liuxiaowei@zju.edu.cn; Tel.: +86-0571-8820-6759

Received: 3 April 2019; Accepted: 16 May 2019; Published: 24 May 2019

Abstract: The paradigm of "divide and conquer" has been well used in Water Distribution Systems (WDSs) zoning planning in recent years. Indeed, Water Network Partitioning (WNP) has played an irreplaceable role in leakage control and pressure management; meanwhile it also has certain drawbacks, such as reduction of the supply reliability of the pipe network system and increased terminal dead water, as a result of the closure of the pipe section. In this paper, an improvement is made to the method proposed by Di Nardo et al. (2013) for optimal location of flow meters and valves. Three improvements to the genetic algorithm are proposed in this work for better and faster optimization in the dividing phase of WNP: preliminary hydraulic analysis which reduces the number of decision variables; modifications to the crossover mechanism to protect the superior individuals in the later stage; and boundary pipe grouping and mutation based on the pipe importance. The objective function considers the master–subordinate relationship when minimizing the number of flow meters and the difference of hydraulic state compared to original WDS. Another objective function of minimizing the deterioration of water quality compared to original WDS is also evaluated. The proposed method is applied for the WNP in a real WDS. Results show that it plays an effective role in the optimization of layout of the flow meters and valves in WNP.

Keywords: water network partition; genetic algorithm; hydraulic; water quality

---

1. Introduction

As one of the most crucial municipal infrastructures, Water Distribution Systems (WDSs) play an irreplaceable role in social and economic development. In older design philosophy of WDSs, we blindly pursued greater pipe redundancy (i.e., complicated looped system) to enhance the system's ability to protect against potential risks (e.g., pipe bursts or fire demand). There was no doubt that it brought higher reliability of supply service but meanwhile we cannot ignore the management inconvenience, especially the difficulty of water loss evaluation in such a complicated WDS, resulting in failure to detect leaks and pipe bursts in time, which were extremely harmful to the sustainable development of the social ecology [1].

Consequently, in recent years, researchers around the world have focused on finding a new approach aimed at simplifying the management of WDSs. The paradigm of "divide and conquer" was hence vigorously promoted, along with a new management model of WDSs, called District Metered Areas (DMAs), which can significantly improve the management level through real-time monitoring of inlet and outlet flows in each subsystem (i.e., water balance analysis); certainly, DMAs play broader roles than water metering. The creation of DMAs can also help the implementation of pressure management [2] with the use of Pressure-Reducing Valves (PRVs), allowing a low level of leakage to be kept in the whole network system; additionally, replacing the PRVs with PATs (Pumps as Turbines) in the water distribution network has proven to be useful for energy recovery [3,4], which has huge application prospects in the future. Furthermore, some studies have shown that the division of DMAs do have a function of protecting the system from water contamination [5] and a better detection of pipe bursting [6].

The two main issues of DMA design are how to shape and dimension the network sub-zones (clustering phase) and where to locate meters and gate valves (dividing phase). Presently, considerable theoretical methods can be used in the clustering phase of DMA design and divide into four categories: graph theory algorithm [7–10], cluster analysis [11–13], community structure [14–17], and multi-agent algorithm [18–20]. The graph theory algorithm is first introduced to solve the WNP problems by Tzatchkov et al. [7]. They used breadth-first search (BFS) and depth-first (DFS) algorithms to find feasible partitioning solutions, and later many graph theoretic heuristic techniques were combined to better address the problems, such as Dijkstra's shortest path algorithm [9] and recursive bisection method [21], but sometimes this method cannot effectively control the number of boundary pipes, resulting in an increase in investment costs; Cluster analysis is a traditionally used method in WDS; the similarity between nodes is used to cluster the sub-zones. Perelman et al. [22] used a bottom-up hierarchical clustering algorithm, which, based on the Euclidean distance between nodes, partitioned the whole network; however, since the Euclidean distance measure does not consider the connectivity of nodes, the internal connectivity of each cluster should be verified; and lately a spectral clustering algorithm [11,12] was widely applied in the WNP issues because of its reliability in minimizing the number of edge-cuts and balancing the number of nodes for each district. Diao et al. [15] first applied the community structure algorithm, based on the concept of 'modularity' to detect the group characteristics of the WDS. This method is also widely used due to its high speed and reliability in decomposition of large-scale, complex systems [23], but the balance between DMAs is not so well, and an improved version of this method is proposed by Giustolisi and Ridolfi [24] which considers the attributes of nodes and pipes. A comprehensive comparison of the methods described above was done by Sela Perelman et al. [22] and Di Nardo et al. [17].

The DMA boundary pipes are determined after the clustering phase, and then the layout of the flow meters (i.e., open pipes) and gate valves (i.e., closed pipes) among the boundary pipes is optimized in the dividing phase to minimize the number of flow meters or the hydraulic performance deterioration [25,26], etc. Simulated Annealing algorithm [27], iterative method [12,15,16], and single-objective or multi-objective optimization based on economic [25], energy and water quality [28,29] issues can be used in this phase. The iterative method is a simple method that suitable for all sizes of WDS [15]. Through the initial analysis of the importance of each pipe (according to diameter or flow rate, etc.), the least important pipe is iteratively closed until normal hydraulic conditions cannot be met. At this point the number of flow meters reaches the minimum. The iterative method is still an empirical method in essence, and it cannot consider the joint effect caused by simultaneous closure of multiple pipes, so there is a certain gap between the results and those obtained by the heuristic algorithm [30]. The heuristic algorithms globally search the layout of the flow meter and valve position based on the hydraulic performance of the water distribution network. However, the heuristic algorithm also has its own drawback, i.e., when the number of decision variables increases, the convergence efficiency drops significantly because the search space grows exponentially.

In this study, the main work is done to improve the optimization performance and efficiency of the genetic algorithm (GA) in the dividing phase of WNP, which consists of selection, crossover, and mutation operations, developed by Goldberg et al. [31]. The contributions of this work are listed below: 1) Changing the dual-objective problem to a single-objective problem by considering the master–subordinate relationship of the two objective function, which improved the computational efficiency; 2) Three improvements integrated in the GA, including preliminary analysis of boundary pipes, modified crossover and mutation mechanism, and the case study verified the superiority of this strategy, especially in terms of stability of output and convergence speed. The objective functions to consider the network resilience and the water quality impact caused by WDS partition were tested independently. As a result, a high similarity in the locations of flow meters and gate valves for these two objective functions are found.

2. Methodology

The framework of the whole partitioning methodology is shown in Figure 1. In the clustering stage, a modularity-based algorithm [32] is applied to deal with the problem of sectorization, and the adjacency matrix A_{ij} of topology structure and total number of pipes m is needed as initial parameters; then the GA is used in the optimization procedure of the dividing stage.

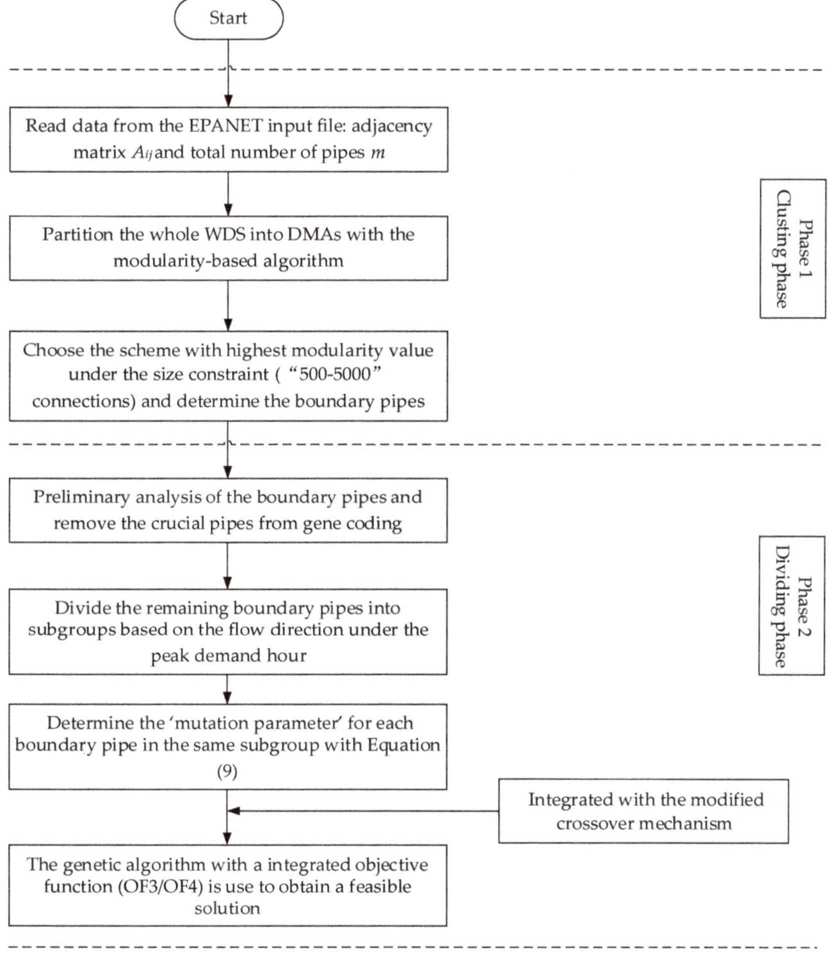

Figure 1. Flowchart of the whole methodology.

2.1. Network Clustering

Similar to many physical and social processes (e.g., business and academic circles) in the world, a water distribution system can also be naturally characterized by a simple topological graph G = G(V,E), where V represents n nodes (consumers, sources, and tanks) and E represents m edges (pipes, valves, and pumps) in a WDS. The approach adopted in this phase owns a bottom-up merge process and is based on the concept of 'modularity', which can quantify the quality of WDS partition [23,32,33], called the Fast-Newman algorithm. The modularity Q is expressed in Equation (1), and $Q \geq 0.3$ indicates a good division into communities [16].

$$Q = \frac{1}{2m} \sum_{vw} \left(A_{vw} - \frac{k_v k_w}{2m} \right) \delta(c_v, c_w) \quad Q \in [-0.5, 1) \tag{1}$$

where m represents the total number of links in the network; v and w are vertexes indexes; k_v (k_w) is the sum of the number of edges connected to vertex v (w); $A_{vw} = 1$ if there is a link connecting the vertexes v and w; $A_{vw} = 0$ otherwise; c_v, c_w is the communities which the vertexes v and w belong to; $\delta(c_v, c_w) = 1$ if vertexes v and w are in the same community, 0 otherwise. The brief steps of the algorithm are described below:

1. At first each node represents a community, and the initial modularity is calculated with Equation (1).
2. Evaluate the modularity change ΔQ_{vw} for each pair of communities v and w with Equation (2), assuming they are combined.
3. Merge the pair of communities v^* and w^* with max (ΔQ_{vw}) (could be negative value) and the modularity of the current structure equals the modularity of the previous structure plus max (ΔQ_{vw}).
4. Step 2 and 3 is repeated until all the nodes merged into one community.

$$\Delta Q_{vw} = \begin{cases} \dfrac{1}{2m} - \dfrac{k_v k_w}{(2m)^2}, & \text{if } v, w \text{ are connected} \\ 0, & \text{otherwise} \end{cases} \tag{2}$$

The partition scheme with highest Q value represents its sub-systems that have stronger internal than external connections, but is not necessarily the best solution because the DMA size needs to be controlled between 500–5000 customer connections according to the guidelines [34]. It becomes much more difficult to control water losses and locate the new bursts when a DMA contains more than 5000 customers [27]; meanwhile more investment will be spent if the size of DMA is too small. Therefore, what we indeed need is the scheme with highest Q value given the premise of satisfying the size constraint. More detail about the Fast-Newman algorithm can be found in previous literature [15,24].

2.2. Network Dividing

For the dividing stage of WNP (Water Network Partitioning), Di Nardo et al. [25] proposed a method based on GA with the following objective function, to protect the hydraulic performance of the original system as much as possible:

$$OF1 = max\left(P_D = \gamma \sum_{i=1}^{n} Q_i H_i \right) \tag{3}$$

where γ is the specific weight of water, Q_i and H_i are the water demand and head at each network node, and P_D is the total node power of the network. Generally, partitioning schemes with more flow meters (i.e., closing fewer pipes) would have a higher objective value P_D, hence the optimal solution is the case that all the boundary pipes are installed with flow meters (i.e., remain open). However, it is not the answer we truly need as more flow meters lead to a less convenient management and higher cost. Therefore Di Nardo et al. [25] tried to solve the above problem by determining the number of flow meters in advance, but it is difficult to decide how many flow meters should be installed as a trade-off between the hydraulic performance and cost, especially for large-scale WDN. Therefore, the number of flow meters should be considered to be one of the objective functions instead of being a predetermined value. Thus, it forms a dual-objective problem as Equation (4):

$$OF2 = \left\{ \begin{array}{l} min\left(N_{fm}\right) \\ max(P_D) \end{array} \right.$$

(4)

At this point, the NSGA-II algorithm [35] is generally used to optimize this dual-objective problem. It is feasible but not the optimal selection—in the previous literature, almost all researchers agree that the number of flow meters should be reduced as much as possible in the dividing phase of WNP under the premise of meeting normal hydraulic conditions (i.e., all demand nodes should meet the minimum service pressure) [15,16]. If possible, the best solution would be each district own just one single inflow meter [25], because fewer flow meters can not only simplify the calculation of the synchronous water balance and improve the reliability of flow monitoring, but also reduce the costs of initial investment and later operation and maintenance costs. Therefore, we would like the second objective function in Equation (4) to be subjected to the first one; in other words, we would give the priority to the choice of the scheme with fewer flow meters, and then consider the role of the second objective function. However, the NSGA-II algorithm cannot embody this relationship between the two objective functions, resulting in a lot of time wasted on optimizing solutions (Pareto frontier).

To address above problem, this paper proposes a new objective function form shown in Equation (5), which can not only combine the two objective functions together, but also embodies the primary and secondary relations of the two objective functions. The purpose of the first term is to minimize the number of flow meters (its value is an integer) and the second term is to search the solution that has less impact on the hydraulic performance compared to original WDN. The value of the second term is between 0 and 1, ensuring that the latter one is subject to the former one.

$$OF3 = min\left(N_{fm} + \frac{P_{D0} - P_D}{P_{D0}}\right)$$

(5)

where N_{fm} is the number of flow meters, P_{D0} is the total node power of the network before WNP and P_D is the total node power after WNP.

The fitness function, Equation (6), is obtained using the boundary construction method, by changing the minimum optimization into the maximum optimization, and the fitness value can be ensured larger than zero.

$$Fit = N_p - OF3 = N_p - N_{fm} - \frac{P_{D0} - P_D}{P_{D0}} = N_{gv} - \frac{P_{D0} - P_D}{P_{D0}}$$

(6)

where N_p is the number of boundary pipes, N_{gv} represents the number of gate valves

To adjust the selective pressure effectively, a rank-based fitness assignment method [36,37] is adopted to better distinguish the fitness between individuals, especially in the later stage of optimization. The specific fitness assignment function is shown in Equation (7)

$$Fit(i) = 2 - SP + \frac{2(SP - 1)(i - 1)}{n - 1}, \ SP\epsilon[1.0, 2.0]$$

(7)

where i is the rank of an individual in the population, SP is the selection pressure and set $SP = 1.5$ in this work, and n is the size of the population. A simple example (the population size n adopts 5) is given to better understand the role of the rank-based fitness assignment method in this work as shown in Table 1. The real fitness values are firstly calculated with Equation (6), then rank the individuals in the following way: the best one has rank n and the worst one has rank 1.

Table 1. A simple example of rank-based fitness assignment method.

Individual Index	Real Fitness Value		Individual Index	Real Fitness Value	Rank		Individual Index	Assigned Fitness Value
1	27.97325		4	27.93649	1		4	0.5
2	28.96611	sort	1	27.97325	2	assign	1	0.75
3	28.94224	→	3	28.94224	3	→	3	1.0
4	27.93649		5	28.95028	4		5	1.25
5	28.95028		2	28.96611	5		2	1.5

Another objective function is adopted in this paper which considers minimizing the deterioration of water quality caused by WNP. The water velocity of pipes is used as an indicator to evaluate the water quality change before and after WNP. Giving the objective function in Equation (8).

$$OF4 = min\left(N_{fm} + \frac{\sum_{i=1}^{p} VO_i \Delta v_i \delta(m_1, m_2)}{\sum_{i=1}^{p} VO_i v_i}\right) \tag{8}$$

where p is the number of pipes; VO is the pipe volume; v is the pipe velocity before WDP; m_1 representing the event that the pipe velocity after WDP is lower than before, m_2 representing the event that the pipe velocity after WDP is lower than 0.1 m/s, and $\delta(m_1, m_2) = 1$ if both events m_1 and m_2 are true at the same time, 0 otherwise; Δv_i is the flow velocity difference before and after WNP.

2.3. Improvements for Better Optimization

The binary coding is used to encode the chromosomes (i.e., individual), one chromosome is composed of a sequence of genes. Gene i assumes 0 if a flow meter installed in i-th pipe; otherwise, the value is 1 if a gate valve installed. As mentioned by other scholars, the efficiency of GA is obviously affected by the number of decision variables (i.e., boundary pipes) in the optimization process of dividing phase [15]. There are two main reasons: first, the increasing number of decision variables inevitably leads to an exponential growth in search space, significantly affecting the optimization efficiency and making the entire optimization process more time-consuming. Second, binary coding cannot truly reflect the relationship between genes [38], which are not really independent of each other (e.g., for a DMA without water source, all the boundary pipes connecting it cannot be closed at the same time otherwise no water will be supplied in this area). Moreover, the common crossover mechanism has great damage to individuals especially for those superior individuals in the later stage of optimization because of the direct break of chromosomes, though in the earlier stage it helps to maintain the diversity of the population. This drawback becomes more apparent when the length of the chromosome increases. To address the above-mentioned issue, three improvements are proposed in this work

Step 1: Reduce the number of decision variables through preliminary analysis

In fact, there are several pipes among the boundary pipes that are not necessary to participate in gene coding, because the closure of any one of them will result in local pressure deficit or even water supply disruption (e. g. main water transmission pipes or a unique inflow pipe for one specific DMA). The gene points corresponding to those pipes must be installed with flow meters can hence be deleted from the chromosome. In this way, the search space is greatly reduced.

The specific method in the procedure is as follows: Each time close one boundary pipe (all other boundary pipes should be open except the selected one) and then run the hydraulic simulation to find the pipes where the minimum pressure is lower than the normal service pressure, i.e., this boundary pipe must be opened to satisfy the minimum pressure and can be removed from the decision variables (the chromosome). All the pipes to be removed from the chromosome can be found by repeating this procedure for all the boundary pipes.

Step 2: Modify the crossover mechanism

A new crossover mechanism is proposed to better protect the individuals in the later stages of optimization; at the same time, it enhances the local search ability of the algorithm. The pseudo-code of new crossover mechanism is described in Figure 2b, where gn represents the current generation, GN represents the total operation generations which should be determined in advance, pc is the first crossover probability with a fixed value of 0.8 meanwhile $pc1$ is the second crossover probability, which is related to the current generation gn. As shown in Figure 2a, mode 1 is the traditional crossover method meanwhile mode 2 is a new crossover method that exchanges the positions of two gene points, one coded 1, another coded 0; both of them are randomly selected (the physical meaning is to change the location of one randomly chosen flow meter meanwhile the total number of flow meters remains unchanged). As the generation (i.e., gn) increases, the probability of the mode 1 becomes smaller and the mode 2 becomes larger because of the declining value of $pc1$. The crossover method of mode 2 can better prevent the superior individuals not being damaged.

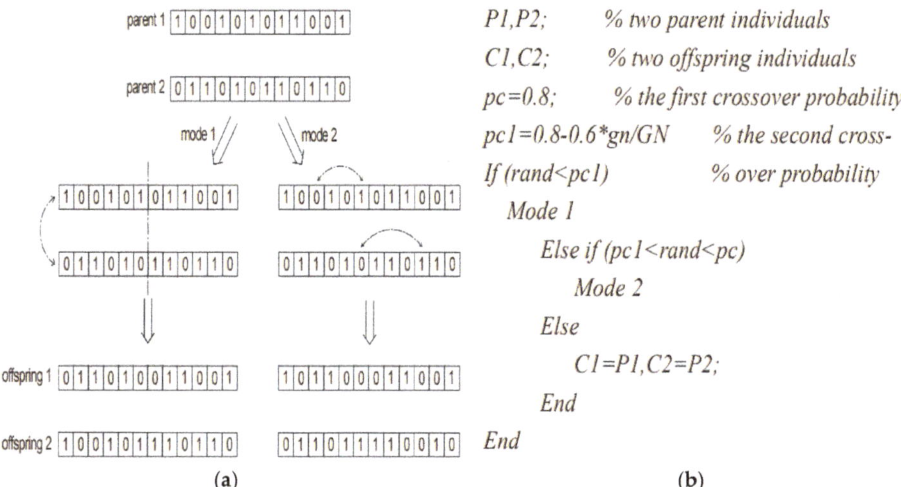

Figure 2. The crossover mechanism: (**a**) schematic diagram; (**b**) the pseudo-code.

Step 3: Improve the mutation parameter

Through assigning 'mutation parameter' to each gene point, the importance level of each boundary pipe can hence be embodied and mutually distinguished in the optimization procedure, effectively accelerating the convergence velocity. At the same time, it will not lose the ability to search for potentially better solution such as the iterative method because of the absolute ordering relationship.

The boundary pipes are first divided into subgroups based on the flow direction under the peak demand hour. To better illustrate the method, an example is shown in Figure 3a; pipe 1, 3, 5 are classified together because they are all the inflow pipes belonging to DMA1 and pipe 2, 6 are grouped together likewise. The pipes classified in the same subgroup play an identical role in the WDS (i.e., supply water for the same DMA), hence they are substitutes for each other and can be classified together for comparison to show the distinction in importance.

Then we determine the 'mutation parameter' for each boundary pipe according to Equation (9). Taking Figure 3a as an example, for DMA1, the flow rates of three inflow pipes 1, 3, 5 are 35 L/s, 136

L/s, 12 L/s respectively, and the corresponding 'mutation parameters' are 0.4290, 0.7049, 0.3661 as calculated by Equation (9).

$$Pa_{i,j} = 0.5 + \frac{0.5\left(Q_{i,j} - Mean_i\right)}{Sum_i} \tag{9}$$

where $Pa_{i,j}$ is the 'mutation parameter' of jth pipe in ith group; $Q_{i,j}$ is the flow rate of jth pipe in ith group; $Mean_i$ is the average flow rate of ith group; Sum_i is the sum of the flow rate of all pipes in ith group. The 0.5 in Equation (9) means the boundary pipes have the same possibility to install flow meters or gate valves if we do not consider any hydraulic characteristics, and the second term in RHS of Equation (9) considers their hydraulic importance (the 0.5 in the second term ensures $Pa_{i,j}$ less than 1).

As we can see, the more important pipe owns larger values of 'mutation parameter' such that it has greater possibility of being an inflow pipe with the flow meter installed (Figure 3b shows the pseudo-code). Taking pipe 3 as an example, it is the most important one of the three and owns the highest value 0.7049. Then, when its corresponding gene point is selected as the mutation point, if a randomly generated number (ranging from 0 to 1) is greater than 0.7049, the gene value becomes 1 (i.e., install the valve), otherwise it becomes 0 (i.e., install the flow meter). In this way the more important pipe has a larger opportunity to become an inflow pipe outside of being closed.

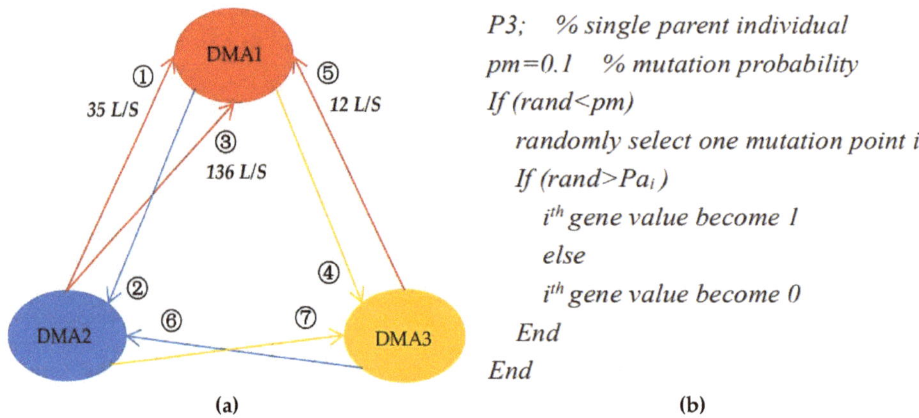

Figure 3. Mutation based on boundary pipes grouping: (**a**) illustration example of grouping; (**b**) the pseudo-code of the mutation mechanism.

3. Results and Discussion

We test the proposed methodology in a medium-size water distribution system located in southern China, H Town [39], which owns one fixed-head reservoir, 898 demand nodes, and 1012 pipes. Under the peak hour demand scenario, a minimum service pressure of 24 m is required-more details can be referred to Table 2.

Table 2. Characteristics of H town.

Property	Value
Number of Nodes	898
Number of Pipes	1012
Number of Reservoirs	1
Total pipe length	518.02 km
Total water demand	1498.9 L/s
Total connections	42826
Average water demand	0.032 L/s/connection

In the clustering stage of WNP, the Fast-Newman algorithm is used to create the sub-zones (results shown in Figure 4). The characteristics of DMAs are shown in Table 3. All DMAs satisfy the size constraint mentioned above and the value of modularity Q is 0.713, a fairly high value because of the relatively obvious community characteristics the WDS owns.

Table 3. Characteristics of H town's DMAs.

DMA Index	Number of Nodes	Water Demand (L/s)	Equivalent Connections
1	48	64.72	1849
2	83	133.94	3827
3	50	98.09	2803
4	50	158.02	4515
5	85	118.71	3392
6	35	60.22	1721
7	53	88.04	2515
8	76	165.33	4724
9	67	143.32	4095
10	40	41.69	1191
11	50	93.42	2669
12	67	48.86	1396
13	77	57.86	1653
14	62	85.6	2446
15	56	140.96	4027

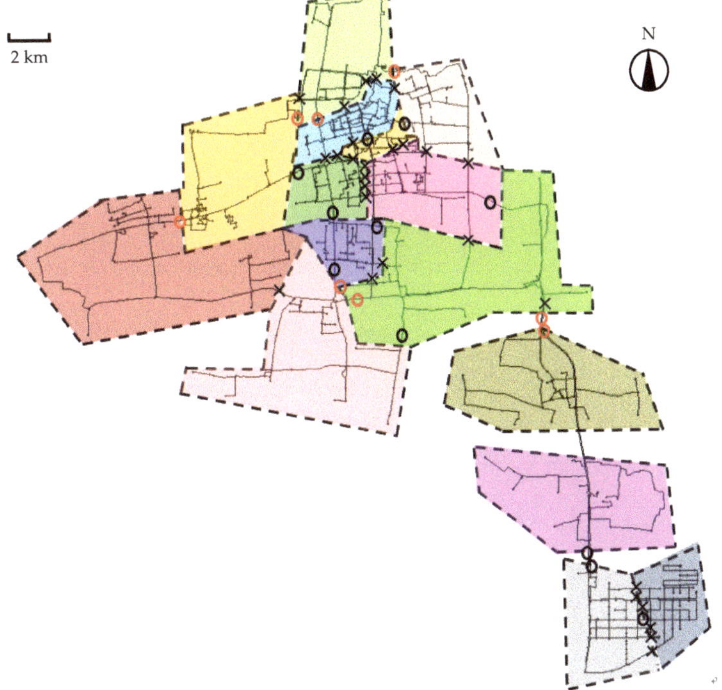

Figure 4. Final DMA configuration of H town.

After the above procedure, a set of boundary pipes containing 48 pipes (corresponding to 48 decision variables) is obtained, and then the layout of the flow meters and gate valves among these boundary pipes is optimized using GA with objective function *OF3* (a population of 50 individuals,

200 generations, *Pcorss* = 0.8, *Pmut* = 0.1). Hydraulic simulation is carried out during the peak water demand and a minimum service pressure of 24 m is required. *MATLAB*, combined with the well-known hydraulic simulator EPANET 2 (Rossman, 2000), is used in this optimization procedure.

The results of five runs (GA without any improvement strategy) are shown in Figure 5a and Table 4. We can clearly see that the results are not good as even the number of minimum flow meters cannot be stably obtained, let alone finding a solution that has the least impact on hydraulic performance under the premise of the minimum number of flow meters. There are non-negligible differences of the locations of flow meters between the solutions (a similar situation exists even if runs 500 generations).

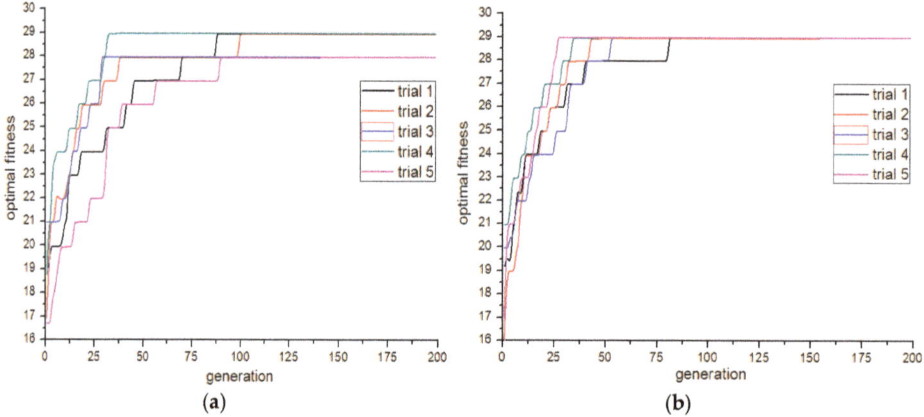

Figure 5. Comparison of optimization results: (**a**) GA without improvements; (**b**) GA with improvements.

Table 4. Results comparison of 5 runs between original GA and improved GA for *OF3*.

GA	Test	Fitness	Pipe Index Installed with Flow Meters	Differences [#]
	trial 1a	27.97325	**45** 152 161 180 219 **221** 269 277 351 359 364 **413** 422 **507** 604 698 755 844 848 873	4
	trial 2a	28.96611	130 152 180 219 269 277 351 359 364 402 422 604 698 755 844 848 873 **882 931**	2
Original	trial 3a	28.94224	130 161 180 **213** 219 269 277 351 359 364 402 422 **488** 604 **749** 755 848 873 **882**	4
	trial 4a	27.93649	152 161 180 **213** 219 **221** 269 **285** 351 359 364 402 422 **488** 604 **749** 755 848 873 **910**	6
	trial 5a	28.95028	131 180 **212 213** 219 269 277 351 359 364 402 422 604 698 **749** 755 848 873 **910**	4
	trial 1b	28.98079	130 152 161 180 219 269 277 312 351 359 364 402 422 604 698 755 844 848 873	0
	trial 2b	28.98073	**45** 152 161 180 219 269 277 312 351 359 364 402 422 604 698 755 844 848 873	1
Improved	trial 3b	28.98079	130 152 161 180 219 269 277 312 351 359 364 402 422 604 698 755 844 848 873	0
	trial 4b	28.98079	130 152 161 180 219 269 277 312 351 359 364 402 422 604 698 755 844 848 873	0
	trial 5b	28.98065	130 152 161 180 219 269 277 312 351 359 364 402 422 698 **749** 755 844 848 873	1

Note: # The difference of pipe index (marked in bold) installed with flow meters, compared with the best solution (trials 1b, 3b, 4b are thought to be the best solution founded so far).

To improve the optimization efficiency, the GA with the improvements described in Section 2.3 is applied in this case study. Table 5 shows the eight pipes excluded by the preliminary hydraulic analysis. The first five pipes are obviously the main transmission pipes, meanwhile the last three are the crucial inflow pipes for the special DMAs. These flow meter install locations are shown in Figure 4 marked in red. The search space is hence reduced from 2^{48} to 2^{40}.

Table 5. The excluded 8 pipes by preliminary hydraulic analysis.

Index	219	269	351	359	364	422	755	848
Diameter	1400	800	1200	1200	1200	400	400	200

The optimization results with improved GA are shown in Figure 5b and Table 4. The hydraulic performance comparison of the ten trials is given in Table 6. The performance indices include minimum pressure P_{min}, mean pressure P_{mean}, resilience index I_r, total node power P_D, the number of flow meters N_{fm} and gate valves N_{gv}. Compared to the original GA, the improved GA converges faster, because the original GA needs at least 100 generations to converge while the improved GA needs 50 generations except the trial 1b needs about 80 generations. The solutions of the improved GA are much better than that of the original GA, since the average fitness increased from 28.55 to 28.98; the average pressure P_{mean} increased from 32.28 m to 33.55 m and the average resilience index I_r increased from 0.6371 to 0.7114. The solutions of the improved GA are more stable than that of the original GA. As shown in the bottom half part of Table 4, trials 1b, 3b, 4b obtain the same result (fitness = 28.98079, the corresponding locations of flow meters and valves are also shown in Figure 4 marked in black), which indicate the best solution found so far. Trial 2b, 5b are just slightly worse than the best solution.

Table 6. Performance comparison of 10 trials (*OF3*).

Scenario	P_{min} (m)	P_{mean} (m)	P_D	I_r	N_{fm}	N_{gv}
Without DMA	29.06	34.49	60977	0.7666	/	/
trial 1a	24.05	33.28	59346	0.6922	20	28
trial 2a	24.12	32.58	58911	0.6724	19	29
trial 3a	24.42	31.81	57455	0.6061	19	29
trial 4a	24.27	31.58	56870	0.5794	20	28
trial 5a	24.32	32.16	58029	0.6353	19	29
average	24.24	32.28	58122	0.6371	19.4	28.6
trial 1b	24.19	33.59	59806	0.7132	19	29
trial 2b	24.21	33.54	59773	0.7129	19	29
trial 3b	24.19	33.59	59806	0.7132	19	29
trial 4b	24.19	33.59	59806	0.7132	19	29
trial 5b	24.08	33.46	59643	0.7043	19	29
average	24.17	33.55	59767	0.7114	19	29

They have only one different flow meter location (marked in red) compared with the best layout scheme, far better than the results obtained by original GA without improvement (the differences are also marked in red), which has four differences on average compared with the best scheme. The worst one has six differences. Taking trial 4a and 4b as an example, even though the scheme of trial 4a owns one more flow meter than the trial 4b, the deterioration of the hydraulic performance of the WDS is unexpectedly more than three times compared with the trial 4b (P_D declines by 6.4% versus 1.9%). The comparison of pressure distribution between the original network and the best WNP scheme is shown in Figure 6a; there is a minor pressure drop compared to the original network and most of the nodes remain the previous pressure level (i.e., 34 m).

To further prove the superiority and practicality of the proposed method, the objective function *OF4* is also tested in this work. The optimization results using GA before and after improvements are shown in Table 7. The performance to the search optimal solution is also improved a lot, much like solving objective function *OF3*. The flow velocity distribution before and after the WNP is shown in Figure 6b (only counts the flow velocity less than 0.1 m/s); we can clearly see that there is a marked increase in water volume when the flow velocity is between 0.01 m/s and 0.06 m/s, but the total amount of water with a flow velocity less than 0.1 m/s is approximately equal, indicating that the velocity distribution becomes more uneven after the WNP.

Through a comparison between two final optimal schemes based on different objective functions, we find there are just two different flow meter locations (*OF3*: pipe index of 277, 698; *OF4*: pipe index of 488, 749). Here we run the optimization procedures independently based on *OF3* or *OF4*. Their solutions gave the same number of the flow meters. Certainly, it is possible to yield a trade-off between the hydraulic performance and water quality through combining the two objective functions together. The two solutions are the two points of the Pareto frontier if we run multi-objective optimization.

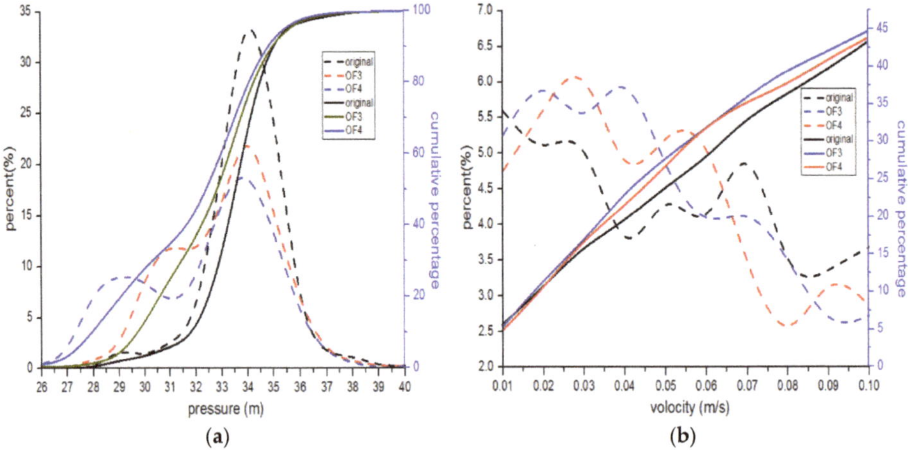

Figure 6. Comparison of pressure and velocity distribution before and after WNP: (**a**) Pressure distribution; (**b**) Velocity distribution with velocity < 0.10 m/s.

Table 7. Results comparison of 5 runs between original GA and improved GA for *OF4*.

GA	Test	Fitness	Pipe Index Installed with Flow Meters	Difference [#]
Original	trial 1a	27.91044	152 161 180 **212 213** 219 **221** 269 **285** 312 351 364 **413** 422 488 **698** 749 755 848 873	6
	trial 2a	28.89954	**43** 130 161 180 **195 212** 219 269 312 351 364 402 422 488 **698** 749 755 848 873	4
	trial 3a	26.94690	**45** 152 161 180 219 **221** 269 **285** 312 351 359 364 **413** 422 **448** 488 604 749 755 848 873	5
	trial 4a	28.93040	**45** 130 152 161 180 219 269 351 359 364 402 422 **698** 749 755 844 848 873 **931**	3
	trial 5a	27.92961	**45** 161 180 219 269 **277 285** 351 359 364 402 422 488 **507** 604 749 755 848 873 **910**	5
Improved	trial 1b	28.93954	130 152 161 180 219 269 312 351 359 364 402 422 488 604 749 755 844 848 873	0
	trial 2b	28.93879	130 152 161 180 219 269 312 351 359 364 402 422 488 749 755 844 848 873 **931**	1
	trial 3b	28.93334	**45** 130 152 161 180 219 **285** 351 359 364 402 422 488 604 749 755 844 848 873	2
	trial 4b	28.93954	130 152 161 180 219 269 312 351 359 364 402 422 488 604 749 755 844 848 873	0
	trial 5b	28.93954	130 152 161 180 219 269 312 351 359 364 402 422 488 604 749 755 844 848 873	0

Note: # The difference of pipe index (marked in bold) installed with flow meters, compared with the best solution (trial 1b is thought to be the best solution founded so far).

4. Conclusions

A new objective function proposed in this article combined with the rank-based fitness assignment method can successfully embody the master–subordinate relationship of the two objective functions and effectively avoid wasting too much time on optimizing unnecessary solutions. The minimum of the number of the flow meter is the master term of objective function while hydraulic performance or water quality deterioration is subordinate term. Three improvements are applied for a better and faster optimization in the dividing phase of WNP, including the reduction of search space through initial hydraulic analysis, modification of the crossover mechanism for a better search ability at the later stage of optimization, and boundary pipe grouping and mutation based on the pipe importance. The proposed method is applied for the WNP in a real WDS. Results show that it plays an effective role in the optimization of layout of the flow meters and valves in WNP.

Overall, in comparison with the iterative method [15,16], which is in essence an empirical method, the heuristic algorithm considers the joint effect caused by the simultaneous closure of multiple pipes (although it is more time-consuming). The multi-objective optimization method usually takes too much time to search the solutions with different number of flow meters. Moreover, the optimization goals should deserve attention. They are usually related to economic feasibility [40–42] or the preservation of the hydraulic reliability of the WDN [11,17]; other aspects such as pressure uniformity and water quality issues are also included. In contrast to Ferrari et al. [21] and Brentan et al. [43], we try not to make the original flow velocity deteriorate, especially for pipes that originally had a low flow velocity. More attention can be paid to better execute pressure management [44] or more quickly detect and control accidental events (e.g., pipe bursts and water contamination) in future research.

Author Contributions: Conceptualization, Y.S. and H.Y.; Formal analysis, H.Y., Y.S. and X.L.; Methodology, H.Y. and S.C.; Supervision, T.Z.; Validation, H.Y.; Writing – original draft, H.Y., Shao Yu and X.L.; Writing – review and editing, Y.S. and T.Z.

Funding: The present research is funded by the National Key Research and Development Program of China (No. 2016YFC0400600), the National Science and Technology Major Projects for Water Pollution Control and Treatment (2017ZX07502003-05, 2017ZX07201-003), the Science and Technology Program of Zhejiang Province (Nos.2017C33174 and 2015C33007), the National Natural Science Foundation of China (No. 51761145022), and the Fundamental Research Funds for the Central Universities (No. 2019FZA4019).

Acknowledgments: The authors greatly thank Yanxi Yu for his discussion and suggestion.

Conflicts of Interest: The authors declare no conflict of interest.

References

1. Ozdemir, O. Water leakage management by district metered areas at water distribution networks. *Environ. Monit. Assess.* **2018**, *190*, 182. [CrossRef]
2. Gomes, R.; Sá Marques, A.; Sousa, J. Estimation of the benefits yielded by pressure management in water distribution systems. *Urban Water J.* **2011**, *8*, 65–77. [CrossRef]
3. Lima, G.M.; Junior, E.L.; Brentan, B.M. Selection of pumps as turbines substituting pressure reducing valves. *Procedia Eng.* **2017**, *186*, 676–683. [CrossRef]
4. Lima, G.M.; Luvizotto, E.; Brentan, B.M.; Ramos, H.M. Leakage control and energy recovery using variable speed pumps as turbines. *J. Water Resour. Plan. Manag.* **2018**, *144*, 04017077. [CrossRef]
5. Di Nardo, A.; Di Natale, M.; Musmarra, D.; Santonastaso, G.F.; Tzatchkov, V.; Alcocer-Yamanaka, V.H. Dual-use value of network partitioning for water system management and protection from malicious contamination. *J. Hydroinform.* **2015**, *17*, 361–376. [CrossRef]
6. Huang, P.; Zhu, N.; Hou, D.; Chen, J.; Xiao, Y.; Yu, J.; Zhang, G.; Zhang, H. Real-time burst detection in district metering areas in water distribution system based on patterns of water demand with supervised learning. *Water* **2018**, *10*, 1765. [CrossRef]
7. Tzatchkov, V.G.; Alcocer-Yamanaka, V.H.; Ortíz, V.B. Graph theory based algorithms for water distribution network sectorization projects. In Proceedings of the 8th Annual Water Distribution Systems Analysis Symposium WDSA, Cincinnati, OH, USA, 27–30 August 2006.
8. Gomes, R.; Marques, A.S.; Sousa, J. Decision support system to divide a large network into suitable district metered areas. *Water Sci. Technol.* **2012**, *65*, 1667–1675. [CrossRef]
9. Alvisi, S.; Franchini, M. A heuristic procedure for the automatic creation of district metered areas in water distribution systems. *Urban Water J.* **2013**, *11*, 137–159. [CrossRef]
10. Di Nardo, A.; Di Natale, M. A heuristic design support methodology based on graph theory for district metering of water supply networks. *Eng. Optim.* **2011**, *43*, 193–211. [CrossRef]
11. Di Nardo, A.; Di Natale, M.; Giudicianni, C.; Greco, R.; Santonastaso, G.F. Weighted spectral clustering for water distribution network partitioning. *Appl. Netw. Sci.* **2017**, *2*, 19. [CrossRef] [PubMed]
12. Liu, J.; Han, R. Spectral clustering and multicriteria decision for design of district metered areas. *J. Water Resour. Plan. Manag.* **2018**, *144*, 04018013. [CrossRef]
13. Perelman, L.; Ostfeld, A. Topological clustering for water distribution systems analysis. *Environ. Model. Softw.* **2011**, *26*, 969–972. [CrossRef]
14. Diao, K.; Farmani, R.; Fu, G.; Astaraie-Imani, M.; Ward, S.; Butler, D. Clustering analysis of water distribution systems: Identifying critical components and community impacts. *Water Sci. Technol.* **2014**, *70*, 1764–1773. [CrossRef]
15. Diao, K.; Zhou, Y.; Rauch, W. Automated creation of district metered area boundaries in water distribution systems. *J. Water Resour. Plan. Manag.* **2013**, *139*, 184–190. [CrossRef]
16. Ciaponi, C.; Murari, E.; Todeschini, S. Modularity-based procedure for partitioning water distribution systems into independent districts. *Water Resour. Manag.* **2016**, *30*, 2021–2036. [CrossRef]

17. Di Nardo, A.; Di Natale, M.; Giudicianni, C.; Musmarra, D.; Santonastaso, G.F.; Simone, A. Water distribution system clustering and partitioning based on social network algorithms. *Procedia Eng.* **2015**, *119*, 196–205. [CrossRef]

18. Di Nardo, A.; Di Natale, M.; Greco, R.; Santonastaso, G.F. Ant algorithm for smart water network partitioning. *Procedia Eng.* **2014**, *70*, 525–534. [CrossRef]

19. Herrera, M.; Izquierdo, J.; Pérez-García, R.; Montalvo, I. Multi-agent adaptive boosting on semi-supervised water supply clusters. *Adv. Eng. Softw.* **2012**, *50*, 131–136. [CrossRef]

20. Izquierdo, J.; Herrera, M.; Montalvo, I.; Pérez-García, R. Division of water supply systems into district metered areas using a multi-agent based approach. In Proceedings of the International Conference on Software and Data Technologies, Sofia, Bulgaria, 26–29 July 2009.

21. Ferrari, G.; Savic, D.; Becciu, G. Graph-theoretic approach and sound engineering principles for design of district metered areas. *J. Water Resour. Plan. Manag.* **2014**, *140*, 04014036. [CrossRef]

22. Sela Perelman, L.; Allen, M.; Preis, A.; Iqbal, M.; Whittle, A.J. Automated sub-zoning of water distribution systems. *Environ. Model. Softw.* **2015**, *65*, 1–14. [CrossRef]

23. Newman, M.E.J. Fast algorithm for detecting community structure in networks. *Phys. Rev. E Stat. Nonlinear Soft Matter Phys.* **2004**, *69*, 066133. [CrossRef]

24. Giustolisi, O.; Ridolfi, L. New modularity-based approach to segmentation of water distribution networks. *J. Hydraul. Eng.* **2014**, *140*, 04014049. [CrossRef]

25. Di Nardo, A.; Di Natale, M.; Santonastaso, G.F.; Venticinque, S. An automated tool for smart water network partitioning. *Water Resour. Manag.* **2013**, *27*, 4493–4508. [CrossRef]

26. Campbell, E.; Izquierdo, J.; Montalvo, I.; Pérez-García, R. A novel water supply network sectorization methodology based on a complete economic analysis, including uncertainties. *Water* **2016**, *8*, 179. [CrossRef]

27. Gomes, R.; Sá Marques, A.; Sousa, J. Identification of the optimal entry points at district metered areas and implementation of pressure management. *Urban Water J.* **2012**, *9*, 365–384. [CrossRef]

28. Di Nardo, A.; Di Natale, M.; Giudicianni, C.; Santonastaso, G.; Tzatchkov, V.; Varela, J. Economic and energy criteria for district meter areas design of water distribution networks. *Water* **2017**, *9*, 463. [CrossRef]

29. Zhang, K.; Yan, H.; Zeng, H.; Xin, K.; Tao, T. A practical multi-objective optimization sectorization method for water distribution network. *Sci. Total Environ.* **2019**, *656*, 1401–1412. [CrossRef] [PubMed]

30. Giudicianni, C.; Nardo, A.D.; Greco, R.; Santonastaso, G.F. Simplified approach for water distribution network dividing. In Proceedings of the WDSA/CCWI Joint Conference, Kingston, ON, Canada, 23–25 July 2018.

31. Goldberg, D.E.; Holland, J.H. Genetic algorithms and machine learning. *Mach. Learn.* **1988**, *3*, 95–99. [CrossRef]

32. Clauset, A.; Newman, M.E.J. Finding community structure in very large networks. *Phys. Rev. E* **2004**, *70*, 066111. [CrossRef] [PubMed]

33. Newman, M.E.J.; Girvan, M. Finding and evaluating community structure in networks. *Phys. Rev. E* **2003**, *69*, 026113. [CrossRef]

34. Morrison, J.; Tooms, S.; Rogers, D. *DMA Management Guidance Notes Version 1*; Water Loss Task Force, IWA: London, UK, 2007; Volume 2, pp. 25–36.

35. Deb, K.; Pratap, A.; Agarwal, S.; Meyarivan, T.A. A fast and elitist multiobjective genetic algorithm: NSGA-II. *IEEE Trans. Evol. Comput.* **2002**, *6*, 182–197. [CrossRef]

36. Abdulal, W.; Al Jadaan, O.; Jabas, A.; Ramachandram, S. An improved rank-based genetic algorithm with limited iterations for grid scheduling. In Proceedings of the IEEE Symposium on Industrial Electronics & Applications, Kuala Lumpur, Malaysia, 4–6 October 2009; IEEE: Piscataway, NJ, USA, 2009.

37. Jadaan, O.A.; Rajamani, L.; Rao, C.R. Ranked based roulette wheel selection method. In Proceedings of the International Symposium on Recent Advances in Mathematics and Its Applications: (ISRAMA 2005), Calcutta Mathematical Society at AE-374, Sector-1, Salt Lake City Kolkata (Calcutta), India, 17–19 December 2005.

38. Macesic, S. Binary-coded and real-coded genetic algorithm in pipeline flow optimization. *Math. Commun.* **1999**, *4*, 35–42.

39. Zhang, Q.Z.; Wu, Z.Y.; Zhao, M.; Qi, J.; Huang, Y.; Zhao, H. Leakage zone identification in large-scale water distribution systems using multiclass support vector machines. *J. Water Resour. Plan. Manag.* **2016**, *142*, 04016042. [CrossRef]

40. De Paola, F.; Fontana, N.; Galdiero, E.; Giugni, M.; Savic, D.; Uberti, G.S.D. Automatic multi-objective sectorization of a water distribution network. *Procedia Eng.* **2014**, *89*, 1200–1207. [CrossRef]

41. De Paola, F.; Fontana, N.; Galdiero, E.; Giugni, M.; Uberti, G.S.D.; Vitaletti, M. Optimal design of district metered areas in water distribution networks. *Procedia Eng.* **2014**, *70*, 449–457. [CrossRef]

42. Galdiero, E.; De Paola, F.; Fontana, N.; Giugni, M.; Savic, D. Decision support system for the optimal design of district metered areas. *J. Hydroinform.* **2016**, *18*, 49–61. [CrossRef]

43. Brentan, B.M.; Campbell, E.; Meirelles, G.L.; Luvizotto, E.; Izquierdo, J. Social network community detection for DMA creation: Criteria analysis through multilevel optimization. *Math. Probl. Eng.* **2017**, *2017*, 9053238. [CrossRef]

44. Al-Washali, T.; Sharma, S.; Al-Nozaily, F.; Haidera, M.; Kennedy, M. Modelling the leakage rate and reduction using minimum night flow analysis in an intermittent supply system. *Water* **2018**, *11*, 48. [CrossRef]

Article

Urban Drainage Network Rehabilitation Considering Storm Tank Installation and Pipe Substitution

Ulrich A. Ngamalieu-Nengoue [1], Pedro L. Iglesias-Rey [1,*], F. Javier Martínez-Solano [1], Daniel Mora-Meliá [2] and Juan G. Saldarriaga Valderrama [3]

[1] Department of Hydraulic Engineering and Environment, Universitat Politècnica de València. Camino de Vera s/n, 46022 Valencia, Spain; ngamaleuulrich@yahoo.fr (U.A.N.-N.); jmsolano@upv.es (F.J.M.-S.)

[2] Department of Engineering and Construction Management, Facultad de Ingeniería, Universidad de Talca, 3340000 Talca, Chile; damora@utalca.cl

[3] Department of Hydraulic Engineering and Environment, Water Supply and Sewerage System Research Center (CIACUA), University of Los Andes, Cra 1 # 18A-11, Bogota D.C., Colombia; jsaldarr@uniandes.edu.co

* Correspondence: piglesia@upv.es; Tel.: +34-96-387-70-00 (ext. 86111)

Received: 21 January 2019; Accepted: 6 March 2019; Published: 12 March 2019

Abstract: The drainage networks of our cities are currently experiencing a growing increase in runoff flows, caused mainly by the waterproofing of the soil and the effects of climate change. Consequently, networks originally designed correctly must endure floods with frequencies much higher than those considered in the design phase. The solution of such a problem is to improve the network. There are several ways to rehabilitate a network: conduit substitution as a former method or current methods such as storm tank installation or combined use of conduit substitution and storm tank installation. To find an optimal solution, deterministic or heuristic optimization methods are used. In this paper, a methodology for the rehabilitation of these drainage networks based on the combined use of the installation of storm tanks and the substitution of some conduits of the system is presented. For this, a cost-optimization method and a pseudo-genetic heuristic algorithm, whose efficiency has been validated in other fields, are applied. The Storm Water Management Model (SWMM) model for hydraulic analysis of drainage and sanitation networks is used. The methodology has been applied to a sector of the drainage network of the city of Bogota in Colombia, showing how the combined use of storm tanks and conduits leads to lower cost rehabilitation solutions.

Keywords: drainage network; climate change; rehabilitation; optimization; SWMM

---

1. Introduction

One of the main problems related to urban drainage systems are the frequent flooding events in urban areas. Various factors can cause these events, such as pipe size, structural failures in the system, objects causing obstructions or an increase in rainfall intensity due to climate change inducing increased runoff flow. Climate change undoubtedly affects the intensity and frequency of meteorological phenomena of present days, including a gradual increase in rainfall intensities in many cities throughout the world, making systems initially well designed begin to fail.

In order to have a better control over the systems and prevent the occurrence of urban flooding events, many mechanisms have been implemented to reduce runoff and increase the capacity of the system. Certainly, one of the most used methods in the last years has been the implementation of storm water detention tanks (STs). These methods refer to the installation of devices in the urban area that can hold back the flow that cannot be evacuated by the system. Later, when the system has regained its transport capacity, the devices return the water to the sewerage system for correct evacuation.

Howard [1] published one of the first works related to STs where it is established that the effectiveness of the STs combined with Wastewater Treatment Plants (WTP) to control runoff depends on STs and WTP's capacity as well as the duration and volume of the precipitation events they must control. Howard's paper also lays out the possibility of using computational tools for the analysis of these devices. His work focused on the usage of probabilistic methods based on precipitation data.

These water storage structures are built to retain a portion of the just-fallen rainwater, a portion that varies according to every storm. Di Toro and Small [2] posited a statistical method based on a probabilistic characterization of both rainwater and runoff that predicts the behavior of the storm water control devices. The long-term behavior of a ST is determined based on its size, operation method and both runoff and precipitation statistical characteristics. Based on these, Di Toro and Small studied the filling, storage and emptying of the retaining structures; flow variations caused by storm water; and first flush's implications regarding water quality.

Early ST sizing methods were based on roughly simplified methods since simulation techniques require a high computational effort in terms of time and memory. Loganathan et al. [3] presented a simplified method that could estimate ST's capacity accounting for previous storms. This method is based on exponential probability density functions for the main hydrological variables involved (runoff volumes, runoff duration, time between events). These functions are used to generate a new statistical distribution that can estimate WTP's capacity and retention volume for a given risk level. One of the main advantages of this method is that it permits us to determine the preliminary volume for an ST. With the same focus, Meredith et al. [4] developed a procedure based on all available historical data for the dimension of a runoff storage structure. The dimensioning process uses the water quality concept because all runoff must pass through the cleaning water process in the WTP. The method was tested in several industrial areas, but it is limited to small areas where rational methods can be applied.

Most of these studies were focused on the treatment of rainwater runoff before the water enters the network. Due to urban area space restrictions, alternative systems for the treatment of runoff water are needed. For this, Takamatsu et al. [5] presented the design of a rainwater storage system as part of the complementary structure of road drainage. They developed a mathematical model based on the hydraulic principals to estimate the efficiency of the pollutant elimination. The main idea was to evaluate the efficiency of a rectangular runoff ST removing suspended solids. In order to validate the mathematical model, a scaled (1/5) network model was built, on which several tests were run to study the influence of different inflows, functioning frequency and pollutant concentrations. The temporal concentration of suspended solids at the exit and the efficiency of the conceptual model were compared. They concluded that there is a correlation between the detention time and the removal efficiency.

Similar studies to Takamatsu et al. have been done to study the behavior of several draining areas under different rainfall scenarios. De Martino et al. [6] proposed the usage of STs as structures that enable controlling the impact of the first-rain contamination, which has the highest pollutant load. As the design of this system depends on a large number of parameters, they consider that the design is not completely defined. In order to study the influence that the rainwater might have, De Martino et al. made several simulations with different rainfall-time series in Campania (Italy). This study made an analysis of the hydrographs and pollutographs generated for several network and tank configurations. The results showed that the rainfall does not have a direct influence in the reduction of pollutant concentration. Based on this, a new methodology for the preliminary design of STs was presented.

In line with analyzing different rainfall amounts, Li et al. [7] analyzed 16 different rain events in three drainage areas. The objective of this study was to analyze the effect of a detention tank's location on the polluted load reduction, mainly particles and metals. Thus, Fu et al. [8] presented a model for optimum distribution of the STs in order to minimize the effect of new residential zones on the discharge water quality. The developed model considered three working scenarios; optimum flow control, minimization of deposit's distribution and a combination of both. A step beyond was made by Vanrolleghem et al. [9], who proposed a real time control system to manage the quality of water poured in the collectors in order to comply with the European Water Framework.

Most of the previous works about the usage of STs are focused on the issue of maximizing the quality of poured water, not to control the potential overflows that may occur due to excessive rainwater, but some studies have been carried out in order to show that STs can also reduce floods. One of the first studies that relate the usage of STs with the rainwater variation due to climate change was done by Andrés-Domenech et al. [10]. The study focused on the effects originating from changes in the rainfall regimes on the efficiency of the actual drainage systems. The proposed analytical statistical model permits us to evaluate the overflow reduction efficiency and the retention tank's volumetric efficiency as a function of the expected climate behavior and urban basins. The tank's efficiency sensitivity is evaluated under the analysis of certain changes in the precipitation. The results show the ability of STs to partially mitigate the resulting effects of climate change. Andrés-Domenech et al. based their work in the same approach as Butler and Schütze [11]: integrate simulation models to obtain an optimum control of the sanitary draining systems. For this purpose, Butler and Schütze developed a model (SYNOPSIS) consisting of a series of sub-models of the sewage network, the treatment plant and the behavior of the natural stream over which the system's evacuation will be made. These sub-models, together with a developed control module, allow the development of water control strategies in order to minimize the impact on the water evacuation. The use of simulation techniques is simple for urban wastewaters. Moreover, the simulation techniques enable water control solutions that were not possible by the classic simplified methods. This work was extended by Fu et al. [8] taking into account the optimization of urban wastewaters as a multi-objective problem. For this, they used a powerful multi-objective genetic model called NSGA II [12] that allows obtaining the Pareto's front for several optimum solutions. For the same topic, Wang et al. [13] proposed a two-stage optimization framework to find an optimal scheme for STs using the storm water management model (SWMM) [14]. As a result, the authors conclude that the use of STs not only reduces flooding, but also the total suspended solids.

Drainage network rehabilitation is a necessary operation to adapt an insufficient network to the new climatologic and environmental conditions. There are several ways to rehabilitate a drainage network with difficulties in the compliance of its design specifications. Traditional ones consist of pipes replaced with larger pipes. A newly developed option is the introduction of STs in the network to store water during the storm, draining it after the extreme events have passed. In this regard, Gaudio et al. [15] proposed a methodology for the hydraulic rehabilitation of urban drainage networks combining a hydraulic analysis model, a data base of rainfall events and a series of synthetic hyetographs. The results statistically analyze the possible places where the flood occurs and the influence that the roughness of the pipes can have on it.

Considering the rehabilitation using pipe replacement and other actions, Sebti et al. [16] proposed an algorithm to analyze the benefits of combining pipe substitution with other solutions installed to reduce the runoff. However, Sebti et al. did not consider hydraulic models to analyze the consequences of rehabilitation works in the network. Instead, they used the simplex algorithm for the optimization. Then, Ugarelli and Di Federico [17] calculated the economic balance between replacing damaged infrastructures or maintaining them for a certain period of time.

The rehabilitation of drainage networks based on the mere substitution of pipes is an approach that has been traditionally carried out. However, this solution is not very applicable in consolidated cities in which the replacement of large parts of the network interferes with other services already installed and generates social problems for the population such as noise, inconvenience or traffic jams. That is why this work proposes a combined use of the installation of STs and the replacement of pipes. This is intended not only to reduce investment costs, but also to minimize the impact that changes may have on citizens.

In short, the objective of this paper is to propose a methodology for the rehabilitation of drainage networks that combines the installation of STs with the replacement of pipes. To do this, the mathematical model of the network is the starting point, using SWWM to conduct the hydraulic analysis and a pseudo-genetic optimization algorithm (PGA) [18] to find the best solutions. The connection between the hydraulic calculation model and the optimization algorithm will be carried out through an adaptation of the SWMM calculation toolkit done by Martínez-Solano et al. [19] in order to reduce the calculation time.

The solution of the optimization problem requires an elevated number of decision variables (DVs) generating a large space of feasible and unfeasible solutions. This search space entails not only a big computational effort, but also may cause the method to fall at local minima, limiting the ability to find the best solution. A reduction in the size of the problem and, subsequently, in the size of the search space might help the convergence of the method. In this paper, a methodology for the reduction of number of DVs based on pre-selection of potential locations of STs and the selection of conduits potentially substitutable is proposed. This is a worthwhile contribution because reducing the size of the search space and improving searching mechanisms are listed as current research challenges in the application of optimization algorithms to real world problems [20,21].

The methodology has been validated in several networks. Although this paper presents the results of application to the E-Chicó sector of the drainage network of Bogotá (Colombia), some model files and additional case studies can be found in the supplementary material.

2. Problem Formulation

In order to formulate the problem of optimizing the rehabilitation of drainage networks based on the combined use of STs and the replacement of pipes, some hypothesis had to be established:

- The computational models for the drainage networks are going to be tested with several rain scenarios, scenarios based on different climate change predictions [22]. Potentially dangerous scenarios are the ones to be considered during the optimization process. The first selected rainfall is a synthetic design rainfall obtained from the IDF curves of 10-year return periods using the alternate block method. The second rainfall is obtained after processing the first through a climate change adaptive scenario. The rehabilitation will be carried out considering only the worst-case scenario.
- The rainfall-runoff transformation model used is the one included in the SWMM model. Specifically, the Curve Number model is used. For this, according to the characteristics of the terrain, the curve number has been specified for each of the sub catchments defined in the model.
- The drainage system models must go through a calibration process, since the analysis must be as accurate as possible. That is, the starting point of the process is a calibrated hydraulic model of the drainage network. The hydraulic model used would be SWMM. Traditionally, this type of simulation is performed considering uniform flow. However, in this case, each configuration is analyzed using the dynamic wave model, because it provides a better representation of floods than the kinetic wave model or uniform model.
- A simplification process is necessary for every model, yet the accuracy of the result must not be compromised. This simplification will highly reduce computational times for every hydraulic simulation.
- The optimization problem will be addressed in monetary units. Thus, the first step would be to find the cost functions that characterize the value of hydraulic variables in monetary units. So, the functions that together form the optimization total cost problem are: pipe replacement cost, ST installation cost and total flood damage cost as introduced by Cunha et al. [23].
- From all described mathematical approaches, it seems heuristics approaches can give the best advantages for the process. Therefore, based on previous experience [24], a PGA method was used.

The main objective of drainage network rehabilitation is to allow the adaptation of the network to new climatologic and environmental condition while fulfilling assigned missions. Due to that importance, it is necessary to define an adequate optimization scenario in order to optimize the process. The formulation of the problem of rehabilitation of a drainage network, combining the installation of ST and the replacement of pipes, is summarized in the flow chart of Figure 1. The first part consists of obtaining the calibrated model of the network, which is the starting point of the process.

This network should represent, as faithfully as possible, the behavior of the network. Afterwards, a simple optimization process is used, with a hydraulic model based on the use of the SWMM Toolkit and a PGA that uses the levels in nodes and pipes and the flooding in nodes to determine the possible diameters of the rehabilitated conduits or the size of the installed STs. In order to define all the details of the optimization algorithm, the following sections describe the DVs, the objective function and the cost functions used.

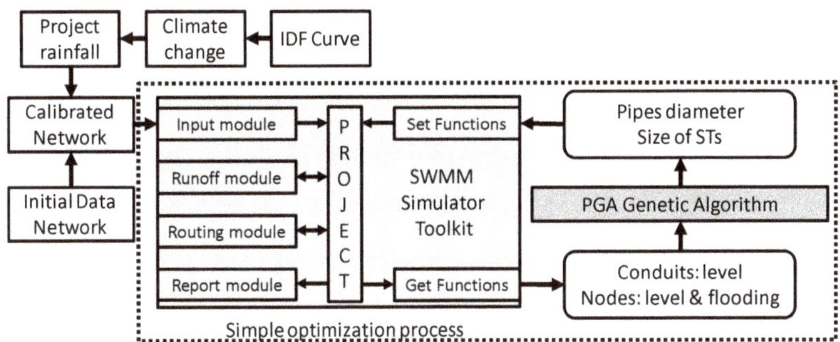

Figure 1. Flow chart of the optimization methodology based on the PGA algorithm and the toolkit of the SWMM model.

2.1. Decision Variables

The process of rehabilitation of a drainage network (Figure 1) involves modifying two types of DVs. On the one hand, there are variables related to pipe diameters, which seek to locate the best combination of sizes to obtain the minimum flooding. The optimization model analyzes the replacement of the pipes based on their transport capacity. The diameter may increase if the original pipe is insufficient to transport the flowing water or decrease if a hydraulic control device [25] should be installed to introduce the same head loss as the calculated pipe's diameter. Consequently, the DV is the size of the pipes, and can vary from 0 (not replace) to a maximum value set before beginning the optimization process. Obviously, if the result associated with a pipe is 0, it is because it is not necessary to replace it for the correct operation of the network. Then, for a better understanding of the optimization methodology, it is convenient to define some parameters related to the pipes. So, N_C is the number of network conduits; m is the number of feasible conduits selected to be replaced, varying between 1 and N_C; and ND is the number of candidate diameters, between ND_0 and ND_{max}.

On the other hand, there are variables related to node storage capacity that seek the minimum volume of STs that reduce floods. The proposed methodology considers the possibility of installing an ST in each node of the network. This involves replacing an existing manhole with an underground ST. The land in which the rehabilitation of the drainage network is developed is mostly urban. Therefore, it is admitted that the depth of excavation is limited. Thus, the maximum depth of ST is what the manhole initially had, so the only parameter needed as DV is the ST cross section. Related to this, every node has a defined storage capacity related to its cross section and the model takes into account some nodes that could be modified into an ST. For this process, every node would have a DV representing the equivalent additional section corresponding to that of the STs in the case it would have been installed in the node location. The DV would vary between zero and a maximum value, predetermined before the optimization process and taking into account the restrictions of the urban geographical space of each node. If the nodes were not to be transformed into a tank, then the variable would have a value of 0, meaning that, it is not adequate to install a ST on the specific node's location, maintaining thus the initial storage capacity of the node. In case the initial hydraulic model has any type of water deposit, the cross section of the deposit might be part of the optimization process.

For every node, whether it is regular or a storage node, its cross section S would be expressed according to the following equation:

$$S = A_S \times z^{B_S} + C_S \tag{1}$$

where A_S, B_S and C_S are characteristic coefficients that adjust the tank's section to different expressions, and z is the water level of the node. In the case of considering tanks of constant section, A represents the cross section while the coefficients B and C are null. However, considering tanks with variable section does not imply a major difficulty in the problem implementation beyond choosing the right DVs.

Again, the definition of some parameters related to the nodes is important for the understanding of the methodology. N_N is the number of network nodes and n is the number of nodes selected to potentially install a ST, varying between 1 and N_N. Each node in which an ST can be installed has the cross section (S) as DV. Since a heuristic optimization model is used, it is necessary to perform a discretization of S. For this reason, a maximum value of the tank cross section (S_{max}) is defined for each node. In this way, N is the number of divisions in which S_{max} is divided. Therefore, N determines the resolution of the section S, varying between N_0 and N_{max}. A simulation performed with the ST cross section divided into N_0 parts is faster than a simulation performed N_{max} divisions. So, to obtain better calculation times, the number of divisions of the ST cross section could be reduced.

Additionally, to introduce tanks with a variable straight section does not imply major difficulty in the problem implementation beyond choosing the right DVs.

As stated before, the size of the problem is a key aspect when trying to optimize real problems. In this work, the optimization algorithm takes into consideration both pipes and nodes. The maximum size of each rehabilitation scenario can be expressed by the following equation:

$$PS_{max} = N_C{}^{ND_{max}} \times N_N{}^{N_{max}} \tag{2}$$

2.2. Objective Function

The objective function of the optimization problem is addressed in monetary units and is represented as the sum of three cost functions:

- The investment cost related to the substitution of each selected pipe of the network.
- The investment cost linked to required volumes of STs to be installed in each solution.
- The damage cost caused by the flooding level in various nodes of the network.

The mathematical expression of the objective function is given by Iglesias-Rey et al. [26]:

$$F = \lambda_1 \sum_{i=1}^{m} C(D_N(i)) \times L_i + \lambda_2 \sum_{j=1}^{n} C(V_{DR}(j)) + \lambda_3 \sum_{k=1}^{N_F} C(V_I(k)) \tag{3}$$

In Equation (3), the first term represents the rehabilitation or replacement cost of the m considered pipes in the network. The second term represents the construction or expansion cost of volume $V_{DR}(j)$ of the n STs installed in the drainage network. This cost concerns the existing STs whose volume will be expanded, and the network nodes where new STs will be installed. The third term represents the total flood damages costs [27] caused by the N_F nodes in which a certain flooding volume $V_I(i)$ appears. All the terms of the objective function have a weight coefficient λ_i, in order to prioritize one term versus another. If λ_i is minimum (eventually null) the term is not considered, but if λ_i is greater, the term is considered in the objective function.

2.2.1. Pipe Replacement Cost Functions

This function represents the cost of rehabilitation of those pipes that are replaced in the optimization process, either because they have poor internal conditions or because they have insufficient transport capacity. In this work, the proposed function is adjusted based on real data from manufacturers, relating the u_{nit} cost of the pipes with their diameter. Finally, the pipe substitution cost is in the form of a second-grade polynomial:

$$C(D_N) = \alpha \times D_N + \beta \times D_N^2 \tag{4}$$

where α and β are coefficients that must be adjusted in each case considering the costs of the project.

2.2.2. Storm Tank Installation Cost Functions

This cost function links the associated investment to the installation of STs in the network. The tanks are installed on-line, simply said in series with the pipes existing in the network. The determination of this cost function is fundamental at the moment to establish the size of the adequate tanks to install. The function defines the tank installation cost based on its storage volume. Here, the idea is to build small and median tanks implying the increase of the storage capacity of the network and the water flow travelling time through the network. The mathematical expression of the cost function is:

$$C(V_{DR}) = A + B \times V_{DR}^C \tag{5}$$

In this expression, $C(V_{DR})$ is the cost associated with the installation of an ST of volume V_{DR} and A, B and C are coefficients that must be adjusted in each case according to the specifics of the project.

2.2.3. Flood Damage Cost functions

Representation of the flood cost function can be based on either total flooding volume or floods based on either total flooding volume or flood level. The first option allows us to eliminate totally the flood, but this option will need more investments. This total water volume can be obtained by summing the flooding volume of each flooding node. The second option allows reducing the flood and the cost of the flood depending on the flooding area. Lee and Kim [27] showed that flooding damage is different from flooding volume and that it was necessary to create a resilience index based on flooding damage because some subareas are immediately damaged by a certain amount of flooding while other subareas are not. They represented flood damage costs as a function of the level reached by water.

Using the Lee and Kim approach, flood damage costs have been determined for a drainage network in Bogota (Colombia) analyzing the replacement costs of the damaged goods. A curve representing the flood cost by m^2 of area as a function of the flood level was obtained for 6 different social stratums, a commercial area and an industrial area [23]. There is also a curve representing the average value of the study area. They can be observed in Figure 2.

The formulation of the proposed problem considers that the resulting flood cost is defined according to land use. Thus, the damage function is mathematically characterized by the expression:

$$C(y) = C_{max} \times \left(1 - e^{-\lambda \frac{y}{y_{max}}}\right)^b \tag{6}$$

In this equation, C_{max} represents the maximum cost, when flood level y_{max} is reached. y is the existing flood level in the specific node; $\lambda = 4.89$ and $b = 2$ are adjustment coefficients of the curve; and the parameter $y_{max} = 1.4$ is the level from which the maximal economic damage is produced. In all the cases, Equation (6) depends totally on the value of C_{max}, which is presented in Table 1, for different land uses and represents the maximum per area unit.

Table 1. C_{max} values for different social strata (Str) linked to land uses.

Land Use	Str. 1	Str. 2	Str. 3	Str. 4	Str. 5	Str. 6	Commercial	Industrial	Average
C_{max} (€/m^2)	142	245	257	584	732	1168	3975	3041	1268

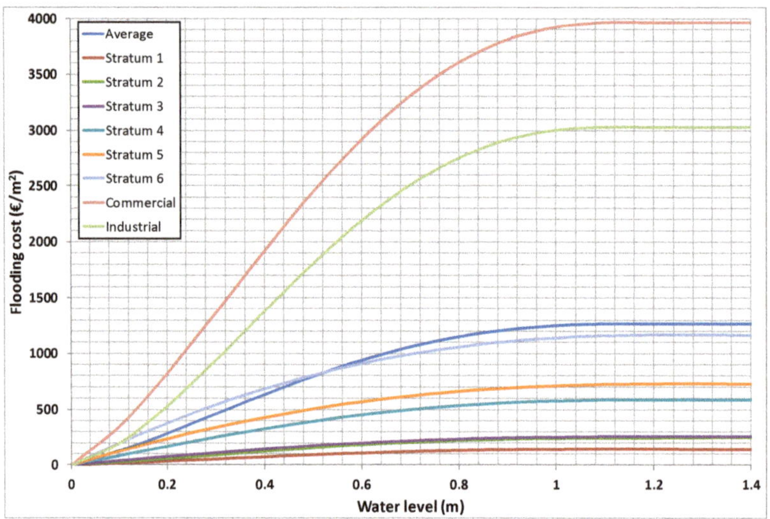

Figure 2. Flooding costs for different social strata linked to land uses.

The cost evaluation $C(y)$ described in Equation (6) required the determination of the flood level in each node. Therefore, the SWMM's ponded area model is used. This model assumes the definition of a ponded area (A_f) in each node. In this way, the flood level is obtained as the relation between the flood volume V_f and the area A_f.

3. Methodology

The optimal design or rehabilitation in drainage networks involves the search for solutions in very large spaces. Accordingly, the probability of finding minimum solutions is very small due to the immensity of SS and the existence of multiple local minima. Developments in the field of genetic algorithms (GAs) have proven to be useful in the optimization of drainage networks.

The GAs test the evolution of a random population via a parallelism that is similar to Darwin´s law of natural selection. Some calibration parameters, including crossover probability, mutation probability and the population size control the optimization process. In this sense, the performance of population-based algorithms is directly related to the balance of exploration and exploitation of the SS. Traditionally, small population sizes have been related to premature convergence, since the population of the algorithm loses diversity too early, converging too early with poor solutions. Conversely, larger population sizes help ensure diversity of individuals, avoiding premature convergence, but the algorithm might waste considerable time exploring regions of the SS without any kind of interest. Consequently, parameter choice should be a trade-off between solution quality and search time. It should be noted that this work uses a modified version of classical GA, PGA, whose complete description can be found in [18].

In the same manner, the stopping criterion is important for population-based algorithms as GAs. Three genetic algorithm stopping conditions are usually found in the literature: the objective function value reaches a certain pre-defined value, a defined absolute number of generations are performed or there is no improvement in the population for X iterations. In this work, the third option was used, i.e., a maximal number of generations G without a change in the objective function value. Considering the characteristics of the problem and based on previous experiences [25], a stopping criterion of 1000 generations without change was selected. As in the calibration of the population size, the choice of stopping criteria must prevent premature convergence, guaranteeing a good exploration and exploitation of the SS.

Additionally, it should be noted that the DVs had to be discretized. On one hand, for STs a maximum area is used, which is a function of the available surface with minimal urban impact and divided into partitions. Therefore, STs can be discretized in N divisions ($N = N_0, \dots, N_{max}$). On the other hand, the number of candidate pipe diameters was ND ($ND = ND_0, \dots, ND_{max}$).

As a previous step, some algorithm tests were performed considering the full SS, i.e., all the nodes and lines of the network. According to protocol, several tests were carried out, in which it was appreciated that the space of solutions is so large that the algorithm barely finds good solutions. The enormous amount of local minimums causes the algorithm to be lost. A smaller solution space allows the optimization algorithm to find minimum values more easily.

So, this work presents a methodology for the rehabilitation of large drainage networks, reducing the SS of the problem. Additionally, this SS reduction will improve the solutions found by the PGA when the entire SS was used. The different options to reduce the SS are the following:

- Reduce the number of nodes (n) in which STs could potentially be installed.
- Reduce the number of lines (m) in which there could potentially be a change in diameter.
- Reduce the discretization N that is made of the section of each of the STs.
- Reduce the number of candidate diameters ND in the pipes.

The four ways to reduce the SS have been organized in a specific methodology. The complete process is summarized in Figure 3, and basically consists of three stages: the pre-location of STs, the pre-selection of conduits and the final optimization. In the following subsections, each of these stages is described in detail.

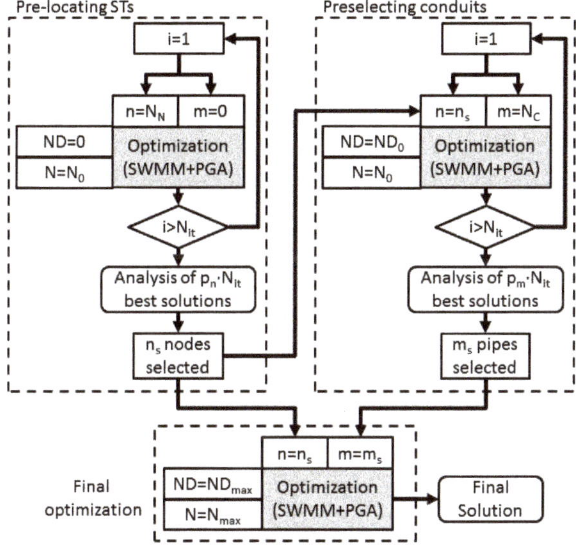

Figure 3. Block diagram of the proposed methodology.

3.1. Pre-locating Storm Tanks

There are some previous studies oriented to the pre-locating of STs [25]. However, in this case methodology described in Figure 1 is used as a basic tool to determine the possible locations of the STs. The first step of the methodology is summarized on the left hand side of Figure 3. Some runs (N_{it}) are performed with all the nodes of the network (n = N_N), but without including the diameters in the optimization process (m = 0). Thus, N_{it} optimizations are made, considering only the cross section

(S) of the tanks as DVs and without modifying the diameters of the network. Since the objective is the reduction of the SS, the discretization of the cross sections of the tanks (S) is carried out with the smallest number of divisions ($N = N_0$). This coarse discretization of each section is carried out since the objective is not to calculate its exact value, but to determine in which nodes the installation of an ST is adequate. That is, the objective of this step is selecting the nodes where STs could be installed in the rehabilitation of the network, i.e., a pre-location of STs.

In this step, the obtained solutions of initial trial runs are ranked according the value of the objective function. Each of these solutions contains a distribution of STs in the network and a dimensioning, even if approximate, of the ST size required in each node. However, the analysis is not focused on the size of the STs but on their location. Therefore, a percentage p_n of the best simulations is selected. The analysis of these simulations allows identifying the nodes where an ST could be installed and a list of n_s possible locations is created. These nodes are selected because they are repeated as the location of an ST in all the p_n selected solutions.

3.2. Locating Lines of Possible Pipe Substitutions

The objective of this second step is to reduce the number of pipes in which the diameter can be modified. Another set of N_{it} optimizations is run. The DVs of each optimization are the cross section of the n_s selected nodes in the previous process and the N_c conduits of the network. For both types of variables the reduction of the SS is applied. Therefore, the discretization of the cross section of the tanks is the coarsest ($N = N_0$) and the range of pipes available is the smallest ($ND = ND_0$). In this step, the aim is to find a pre-location of pipes to be substituted.

Analogous to what happened in the previous step, an analysis of the solutions with the best value of the objective function is carried out. This analysis is not focused on the section of the tanks or on the diameter of the ducts. The analysis is centered on locating the conduits whose dimensions have been modified with respect to the initial situation.

Finally, a percentage p_m of the best solutions are selected. The conduits selected are those that appear repeated in the solutions defined by the percentage p_m. Analyzing these solutions, the list of m_s pipes whose replacement is repeated in the p_m best solutions is selected.

3.3. Final Optimization, Location and Optimization of Storm Tanks and Pipe Diameters

The last step considers the results of the two previous steps: the pre-location of the STs and the location of the pipes that could potentially be rehabilitated. A simulation is defined with the n_s selected nodes and the m_s selected lines. Although the number of DVs is smaller, the exploration of each of these variables must now be greater. Therefore, the STs are discretized to the maximum ($N = N_{max}$) and the list of candidate diameters for the conduits is also the largest ($ND = ND_{max}$). This final optimization determines the location and size of the STs to be installed and the diameters of the pipes to be rehabilitated.

In short, the reduction of SS in a problem with continuous and discrete variables has been based on two aspects: the reduction of the number of DVs and the level of detail of each of the variables. During the first two stages, the number of DVs is reduced by two with a lower level of exploration of each variable. In the final simulation, a smaller number of variables is used, but with a higher level of exploration. This reduction of the SS allows obtaining solutions to problems with large SS and multiple local minimums.

4. Case Study

The drainage network is called E-Chicó. It is divided into 35 hydrological sub-catchments over an area of 51 hectares. All the conduits are circular with diameters between 300 and 1400 mm, with a total length around 5000 m. The network works completely by gravity with a maximum difference of 39.28 m.

The diagnosis of the network and the evaluation of the possible solutions have been carried out using projected rain obtained by means of the alternative blocks method and the IDF curve. This process has been done with the original IDF curve (actual rain), the one with which the network was designed, and the one obtained after applying several climate change models [22]. As can be seen (Figure 4), the consideration of the climate change effects on the study area implies an increase in the intensity of rainfall above 45%.

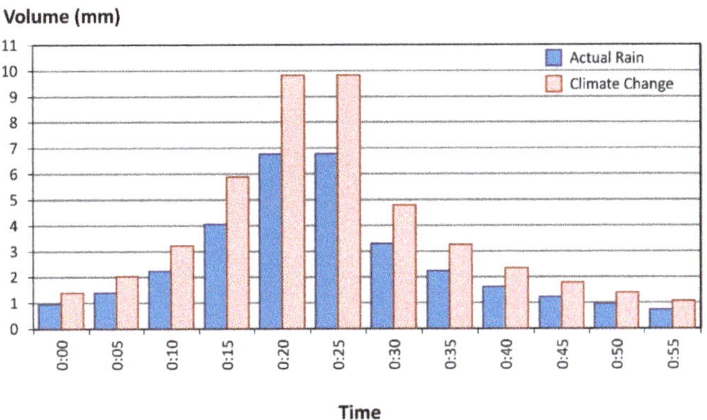

Figure 4. Alternates blocks rains: Actual and Climate Change.

The results of the first simulation considering the current IDF curve barely cause flooding in any node of the network. On the other hand, if the network is analyzed with the IDF curve obtained from climate change analysis, flooding occurs in 11 nodes of the network with a volume of 3833 m^3, representing 18% of the total runoff of the network (21,233 m^3). Figure 5 shows the nodes in which the main floods occur (flood level over 10 cm), indicating the flood volume V; the maximum level y reached by the water in the node and the cost associated with flood damage. Table 2 shows the detail of the flooded nodes: the flood volume, area and level, and the damage cost obtained from Equation (6). The nodes shown in Figure 5 are highlighted in bold in Table 2.

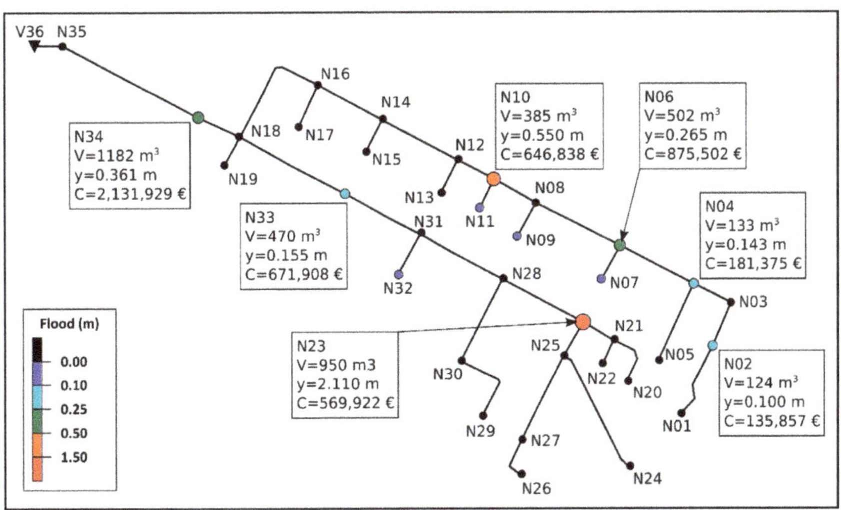

Figure 5. Representation of E-Chicó Flooding Nodes during the rainfall event.

Table 2. Data of the flooded nodes. .

Node.	Flood Volume (m^3)	Flood Area (m^2)	y_{max} (m)	C(€)
N02	123.56	1240	0.100	135,857
N04	132.56	930	0.143	181,375
N06	501.79	1890	0.265	875,502
N07	23.95	1250	0.019	6644
N09	1.82	1130	0.002	45
N10	385.12	700	0.550	646,838
N11	25.83	820	0.032	11,288
N23	949.54	450	2.110	569,922
N32	36.65	1500	0.024	12,727
N33	469.82	3030	0.155	671,908
N34	1181.87	3270	0.361	2,131,929
TOTAL	3832.51			5,244,034

The total cost of flood damage in the study area amounts to 5.24 million euros, which suggests the importance of rehabilitating this sector of the network and applying the proposed methodology.

4.1. Application of the Drainage Network Rehabilitation Methodology to E Chico

In the first place, the validity of the methodology described in Figure 1 will be tested. Taking as a starting point the simulation whose results are obtained in Figure 3, three different simulation scenarios are considered:

- Scenario 1: Rehabilitation of the network based only on the modification of conduits of the network and substituted them by another with different diameter. This scenario has 35 DVs, as all the conduits are potentially changed.
- Scenario 2: Rehabilitation of the network by installing only STs. This scenario also has 35 DVs, corresponding to the 35 potential nodes in which STs can be installed. The maximum section to be installed in the tank potentially installable is defined in each node. Subsequently, the optimization method selects the cross section according to the discretization (N) of this variable. It should be noted that the section minimum value corresponds to a diameter of 1.2 meters, corresponding to the cross-sectional value of a manhole.
- Scenario 3: Rehabilitation of the network combining the installation of conduits and STs. The number of DVs is 70.

In order to be able to simulate the previous scenarios, all the data described in the formulation of the problem should be defined. That is, it is necessary to specify the values of the installation cost functions of the new conduits and the investment cost for the construction of the new STs to be installed at the nodes. This involves determining the values of the parameters α and β of Equation (4) and parameters A, B and C of Equation (5). In this case, price databases have been used for the area where the network is located (Colombia). In this way, the values of the characteristic parameters of Equations (4) and (5) are shown in Table 3.

Table 3. Coefficients for pipes and STs cost curves.

α	B	A	B	C
40.69	208.06	16923	318.4	0.65

At the same time, simulation of scenarios 1 and 3 requires defining a full range of pipe diameters. The full range of diameters used in this case is shown in Table 4. This table assumes a value of the parameter $ND = ND_{max} = 25$, since it must be combined with an additional state corresponding to the event of leaving the drainage line as it is in the model; that is, without rehabilitation. Table 4 also shows the installation cost of each diameter. These costs are obtained from the application of Equation (4) with the coefficients defined in Table 3.

Table 4. Full range ($ND = ND_{max}$) of commercial diameters used in the example.

D (mm)	300	350	400	450	500	600	700	800	900	1000	1100	1200
C (€/m)	30.93	39.73	49.56	60.44	72.36	99.31	130.43	165.71	205.15	248.75	296.51	348.43
D (mm)	1300	1400	1500	1600	1800	1900	2000	2200	2400	2600	2800	3000
C (€/m)	404.51	464.76	529.16	597.73	747.35	828.4	913.61	1096.52	1296.07	1512.27	1745.11	1994.6

The results of the optimization of the first three scenarios will allow comparing the effect of carrying out the rehabilitation of the drainage network by different methods. The results will show the benefits of rehabilitation through the combined use of STs and the replacement of pipes. One of the main conclusions obtained from this first analysis is the time necessary to complete the simulations due to the high number of DVs and the wide of solutions space. In addition, after performing numerous simulations, a large dispersion in the results was also observed. Additionally, it could be concluded that the combination of diameter changes and ST installation gave better results than the optimization of any of them separately.

For the Scenario 3 corresponding to the rehabilitation of the whole network, some previous simulations have been performed with the following characteristics: $n = 35$, $m = 35$, $N = 40$ and $ND = 25$. The best objective function value obtained was 268,292 €, corresponding to the substitution of 21 pipes and the installation of 16 STs. Afterwards, a sensitivity analysis with different population sizes and different maximum number of generations was done, but no improvement was obtained.

So, the presented methodology has undergone an improvement process in order to obtain optimal results without the need to significantly modify the parameters of the genetic algorithm used to perform the simulations. Since these parameters are estimated based on the number of DVs and the complexity of the problem, the objective was now focused on reducing the number of nodes and lines that can be modified.

4.2. Application of the Solution Space Reduction Methodology to E-Chicó

The application of the proposed methodology (Figure 3) to the E-Chicó network begins with the process of *pre-locating STs*. The parameters used in this process are:

- The number of simulations defined is one hundred ($N_{it} = 100$).
- The discretization of the ST area is reduced to its minimum value ($N = N_0 = 10$).
- Only the sections of the tanks potentially to be installed in the nodes of the network are considered as DVs ($n = N_N = 35$).
- No conduit can be modified during the process ($m = 0$).
- The basic parameters of the PGA algorithm, population size (N_{pop}) and the end criterion based on a number of generations (N_{gen}) without change, are fixed at 100.

Once all the simulations have been carried out, the percentage p_n of solutions with the best value in the objective function is selected. In short, the 10 best solutions are selected. In each one, it is analyzed in which nodes an ST is installed. This generates a list of pre-locations of STs in the network. In this case, this list contains a total of 15 possible locations of the STs. In Figure 6, the shaded cells represent the selected nodes.

N01	N02	N03	N04	N05	N06	N07
N08	N09	N10	N11	N12	N13	N14
N15	N16	N17	N18	N19	N20	N21
N22	N23	N24	N25	N26	N27	N28
N29	N30	N31	N32	N33	N34	N35

Figure 6. Selected nodes as STs' pre-location.

In order to validate the methodology of pre-localization of STs, the previous process has been repeated, but by expanding the discretization of the tank area to the maximum value ($N = N_{max} = 40$). The results obtained lead to the same list of selected nodes as indicated in Figure 6.

The second stage of the SS reduction process is the pre-selection of conduits. For this, the pre-location of STs from the previous stage is used and a reduction in the number of potentially substitutable pipes is sought. The parameters used in this process are:

- The number of simulations is the same as in the previous stage ($N_{it} = 100$).
- The DVs are the areas of the ns nodes selected in the first stage and the diameters of all the conduits ($n = N_C = 35$) of the network.
- The discretization of the area of the STs is kept at the minimum value, as it happened with the previous stage of the process.
- The basic parameters of the PGA algorithm are the same as in the previous phase ($N_{pop} = 100$, $N_{gen} = 100$).
- Instead of using the full range of diameters (Table 4), a range of reduced diameters is used, the details of which can be seen in Table 5. This table shows the diameters D of the reduced range and the unit costs (C) obtained from the application of the cost function (4) with the parameters of the Table 3. Note that although Table 5 only has 9 values, the number of options for the DV is $ND = 10$. This additional value corresponds to the option of not taking any action on the pipe.

Table 5. Reduced diameter range ($ND = ND_0 = 10$).

D (mm)	300	400	600	800	1000	1200	1500	1800	2000
C (€/m)	30.93	49.56	99.31	165.7	248.74	348.43	529.16	747.35	913.61

Once the simulations have been carried out, it has been decided to define two different conduit pre-selections. One is considering a probability of 10% of the best solutions ($p_{m1} = 10\%$) and another one considering only 5% of solutions with a lower value of the objective function ($p_{m2} = 5\%$). The preselected lines in each case are shown in Figure 7. In the left part of the figure, the values corresponding to 10% are collected, highlighting the preselected lines by gray color. On the right hand side, the pipes selected for 5% of the best solutions obtained in this process are represented in an analogous way.

P01	P02	P03	P04	P05	P06	P07
P08	P09	P10	P11	P12	P13	P14
P15	P16	P17	P18	P19	P20	P21
P22	P23	P24	P25	P26	P27	P28
P29	P30	P31	P32	P33	P34	P35

$p_{m1} = 10\%$

P01	P02	P03	P04	P05	P06	P07
P08	P09	P10	P11	P12	P13	P14
P15	P16	P17	P18	P19	P20	P21
P22	P23	P24	P25	P26	P27	P28
P29	P30	P31	P32	P33	P34	P35

$p_{m2} = 5\%$

Figure 7. Selected lines to potentially be replaced for $p_{m1} = 10\%$ and $p_{m2} = 5\%$.

The pre-selection of ducts has been tested to validate performing the same simulations but using the full range of diameters ($ND = ND_{max} = 25$). The results obtained, in terms of the lines that would be pre-selected, are the same as those obtained in the case of using only the reduced range ($ND = ND_0 = 10$).

In short, the process of conduits pre-selection leads to select 15 pipes in case of selecting 10% of the best solutions or only 8 lines when the 5% of best simulations is considered. At this moment, the final stage of the process is the final optimization. In this phase, the reduction of the SS is used: 15 possible locations of STs and 8 or 15 possible lines to be rehabilitated. Therefore, the analysis of the network requires the definition of two new scenarios:

- Scenario 4: Rehabilitation of the network combining the possible installation of STs in the 15 selected nodes and the 15 conduits that can be substituted.
- Scenario 5: It is the same scenario as the previous one (scenario 4), with the difference that the number of potentially replaceable conduits is only 8.

4.3. Results Analysis

The results obtained from the optimization in the five scenarios considered are shown in Table 6. In each scenario, the following values appear: the number of DVs (nodes in which STs can potentially be installed and lines where their diameter could potentially be modified), the value of the objective function (divided into flood costs, investment costs in STs and investment costs in pipes), the number of elements to install of each type (STs and conduits) and the size of the SS of the problem. The scenario that offers the worst results (scenario 1) would be classic rehabilitation based solely on the replacement of pipes. The solution based on the use of STs has better results than the rehabilitation based on pipe substitution (scenario 2). However, the combined use of pipe substitution and installation of STs is shown as the best option of all (scenario 3).

Table 6. Result summary of the E Chico Optimization Process.

Scenario	No. DVs		Objective Function	Terms in the Objective Function			No. Elements in the Solution	
	Nodes	Lines		Floods	STs	Pipes	STs	Pipes
1	0	35	791,214	24,753	0	766,461	0	21
2	35	0	273,455	5392	268,063	0	6	0
3	35	35	268,292	20,238	230,087	17,968	4	4
4	15	15	245,547	8353	213,133	24,061	4	5
5	15	8	213,981	12,701	186,353	14,927	3	3

Comparing the solutions that use joint rehabilitation pipes-STs, it is observed that the reduction of the SS significantly improves the solutions. That is, pre-location of STs and pre-selection of conductions reduce the SS in an amount that allows a better exploration. Therefore, according to Table 6, the best solution is the one proposed by scenario 5, which considers the installation of 3 STs and the replacement of 3 conduits. As it is observed, the reduction of the SS between scenario 3 and scenarios 4 and 5 is many orders of magnitude, which helps to explain the improvement in the solutions.

The methodology used assesses the flood economically. Therefore, the solutions of scenarios 4 and 5 present flood costs (8353 and 12,701 respectively). These flood costs correspond to nodes whose flood level does not exceed 1.5 cm. However, from the point of view of the function defined in Equation (6), any level of flooding has an associated cost. Additionally, Tables 7 and 8 present the results of the flood nodes for scenarios 4 and 5. As can be seen, the cost of flooding is not zero. However, both flood volumes (52.77 and 75.78 m^3) and flood levels are very low. Therefore, from a practical point of view it can be considered that the solutions are acceptable.

Table 7. Flooding results in scenario 4.

Node	Flood Volume (m^3)	Flood Area (m^2)	y_{max} (m)	C (€)
N10	3.77	700	0.013	307.94
N32	12.96	1500	0.010	1680.29
N33	36.04	3030	0.009	6364.57
TOTAL	52.77			8352.80

Table 8. Flooding results in scenario 5.

Node	Flood Volume (m^3)	Flood Area (m^2)	y_{max} (m)	C (€)
N02	12.82	1240	0.013	3024.66
N03	10.82	1080	0.010	1619.42
N32	12.96	1500	0.009	1681.56
N33	36.08	3030	0.012	6775.01
TOTAL	75.78			12,700.64

Figure 8 represents the location of the STs obtained in scenario 5, as well as their size. Also, the flooded nodes are represented (blue nodes). Additionally, the figure represents the lines that have been necessary to modify and the size of the new conduits. It should be noted that in the case of conduits T04 and T10, the solution obtained involves the installation of a diameter smaller than the original one. This clearly indicates the need to install in these sections a resistant element (a gate or an orifice) that introduces a head loss equivalent to the one that involves the installation of the new smaller diameter.

The installation of control devices in drainage networks allows the accumulation of water in certain points of the network, decreasing the time of concentration downstream, and thereby reducing flooding. This is why this "hydraulic control" is one of the techniques that is often used to improve the efficiency of STs. In this case, the appearance of solutions, such as those of conduits T04 and T10, supposes the need to install a gate or an orifice at the outlet of the STs that introduces a certain head loss. This head loss must be such that the equivalent capacity of the original pipeline together with the control device used is equal to the transport capacity of the calculated pipeline.

In short, two of the three STs installed should have hydraulic control. Therefore, these tanks must have a control element at the exit that contains the avenue of water in the tanks and reduces the water travel time towards the downstream sections. If Figure 8 is compared with Figure 5, the installation of the STs together with the hydraulic control elements allows not only a reduction of flooding upstream, but also control of flooding that occurred in nodes downstream from these STs (i.e., nodes N31 and N34).

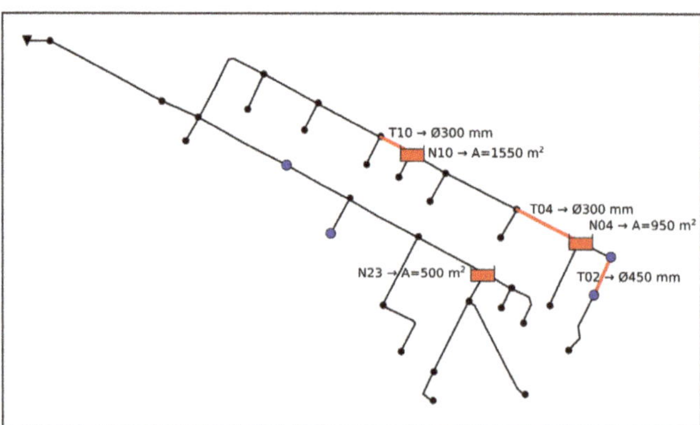

Figure 8. Representation of STs installed and pipes to replace according to scenario 5 results.

Simulations with a larger number of DVs are usually associated with a longer calculation time. However, the real difficulty of a problem is expressed by its size (PS_{max}) according to Equation (2). Therefore, in Table 9 the size of the problem of each scenario is presented. The table also includes a measurement of the calculation time, expressed as the number of times it is necessary to evaluate the objective function.

As can be seen, the larger the size of the problem, the greater the number of evaluations of the objective function required to find the solution. It is for this reason that the use of an SS reduction technique and reduction of the size of the problem has been effective.

Table 9. Size problem and computation time of the different scenarios.

Scenario	No. DVs	Problem Size	No. Evaluations of the Objective Function
1	35	4.0×10^{38}	51,200
2	35	5.8×10^{61}	65,485
3	70	2.3×10^{100}	709,000
4	30	2.8×10^{76}	205,300
5	23	4.1×10^{69}	151,300

The reduction in the number of DVs allows a better exploration of the SS, which means the smaller the number of decision variables, the better the solution found. However, reducing the number of DVs entails another effect. Simulations performed with large numbers of DVs cause a large dispersion of solutions, while solutions obtained with fewer DVs offer a much narrower range of solutions. Thus, Figure 9 represents a comparison of the solution variability range offered in scenarios 3, 4 and 5. It shows on the X-axis the relation between the value of the objective function in a simulation and the minimum global value for this scenario. On the Y-axis is represented the frequency of the different solutions obtained with respect to the minimum value obtained. That is, the more vertical the curve, the greater concentration of solutions. On the contrary, if the slope of the curve is smaller, it indicates that there is a greater dispersion of the solutions. Suppose that a value of 1.5 is set as a reference (the solution obtained has a maximum extra cost of 50% over the optimal solution). In that case, only 37% of the solutions obtained with scenario 3 are below that level. In the case of scenario 4, 66% of solutions present a cost smaller than 1.5 times the minimum value of the objective function. For scenario 5, this percentage is even greater (81%). In short, the reduction of the SS not only allows obtaining better solutions but at the same time it makes the range of good solutions (solutions close to the minimum) greater.

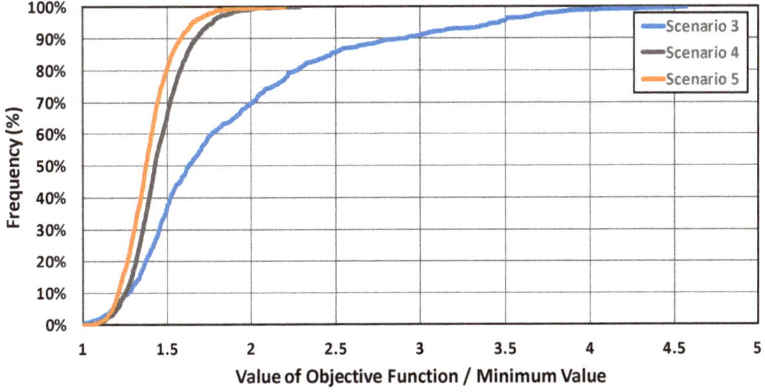

Figure 9. Comparison between the whole problem results and methodology application results.

4.4. Sensitivity Analysis of the SS Reduction

The main uncertainty is whether the SS reduction methodology could affect the final solution. It must be borne in mind that STs pre-localization and conduits pre-selection were performed using the PGA algorithm with a population size $N_{pob} = 100$ and a termination criterion $N_{gen} = 100$. In order to validate the selection, a sensitivity analysis of the selection process of STs and conduits has been carried out. The process of reducing the SS was repeated with different values of N_{pop} and N_{gen}. The different values of the population size were 35, 50, 75, 100 and 300 elements, while the values used as finalization criteria were 50, 100 and 150 generations without a change in the value of the objective function. The results obtained are shown in the following figures. In Figure 10, the nodes that could potentially be a ST location are collected for each combination of values of N_{pop} and N_{gen}. On the other hand, Figure 11 shows the lines potentially replaceable. In both cases, the node or line selected has been indicated with an X in the corresponding cell.

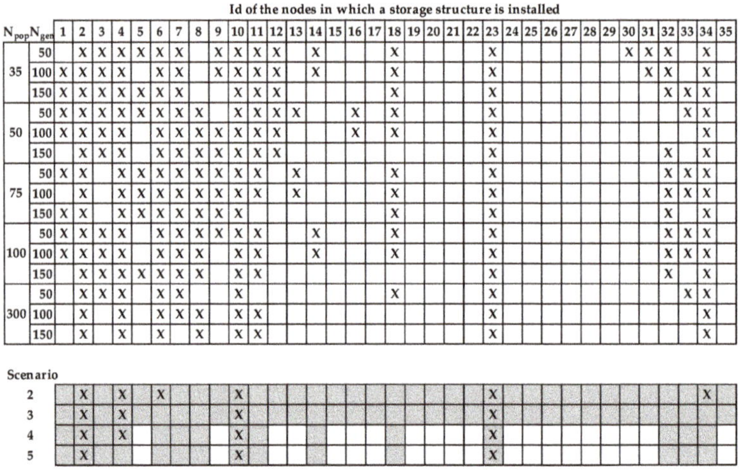

Id of the nodes in which a storage structure is installed

N_{pop}	N_{gen}	1	2	3	4	5	6	7	8	9	10	11	12	13	14	15	16	17	18	19	20	21	22	23	24	25	26	27	28	29	30	31	32	33	34	35
35	50		X	X	X	X	X	X		X	X	X	X		X				X					X							X	X	X		X	
	100	X	X	X	X		X	X		X	X	X	X		X				X					X								X	X		X	
	150	X	X	X	X	X	X	X			X	X	X						X					X									X	X	X	
50	50	X	X	X	X	X	X	X	X		X	X	X	X			X		X					X										X	X	
	100	X	X	X	X		X	X	X	X	X	X	X				X		X					X											X	
	150		X	X	X		X	X	X	X	X	X	X											X									X		X	
75	50	X	X		X	X	X	X	X	X	X	X		X					X					X									X	X	X	
	100		X		X	X	X	X	X	X	X	X		X					X					X									X	X	X	
	150	X	X		X	X	X	X	X	X	X								X					X									X		X	
100	50	X	X	X	X		X	X	X	X	X	X			X				X					X									X	X	X	
	100	X	X	X	X		X	X	X		X	X			X				X					X									X	X	X	
	150		X	X	X	X	X	X	X		X	X												X									X		X	
300	50		X	X	X		X	X			X								X					X										X	X	
	100		X		X		X	X	X		X	X												X											X	
	150		X		X		X		X		X	X												X											X	

Scenario

Scenario	1	2	3	4	5	6	7	8	9	10	11	12	13	14	15	16	17	18	19	20	21	22	23	24	25	26	27	28	29	30	31	32	33	34	35
2		X		X		X				X								X																X	
3		X		X						X								X																	
4		X		X						X								X																	
5		X								X								X																	

Figure 10. Sensitivity analysis results for pre-located STs with different N_{pop} and N_{Gen}.

Id of the conduit in which a storage structure is installed

N_{pop}	N_{gen}	1	2	3	4	5	6	7	8	9	10	11	12	13	14	15	16	17	18	19	20	21	22	23	24	25	26	27	28	29	30	31	32	33	34	35
35	50		X	X	X	X	X		X	X		X	X	X	X			X		X		X				X						X	X			
	100	X	X	X	X	X	X	X	X	X	X	X														X							X			X
	150		X	X	X	X				X	X		X		X						X					X		X					X			X
50	50		X	X	X		X	X	X	X	X			X			X		X		X	X				X							X			
	100		X	X	X		X	X	X	X	X										X					X										X
	150		X	X	X		X	X	X		X	X	X	X			X				X			X		X	X			X			X			
75	50		X	X	X		X	X	X		X	X	X	X				X		X		X				X	X						X			X
	100		X	X	X	X	X	X	X	X	X			X	X			X		X			X			X							X			X
	150		X	X	X	X	X	X	X	X	X			X	X	X										X							X			X
100	50		X	X	X	X	X	X	X	X	X	X		X					X			X	X			X			X				X			X
	100		X	X	X		X	X		X	X	X		X		X										X										X
	150		X	X	X		X	X	X	X	X		X		X		X					X				X	X									X
300	50		X	X	X		X		X	X	X	X		X				X		X	X		X			X			X			X				
	100		X	X	X		X	X	X	X	X	X	X					X		X			X			X										
	150		X		X		X	X	X	X	X		X					X	X	X	X	X				X	X									X

Scenario

Scenario	1	2	3	4	5	6	7	8	9	10	11	12	13	14	15	16	17	18	19	20	21	22	23	24	25	26	27	28	29	30	31	32	33	34	35
1		X	X	X		X		X	X	X		X		X	X		X	X							X		X	X	X	X		X		X	X
3			X							X																	X					X			
4		X	X							X															X							X			
5		X	X							X															X										

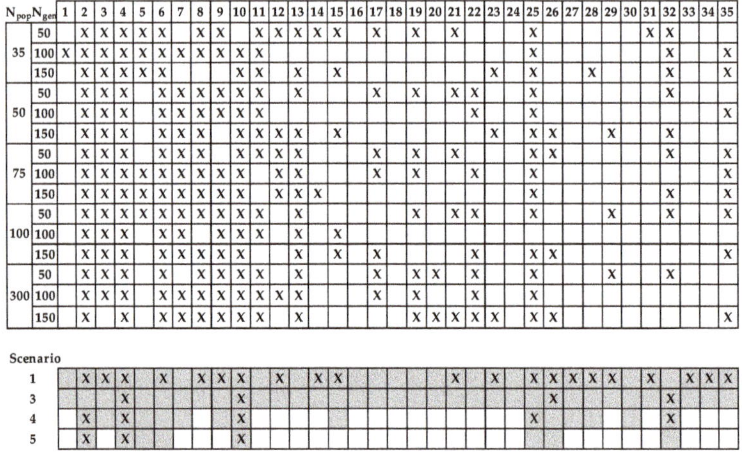

Figure 11. Sensitivity analysis results for pre-selecting conduits with different N_{pop} and N_{Gen}.

The sensitivity analysis of Figures 10 and 11 validates the SS reduction methodology based on the pre-location of STs and the pre-selection of conduits. In the lower part of Figure 10, the DVs used in each scenario are indicated by a shaded cell, and X indicates the nodes that finally provide a solution for the installation of a ST. The nodes that finally appear as solutions in the scenarios are those that had been selected mostly in the process of pre-selection of STs. In short, regardless the value of N_{pop} and N_{gen} the preselected nodes would have been almost the same, and in any case those that appear in the final solutions of scenarios 3, 4 and 5 would have always been selected.

Something similar happens in the case of pre-selection of conduits. Figure 11 shows the results of the sensitivity analysis and the results of scenarios 1, 3, 4 and 5. As in Figure 10, the DVs used are indicated as shaded cells, and the lines that are finally replaced are indicated with an X. The results show that the lines selected for different values of N_{pop} and N_{gen} are almost the same. That is, selected lines are those that are finally found in the final solutions of scenarios 3, 4 and 5. Moreover, N_{pop} and N_{gen} parameters have no influence on the pre-location of STs or on the pre-selection of pipes. Therefore, the SS reduction methodology can be considered reliable.

5. Conclusions

The proposed methodology for the rehabilitation of drainage networks, based on the combined use of STs and the replacement of conduits, has been shown to be effective in solving the problems of insufficient drainage networks. Despite the effectiveness of the presented methodology, a certain disparity is detected in the obtained solutions due to the enormous size of the SS. Therefore, a complementary strategy that allows the reduction of SS is presented. Once the methodologies area applied to the case study, it is possible to extract the following conclusions:

- The increase in rainfall intensities caused by climate change causes originally well-designed networks to present flooding problems. The use of STs has been shown as an effective technique to solve this problem. However, the effectiveness of this method is even greater when it is combined with the rehabilitation of some pipes of the drainage network. The results shown by scenarios 1, 2 and 3 presented above show how the combined action of STs and pipe renewal (scenario 3) is much more effective than the isolated action of any of the other two actions (scenarios 1 and 2).

- The number of DVs required for the rehabilitation of a drainage network can be very high. This may cause (Figure 9) the optimization model to give solutions that are quite dispersed and sometimes far from the optimal values required. It is well known that the importance of meta-heuristic algorithms does not lie in the optimality of their solutions, but in the ability to obtain large sets of good solutions. That is why the disparity of solutions observed with large numbers of DVs can become a problem.

- An alternative methodology has been proposed for the reduction of the SS that permit us to locate solutions that are better and less dispersed than those obtained with the initial model. This methodology is based on the pre-location of STs and the pre-selection of possible conduits. The methodology uses the PGA developed, reducing the SS by using a smaller discretization of the size of the STs, a smaller range of diameters of pipes, and a smaller number of elements (tanks and pipes). The final result has proven to be effective for the cases analyzed, not only because it reduces SS and computing times, but also because it obtains better solutions (scenario 4 and 5). Likewise, the solutions obtained with the scenarios that use a reduced SS (scenarios 4 and 5) are much more concentrated and less dispersed (Figure 9) than those obtained initially.

- The SS reduction methodology in the case study has been shown to be reliable and stable since variations in the N_{pop} and N_{gen} parameters of the PGA model hardly modify the preselected lines and nodes.

The best solutions suggest the suitability of installing control devices as a strategy to improve the operation of drainage networks. These control elements are static: either orifices or gates in a fixed position. In the solutions of scenarios 4 and 5, the installation of STs is complemented with the presence of this devices at the exit of the ST. This additional hydraulic resistance has been represented by a conduit with a diameter lower than the original one. As indicated previously, this type of operation helps to increase the concentration times in the critical nodes of the network.

The methodology based on the use of STs is adequate for the rehabilitation of drainage networks. The optimization model and the SS reduction methodology allow obtaining good solutions to a really complex problem. The resolution methodology used in the E-Chicó network can be extended to any type of drainage network, since the approach of the method has been so general to be easily extrapolated to other cases.

The extrapolation of the method to other networks generates some interesting uncertainties, such as the possible behavior of drainage networks with pumping stations. In this case, where the network has already installed pumping systems, the start and stop levels of the pumps and the size of the suction tank can be defined as DV. In the case of installing new pumping systems, the DVs would be the flows to be evacuated by each pump. In this case, it would be necessary to include in the objective function the investment cost associated with the construction of the new pumping equipment. In any case, it seems that the exposed methodology would be applicable to this situation.

It should be noted that the best solutions do not always eliminate the presence of floods in the network. In the case that solutions without flood are desirable, there are two possibilities. The first one is to allocate higher costs to the presence of floods that shift the optimum to solutions where the installation of STs or new conduits is profitable. The second option would be to run the problem statement from a multi-criteria point of view. With this type of approach, one could determine the level of investment required for each level of expected flood damage.

Supplementary Materials: The following are available online at http://www.mdpi.com/2073-4441/11/3/515/s1.

Author Contributions: All authors contributed extensively to the work presented in this paper. U.A.N.N P.L.I.-R. and F.J.M. contributed to the research, the modeling, the data analysis, and the writing of the paper. D.M.-M. contributed in the development of some cost functions and the manuscript review. J.G.S.V. collaborated to the modeling part of the E-Chicó drainage network used in the case study.

Funding: This research was funded by the Program Fondecyt Regular, grant number 1180660.

Acknowledgments: This work was supported by the Program Fondecyt Regular (Project 1180660) of the Comision Nacional de Investigación Científica y Tecnológica (Conicyt), Chile.

Conflicts of Interest: The authors declare no conflict of interest.

Abbreviations

The following abbreviations are used in this manuscript:

DV	Decision variable
GA	Genetic Algorithm
IDF	Intensity, Duration, Frequency
NSGA	Non-dominated Sorting Genetic Algorithm
PGA	Pseudo-Genetic Algorithm
SS	Search Space, Solution Space
ST	Storm Tanks
SWMM	Storm Water Management Model
WTP	Wastewater Treatment Plant

Notation

α, β	characteristic coefficients for the cost function related to the installation of new conduits
A, B	characteristic coefficients for the cost function related to the installation of a new ST of volumen V_{DR}
A_f	ponded area of a node to represent the flood level
A_S, B_S, C_S	characteristic coefficients of tank cross section equation
b	adjusts coefficients of the damage cost curve
$C(DN(i))$	ccost related to the installation of the diameter i
$C(V_{DR}(j))$	cost related to the construction or expansion of the volumen $V_{DR}(j)$
$C(y)$	damage cost related to a flood level y
C_{max}	maximum economic damage cost, when flood level y_{max} is reached
G	number of generations of the PGA
L_i	length of the conduit i
λ	adjusts coefficients of the damage cost curve
λ_i	Lagrange multipliers of the objective function
m	number of feasible conduits selected to be replaced, varying between 1 and N_C
m_s	number of possible pipe locations created after the pre-selecting pipes location step
n	number of nodes selected to potentially install a ST, varying between 1 and N_N.
N	number of divisions for the cross section S, varying between N_0 and N_{max}
N_C	number of conduits of the network
N_D	number of candidate diameters, between ND_0 and ND_{max}
N_{gen}	stop criteria: maximal number of generations G without change in the objective function value
N_{it}	number of optimizations (runs) of the PGA algorithm to develop one of the steps of the SS reduction method
N_N	number of nodes of the network
N_{pop}	population size of the PGA
n_s	number of possible ST locations after the pre-locating STs step
p_m	percentage of the best solutions selected in the pre-selecting pipes location step
p_n	percentage of the best solutions selected in the pre-locating STs step
PS_{max}	maximum size of the optimization problem
S	tank's cross section
S_{max}	maximum value of the tank cross section
$V_{DR}(j)$	volume of the j-th ST
V_f	volume of water flooded in a node
y	existing flood level on the specific node
y_{max}	level from which the maximal economic damage is produced
z	water level in a storm tank

References

1. Howard, C.D.D. Theory of Storage and Treatment-Plant Overflows. *J. Environ. Eng. Div.* **1976**, *102*, 709–722.
2. di Toro, D.M.; Small, M.J. Stormwater Interception and Storage. *J. Environ. Eng. Div.* **1979**, *105*, 43–54.
3. Loganathan, V.G.; Deniur, J.W.; Segarra, R.I. Planning Detention Storage for Stormwater Management. *J. Water Resour. Plan. Manag.* **1985**, *111*, 382–398. [CrossRef]
4. Meredith, D.D.; Middleton, A.C.; Smith, J.R. Design of Detention Basins for Industrial Sites. *J. Water Resour. Plan. Manag.* **1990**, *116*, 586–591. [CrossRef]
5. Takamatsu, M. Hydraulic model for sedimentation in storm-water detention basins. *J. Environ.* **2009**, *136*, 527–534. [CrossRef]
6. de Martino, G.; de Paola, F.; Fontana, N.; Marini, G.; Ranucci, A. Pollution Reduction in Receivers: Storm-Water Tanks. *J. Urban Plan. Dev.* **2011**, *137*, 29–38. [CrossRef]
7. Li, Y.; Kang, J.-H.; Lau, S.-L.; Kayhanian, M.; Stenstrom, M.K. Optimization of Settling Tank Design to Remove Particles and Metals. *J. Environ. Eng.* **2008**, *134*, 885–894. [CrossRef]

8. Fu, G.; Butler, D.; Khu, S.T. Multiple objective optimal control of integrated urban wastewater systems. *Environ. Model. Softw.* **2008**, *23*, 225–234. [CrossRef]

9. Vanrolleghem, P.; Benedetti, L.; Meirlaen, J. Modelling and real-time control of the integrated urban wastewater system. *Environ. Model. Softw.* **2005**, *20*, 427–442. [CrossRef]

10. Andrés-Doménech, I.; Montanari, A.; Marco, J.B. Efficiency of Storm Detention Tanks for Urban Drainage Systems under Climate Variability. *J. Water Resour. Plan. Manag.* **2012**, *138*, 36–46. [CrossRef]

11. Butler, D.; Schütze, M. Integrating simulation models with a view to optimal control of urban wastewater systems. *Environ. Model. Softw.* **2005**, *20*, 415–426. [CrossRef]

12. Deb, K.; Pratap, A.; Agarwal, S.; Meyarivan, T. A fast and elitist multiobjective genetic algorithm: NSGA-II. *IEEE Trans. Evol. Comput.* **2002**, *6*, 182–197. [CrossRef]

13. Wang, M.; Sun, Y.; Sweetapple, C. Optimization of storage tank locations in an urban stormwater drainage system using a two-stage approach. *J. Environ. Manag.* **2017**, *204*, 31–38. [CrossRef] [PubMed]

14. Rossman, L.A. *Storm Water Management Model User's Manual*; U.S. Environmental Protection Agency: Cincinnati, OH, USA, 2015.

15. Gaudio, R.; Penna, N.; Viteritti, V. A combined methodology for the hydraulic rehabilitation of urban drainage networks. *Urban Water J.* **2016**, *13*, 644–656. [CrossRef]

16. Sebti, A.; Bennis, S.; Fuamba, M. Cost Optimization of Hydraulic and Structural Rehabilitation of Urban Drainage Network. *J. Infrastruct. Syst.* **2014**, *20*, 04014009. [CrossRef]

17. Ugarelli, R.; di Federico, V. Optimal Scheduling of Replacement and Rehabilitation in Wastewater Pipeline Networks. *J. Water Resour. Plan. Manag.* **2010**, *136*, 348–356. [CrossRef]

18. Mora-Melia, D.; Iglesias-Rey, P.L.; Martinez-Solano, F.J.; Fuertes-Miquel, V.S. Design of Water Distribution Networks using a Pseudo-Genetic Algorithm and Sensitivity of Genetic Operators. *Water Resour. Manag.* **2013**, *27*, 4149–4162. [CrossRef]

19. Martínez-Solano, F.J.; Iglesias-Rey, P.L.; Saldarriaga, J.G.; Vallejo, D. Creation of an SWMM Toolkit for Its Application in Urban Drainage Networks Optimization. *Water* **2016**, *8*, 259. [CrossRef]

20. Maier, H.R.; Kapelan, Z.; Kasprzyk, J.; Kollat, J.; Matott, L.S.; Cunha, M.C.; Dandy, G.C.; Gibbs, M.S.; Keedwell, E.; Marchi, A.; et al. Evolutionary algorithms and other metaheuristics in water resources: Current status, research challenges and future directions. *Environ. Model. Softw.* **2014**, *62*, 271–299. [CrossRef]

21. Mala-Jetmarova, H.; Sultanova, N.; Savic, D. Lost in optimisation of water distribution systems? A literature review of system operation. *Environ. Model. Softw.* **2017**, *93*, 209–254. [CrossRef]

22. Gulizia, C.; Camilloni, I. Comparative analysis of the ability of a set of CMIP3 and CMIP5 global climate models to represent precipitation in South America. *Int. J. Climatol.* **2015**, *35*, 583–595. [CrossRef]

23. Cunha, M.C.; Zeferino, J.A.; Simões, N.E.; Saldarriaga, J.G. Optimal location and sizing of storage units in a drainage system. *Environ. Model. Softw.* **2016**, *83*, 155–166. [CrossRef]

24. Mora-Melia, D.; Iglesias-Rey, P.L.; Martinez-Solano, F.J.; Ballesteros-Pérez, P. Efficiency of Evolutionary Algorithms in Water Network Pipe Sizing. *Water Resour. Manag.* **2015**, *29*, 4817–4831. [CrossRef]

25. Leitão, J.P.; Carbajal, J.P.; Rieckermann, J.; Simões, N.E.; Marques, A.S.; de Sousa, L.M. Identifying the best locations to install flow control devices in sewer networks to enable in-sewer storage. *J. Hydrol.* **2018**, *556*, 371–383. [CrossRef]

26. Iglesias-Rey, P.L.; Martínez-Solano, F.J.; Saldarriaga, J.G.; Navarro-Planas, V.R. Pseudo-genetic Model Optimization for Rehabilitation of Urban Storm-water Drainage Networks. *Procedia Eng.* **2017**, *186*, 617–625. [CrossRef]

27. Lee, E.; Kim, J. Development of Resilience Index Based on Flooding Damage in Urban Areas. *Water* **2017**, *9*, 428. [CrossRef]

Article

Multi-Objective Optimization for Urban Drainage or Sewer Networks Rehabilitation through Pipes Substitution and Storage Tanks Installation

Ulrich A. Ngamalieu-Nengoue [1], F. Javier Martínez-Solano [1,*], Pedro L. Iglesias-Rey [1] and Daniel Mora-Meliá [2]

[1] Department of Hydraulic Engineering and Environment, Universitat Politècnica de València, Camino de Vera s/n. 46022 Valencia, Spain; ngamaleuulrich@yahoo.fr (U.A.N.-N.); piglesia@upv.es (P.L.I.-R.)
[2] Department de Ingeniería y Gestión de la Construcción. Facultad de Ingeniería, Universidad de Talca, Camino de los Niches km 1, Curicó 3340000, Chile; damora@utalca.cl
* Correspondence: jmsolano@upv.es; Tel.: +34-96-387-7610

Received: 7 March 2019; Accepted: 29 April 2019; Published: 3 May 2019

Abstract: Drainage networks are civil constructions which do not generally attract the attention of decision-makers. However, they are of crucial importance for cities; this can be seen when a city faces floods resulting in extensive and expensive damage. The increase of rain intensity due to climate change may cause deficiencies in drainage networks built for certain defined flows which are incapable of coping with sudden increases, leading to floods. This problem can be solved using different strategies; one is the adaptation of the network through rehabilitation. A way to adapt the traditional network approach consists of substituting some pipes for others with greater diameters. More recently, the installation of storm tanks makes it possible to temporarily store excess water. Either of these solutions can be expensive, and an economic analysis must be done. Recent studies have related flooding with damage costs. In this work, a novel solution combining both approaches (pipes and tanks) is studied. A multi-objective optimization algorithm based on the NSGA-II is proposed for the rehabilitation of urban drainage networks through the substitution of pipes and the installation of storage tanks. Installation costs will be offset by damage costs associated with flooding. As a result, a set of optimal solutions that can be implemented based on the objectives to be achieved by municipalities or decisions makers. The methodology is finally applied to a real network located in the city of Bogotá, Colombia.

Keywords: drainage networks; flooding; rehabilitation; multi-objective optimization; SWMM

1. Introduction

Over time, sewer networks present limitations in terms of carrying out the purpose to which they were assigned at the time of their construction. Unfortunately, their drainage capacity is limited to a few historic rainfall events or storms. Rainfalls with greater intensities may produce floods. For example, the total damage associated with floods in Spain between the years 1971 and 2017 reached 6 billion euros. This amount is more than 60% of the total amount of damages paid by insurance companies [1]. Flooding in urban areas has become increasingly common for different reasons. Kordana [2] categorizes the factors affecting the capacities of sewer networks as political, economic, social, technological, legal or environmental. Anthropogenic factors, linked to excessive urbanization, are partly responsible; some other reasons are related to climate change [3]. Extreme rainfalls events are important parameters affecting various natural and socio-economic systems. The evolution of rainfall also affects the efficiency of drainage networks. Starzec et al. [4] suggest that omitting the evolution of storms can led to undersizing the network. Mailhot and Duchesne [5] suggest that the probability of the occurrence of intense rainfall will increase in the future due to the effects of greenhouse gases. Gulizia and Camilloni [6] made a comparison among different global climate models which had been applied to South America. These models showed that an increase of up to 36% in rainfall events is expected in some regions. Ma et al. [7] studied the effects and frequency of flash floods in China. They arrived at the conclusion that flash floods are still the main cause of deaths in flood disasters. This situation is becoming alarming, since it is to be expected that urban drainage networks designed for past conditions will not function effectively in the future [8]. It should be noted that climate change is suspected as the cause of floods due to its effects on rainfall intensity and frequency.

However, not only climate change affects drainage and sewer networks performance; aging, structural collapses and exfiltration (leaking) might affect them too [5]. Moselhi et al. [9] presented a methodology based on neural networks to automatically detect and classify defects in drainage networks pipes. The aim of their methodology was to implement appropriate actions to upgrade the network.

Managers and others responsible for the operation of drainage networks must implement effective measures to ensure that infrastructure will be ready for these changes [8]. Several options are available for these managers: Real Time Control (RTC), reparation or rehabilitation. Driessen et al. [10] define five different complementary approaches to improve the resilience of hydraulic infrastructure: prevention, defense, mitigation, preparation and recovery. Among these strategies, the most traditional is flood defense, based on the idea of *"keep water away from people"*. However, these traditional methods have shown limitations, since they cannot guarantee the cost-benefit effectiveness of the proposed solution.

Over the years, several studies on pipe replacement have been carried out to achieve optimal rehabilitations of defective sewer networks in order to prevent floods and environmental threat while minimizing investments costs. Reyna et al. [11] have proposed some solutions to the challenges which are encountered in the process of finding new strategies for sewer network rehabilitation methodologies which seek to maintain the cost-effectiveness. Abraham et al. [12] proposed an integrated management methodology considering Markovian probability-based models combined with deterministic models to predict structural failures in sewer networks. Based on these models, they proposed a priority rehabilitation plan which would reduce investment costs. The methodology first discretizes the network in small systems which are identified based on their structural characteristics. However, all these approaches become ineffective when the effects of climate change are taken into account.

Traditionally, the design of urban drainage systems is based on statistical analyses of past events. Keeping in mind that an increase in the intensity and frequency of extreme rainfall events will most probably result in more frequent flooding, Mailhot and Duchesne [5] proposed a revision of the design criteria of urban drainage infrastructure which takes into account uncertainties related to climate change. Gaudio et al. [1] proposed a combined methodology for the hydraulic rehabilitation of urban drainage networks. In contrast to traditional methods, they concluded that not only intensity, but also rainfall volume, influence the hydraulic design. They combined the observed rainfall data and synthetic hyetographs to deduce critical hydraulic conditions in terms of overflow volumes rather than rainfall volumes. All these studies conclude that rainfall volumes might be even more important than rainfall intensities as a design criterion.

During the last decade, a new trend in drainage network rehabilitation has been developed and implemented: the Best Management Practices (BMP). BMPs consist of reducing flooding at low cost using source controls and other devices to increase infiltration and reduce runoff. Some interesting studies have been carried out to popularize the technique. Sebti et al. [13] developed an optimization model based on Linear Programming for the optimal selection and placement of structural measures. They considered four types of structural BMPs: retention pond, green roofs, infiltration tranches and vegetated depressions. They tried to minimize the total cost of BMPs, limiting the combined sewer overflow as a constraint. Zahmatkesh et al. [14] used a global climate model (CMIP5) to generate rainfall scenarios that were used as the input for hydraulic analyses. As a result, the impact of climate change on the Bronx River watershed (New York City) was analyzed. Considering the impact of climate change on watershed runoff, the potential for Low-Impact Development (LID) controls to mitigate the effects was investigated. Stormwater runoff and LID controls were modeled using the Storm Water Management Model (SWMM) [15]. The results obtained by Mora et al. [16] showed that green roofs are efficient for small to moderate rainfalls, but that their implementation could prevent the flooding due to extreme rainfall events.

As stated by Driessen et al. [10], any strategy which seeks to reduce floods has some costs. Ugarelli and Di Federico [17] presented a cost-based methodology for drainage networks service level upgrading. This approach unifies both upgrading actions and damage costs, that is, costs associated with the risk of failure. A comparison between maintenance costs, renovation costs and rehabilitation costs was first performed in order to select the optimal course of action taking into account economical and engineering factors. Ngamalieu-Nengoue et al. [18] performed an analysis of costs related to pipe renewal and the installation of detention storm tank. They presented an optimization methodology to obtain the best solutions for drainage network rehabilitation linking SWMM as the hydraulic solver and Genetic Algorithm as the optimization model. The methodology makes it possible to define the pipes that should be changed and the places in which detention tanks should be installed. In this process, floods were considered unacceptable, and solutions had to ensure that no such event could occur.

Floods produce damages, and these damages have costs. Exact flood damage costs cannot be known until the capacity of the drainage network is exceeded. However, several studies have shown that flood damage costs depend on the area where flooding occurs and on the depth of the water. Lee and Kim [19] demonstrated that flooding volume in urban areas was not linearly proportional to flooding damage. There are two ways to represent flood cost damages [20]: (i) proportional to the flood volume at nodes, and (ii) proportional to the level of flooding. In the latter case, the flooding area in each node must be defined. The study concluded that flooding costs are a function of the water level reached. Both approaches were confirmed by Lee and Kim [19], who pointed out that flooding damage is different from flooding volume, because each subarea has different components.

Facing this growing risk, urban drainage management is moving towards a flood risk management approach related to city resilience, that is, the capacity to continue functioning even under hazardous conditions. Fadel et al. [21] presented a risk-based method for introducing protection measures. The standard approach uses comparisons of damage costs with and without measures as a decision-making tool. These authors include the concept of risk with an associated probability. They applied this methodology to different rainfall scenarios ranging from 2 to 200 years of return period, and to three protection scenarios: no measure, land zoning and levee. The results showed that land zoning, i.e., an adaptation measure was a better solution than levees i.e., a structural measure. They concluded their study by recommending the use of flood costs for their risk assessment, as it proved its cost-effectivity.

Due to their unpredictable nature, flood damage costs cannot be harmonized with construction costs. Investments on sewer network adaptation are made, but damages associated with floods depend on the probability of their occurrence. These costs are different in nature, and a multi objective optimization algorithm is advisable to establish a relationship between both functions.

The incorporation of detention tanks in drainage networks might be considered as a form of implementation of BMP. This practice is becoming popular and is being implemented in the design of sewer networks to avoid floods and the contamination of receiving water bodies. The use of detention tanks has increased over the last decade [22]. Detention tank volumes are calculated depending of rainfall levels. Starzec [23] proposed a method to determine the required volume of these devices as a function of the time to the maximum flow, instead of using rainfall duration. Pochwat and Słyś [24] proposed the use of artificial neural networks as a tool for the estimation of the duration of rainfall events. Other authors [18,25] use heuristic methods to determine the optimal size of detention tanks.

At first, in-line detention tanks were used to control pollution in both storm and waste water networks. De Martino et al. [26] compared three different configurations of these tanks using efficiency in pollutant removal as ranking criteria. Andrés-Doménech et al. [27] studied the resilience of storm water detention tank efficiency levels with respect to changes in rainfall forcing. Wang et al. [28] proposed a rehabilitation methodology connecting SWMM to a multi-objective framework to find optimal locations for storage tank installation, thereby showing that storage tanks reduce not only TSS, but also flooding. Cunha et al. [29] showed that the location, dimensions and flow control capabilities of storage tanks strongly affect their efficiency. In the work of Cunha et al., only storage tanks were implemented as a strategy for flood reduction; Simulated Annealing was used as the optimization model. As an alternative solution, Dziopak [30] or Słyś [31] proposed the use of control devices, allowing storage in the channel themselves.

Hence, in previous works, some actions were taken and optimized to eliminate the effects of floods. Concretely, Cunha et al. [29] used orifices as a hydraulic control in storage tanks, while Iglesias-Rey [20] combined storage tanks with increasing sewer transport capacities. Both works were done using evolutionary algorithms with a single objective optimization model. In this paper, a novel multi-objective approach combining not only pipe renewal but also storm tank size and location is developed. Furthermore, the model takes into account the economic effects of floods. Consequently, this paper is a worthwhile contribution because, as discussed previously, no previously published works have combined these three items.

Since the damage associated with flooding depends on the probability of rainfall, a multi-objective approach is assumed. Two different cost functions will be defined: investment costs for pipes and tanks, and damage costs associated with the flood itself. A multi-objective optimization algorithm was built linking an adapted Non-dominated Sorting Genetic Algorithm (NSGA-II) [32] with a SWMM programmer toolkit [33] to perform the rehabilitation of drainage networks by combining the use of pipe substitutions and the installation of storage tanks. Flooding damages are quantified in term of money based on the water level of the flood. As a result, a set of Pareto fronts were obtained relating both types of costs. These solutions can be used by network managers to make decisions concerning the rehabilitation plans and investments within the context of budget limitations.

Finally, as a study, this methodology will be applied to the E-Chicó area of Bogotá, Colombia.

2. Methodology

2.1. Problem Formulation

Urban drainage systems are designed to cope with predicted storms. When these systems experience extreme rainfall, the excess of water accumulates out of the network and floods occur. Flooding may cause important damage in cities, and this damage can be converted into cost. Hence, floods may be represented as a damage function. Several corrective actions might be taken to avoid flooding. Among them, increasing the size of pipes or installing retention tanks stand out as obvious choices. These actions imply an investment and have associated costs. The problem consists of harmonizing the installation costs of these corrective actions with the damage costs which would arise as a consequence of flooding.

The relationship between both types of costs presents two aspects. On the one hand, there is a hydraulic connection. Investments imply increasing transport capacity of the network and, consequently, reducing floods. On the other hand, flood damage costs depend on extreme rainfall probability. In other words, investments are made based on the probability of a storm occurring, while damage may occur with some level of probability.

To relate both concepts, a mathematical model of the drainage system was combined with a multi-objective optimization algorithm. Let a solution be a set of corrective actions adopted for the network. For every proposed solution, the hydraulic solver calculates the behavior of the network, focusing on the presence of flood. Then, the level of fit is calculated with both installation costs and flood damage. Finally, the optimization algorithm ranks every feasible solution to produce the next generation of solutions until the termination criterion is reached. Coupling between the optimization algorithm and the hydraulic model was achieved using a connection library, as described in [33]. An outline of the process is shown in Figure 1.

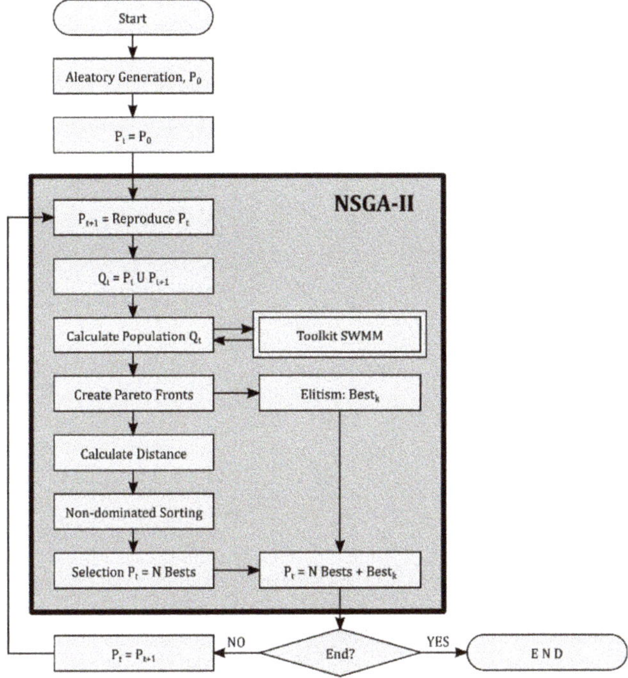

Figure 1. Flow chart of NSGA II performed in the methodology.

The methodology assumes the following hypotheses:

- The design storm is supposed to be static and the same for the whole network. Some authors [4] state that static rainfall may produce undersized detention tanks. The design storm and its evolution are beyond the scope of this work.
- The hydraulic model will consider that the network receives water flow directly into the different nodes. Therefore, the rainfall-runoff transformation model is performed independently.
- SWMM [15] will be used as analysis tool. A calibrated mathematical model of the drainage network is used to provide a precise view of the operation scenario.
- The mathematical model of the network should be simplified to reduce the computational time.
- The corrective actions are the installation of storage tanks and the change in pipe diameter. The unknown variables are the volume of these tanks and the new diameter of the pipes.
- The detention tanks are considered to be single chamber, in-line tanks without control devices. The invert elevation of the tanks is assumed to be the same as that of the node in which they will be installed.
- The optimization problem will be addressed in term of costs. So, cost functions for both investment and damage must be defined. However, damage is associated with a probability and investment is not, i.e., both functions are treated separately.
- The decision variables are nodes potentially becoming storage tanks and conduits where dimensions may be changed.

2.2. Objective Functions

The optimization process is based on the minimization of an objective function. In this case, two objectives are in conflict in the proposed objective function, i.e., on the one hand, the investment costs and, on the other, the flood damages costs. More investments reduce flood damage costs and vice versa.

The investments costs (F_1) include the renewal of pipes and the installation of tanks. Damage function (F_2) relates flooding with the costs associated with it. Both functions express hydraulic variables values in monetary units and were presented in detail in [20]. These functions are expressed by Equations (1) and (2):

$$F_1 = \sum_{i=1}^{N_C} C_C(D_i) \times L_i + \sum_{j=1}^{N_T} C_T(V_j) \tag{1}$$

$$F_2 = \sum_{k=1}^{N_N} C_I(y_k) \times A_k \tag{2}$$

In Equations (1) and (2), N_C represents the total number of conduits in the network, N_N represents the total number of nodes in the network and N_T represents the total number of storage tanks installed in the network.

The cost of pipe substitution, $C_C(D_i)$, was obtained from real data supplied by pipe manufacturers. With these data, a mathematical formulation was carried out to express economically the cost of pipe substitution in euros per meter, depending on the diameter of the pipe to be installed:

$$C_C(D_i) = A \times D_i + B \times D_i^2 \tag{3}$$

The cost of storage tank installation, $C_T(V_j)$, is associated with the storage volume of a tank (V_j) that may be necessary to install on one node location of the network to absorb the excess water that cannot be evacuated normally through the drainage network. This function is composed of a fix term which represents the minimum costs associated with the construction (C_{min}) and a variable term which depends on the total volume through a constant (C_{var}) and an exponent (n):

$$C_T(V_j) = C_{min} + C_{var} \times V_j^n \tag{4}$$

The flood damage cost function represents damages caused by the flooding. Some authors consider flooding as a volume ([34]); however, others present flooding as the highest depth reached by the water (y) out of the network. Lee and Kim [19] showed that flood damage is different from flood volume. They proposed a resilience index based on flood damage because some subareas are immediately damaged by a certain amount of flooding, while others are not. They represented flood damage costs as a function of the depth reached by water. Following their example, the damage function was expressed as a function of the depth y of the flooding:

$$C_I(y_k) = C_{max} \times \left(1 - e^{-k \times \frac{y_k}{y_{max}}}\right)^2. \tag{5}$$

In Equation (5), C_{max} represents the maximum cost per square meter that a flood might cause. For a certain depth (y_{max}), the damage is considered as irreparable; therefore, the function stops growing and the cost will reach the maximum value. Coefficient k is based on historical data of damages caused by flooding.

2.3. Optimization Algorithm

In the problem presented above, the two objective functions are in conflict with each other, that is, improvement of one may worsen the other. A solution is dominated if another solution improves all its objective functions. A solution is non-dominated if no other solution dominates it. A set of all non-dominated solutions is called a Pareto front. The Pareto front might be formed by an infinite number of non-dominated solutions. In this case, the NSGA-II algorithm will be used.

NSGA-II was first introduced by Deb et al. [32]. This method adopts a fast non-dominated sorting approach to rank solutions through an implicit elitist selection method based on the concepts of Pareto dominance and crowding distance. If all solutions in a Pareto front are sorted according to the different objectives, the crowding distance of a solution for an objective is the average distance of its two neighboring solutions, as shown in Figure 2. Every objective has its own crowding distance which is a solution that may be computed by combining the entire individual crowding distance values in each objective function [35]. The crowding distance value of a solution is an estimate of the density of solutions surrounding that solution.

Additionally, an additional elitism aspect was added for this work. For every generation, the best individual in every objective was selected for the next generation. That is, according to Figure 2, individuals P_1 and P_N were selected for the next generation. Usually, the stopping criterion for multi-objective algorithms is the number of simulations or generations. In this case, the evolution of the best values for every objective was also used as stopping criterion. If, after a certain number of generations, none of the values of the objectives ($F_2(P_1)$ and $F_1(P_N)$) are improved, then the algorithm stops. Another measure for assessing the algorithm was the evolution of the crowding distance. It is desirable that solutions in a Pareto front be equally distributed. In this sense, the standard deviation of the crowded distance in the first Pareto front was also taken as a measure of the quality of the front. The smaller the standard deviation, the better the solution.

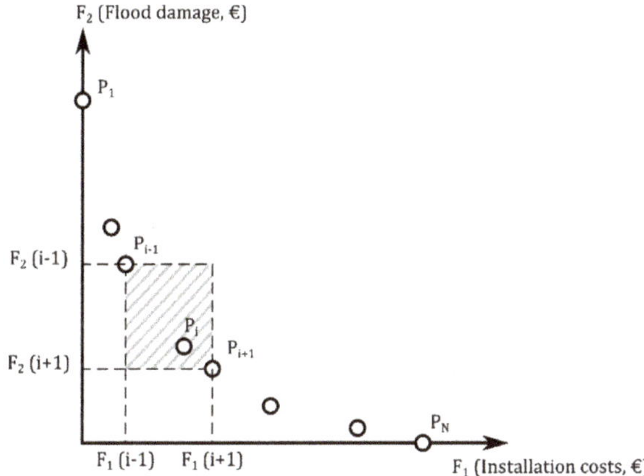

Figure 2. Main features of a Pareto front.

3. Case Study

In order to test the methodology mentioned above, a sub-catchment of the drainage network of Bogotá city, Colombia, was used. The proposed algorithm was applied to the rehabilitation of drainage or sewer networks through the substitution of pipes and the installation of storage tanks. The part of the drainage network studied is generically known as E-Chicó district. E-Chicó is divided into 35 hydrological sub-catchments expanded over a surface of 51 ha. The network is composed of 35 circular conduits with diameters varying from 300 to 1400 mm, and 35 connecting nodes. The total length of the network is around 5000 m. The network works completely by gravity, since the terrain profile is favorable to the drainage of rainwater. The height difference between the highest and the lowest points is 39.28 m. Figure 3 shows the E-Chicó drainage network. The complete description of the network used as a case study can be found as supplementary material of this article.

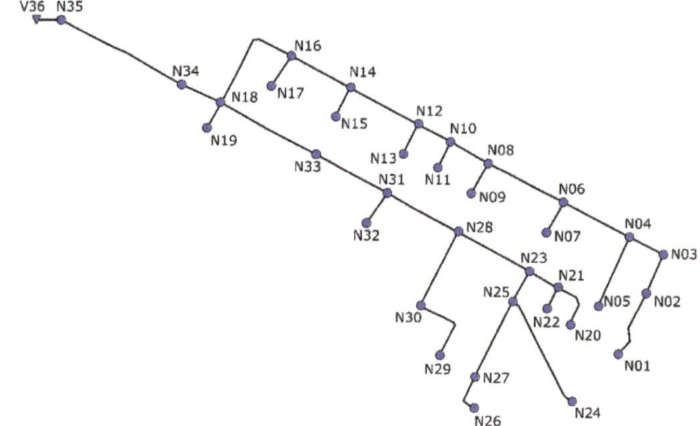

Figure 3. Representation of E-Chicó drainage network.

For diagnostic and further evaluation of possible solutions, a design storm was used based on the intensity–duration–frequency (IDF) curve for a return period of ten years. The IDF curve was obtained after the application of a climate change scenario [6]. Then, a design storm was calculated using the alternating block method with a time interval of 5 min and a minimum duration of 10 min. Both, the IDF curve (left) and the design storm (right) are presented in Figure 4.

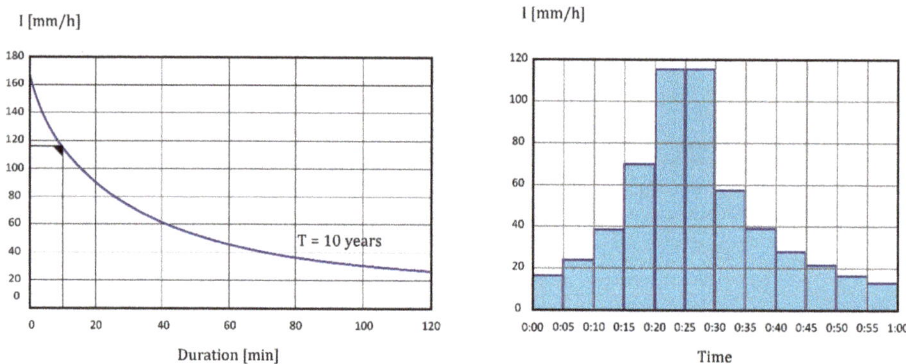

Figure 4. Rainfall used in the case study (**left**: intensity–duration–frequency (IDF) curve for a return period of 10 years, **right**: design storm).

The cost functions used for the case study were based on real data, according to Ngamalieu-Nengoue et al. [18]. Equations (3)–(5) will become, respectively:

$$C_C(D_i) = 40.69 \times D_i + 208.06 \times D_i^2 \tag{6}$$

$$C_T(V_j) = 16923 + 318.4 \times V_j^{0.65} \tag{7}$$

$$C_I(y_k) = 1268 \times \left(1 - e^{-4.89 \times \frac{y_k}{1.4}}\right)^2 \tag{8}$$

A first simulation provided the behavior of the network without any intervention. In this preliminary analysis, the network presented a flooding total volume of 3832 m^3, which represents 17.6% of the generated runoff (21,766 m^3). Using the damage function described in Equation (5), it was calculated that this flood would cost 5.24 million euros. Furthermore, more than 30% of the nodes showed flooding (11 out of 35 nodes). In summary, the preliminary analysis of the network shows that it is unable to drain the selected project rain. Therefore, the E-Chicó drainage network was considered adequate as a model to which the multi-objective optimization NSGA-II rehabilitation drainage networks could be applied.

Three different rehabilitation scenarios were performed depending on the selection of the decision variables:

- Scenario 1 (35C). All the conduits were selected as suitable for rehabilitation.
- Scenario 2 (35T). All the nodes were selected as possible locations for the installation of storage tanks.
- Scenario 3 (35C + 35T). A combination of Scenarios 1 and 2. All the conduits and all the nodes were selected as decision variables.

For all 3 scenarios, the crossover probability was fixed at 80%, while the mutation coefficients were calculated as the inverse of the number of decision variables, as suggested by [36].

4. Results

For every scenario, 10 different population sizes and six different values for the number of iterations were used. Apart from the results of the solutions in the Pareto fronts, some other indexes were gathered to assess the algorithm. The simulation time varied from 1–15 days, depending on the scenario and the selection of the parameters. For example, the rehabilitation of the whole network, corresponding to the Scenario 3 (35T + 35C) with a population size of 1000 individuals and a maximum number of generations 15,000, provided results after 15 days. Next, the results of these simulations are presented.

Figure 5 represents the results obtained for the 3 different scenarios presented above, that is, the substitution of pipes (35C), the installation of storage tanks in some locations (35T), and the combination of both alternatives (pipes substitution and storage tanks installation, 35T + 35C). These simulations were performed with a population size of 200 individuals and number of generations of 10,000 for all scenarios. Figure 5 also shows that Scenario 3 presents better results than Scenarios 1 and 2; even though this combined scenario represents a bigger problem (70 decision variables versus 35 of Scenarios 1 and 2), the combination of the two different strategies led to better results.

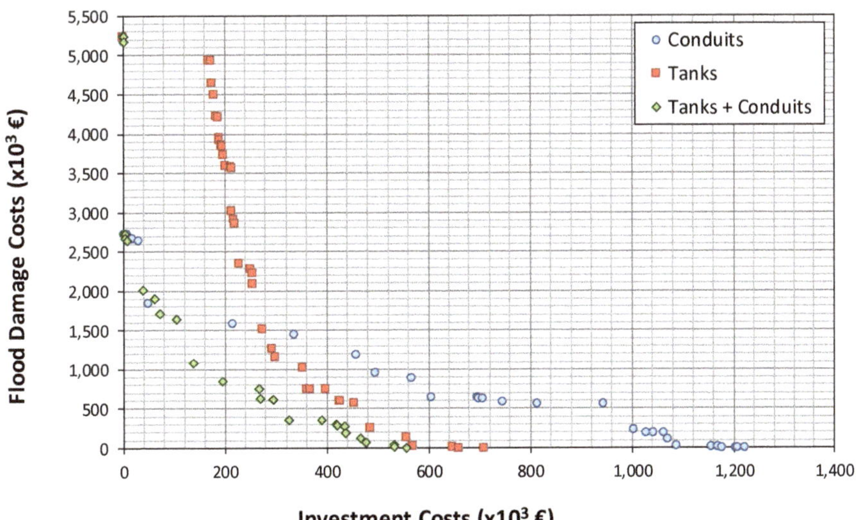

Figure 5. Pareto front representation of 3 different scenarios.

The NSGA-II presents a set of feasible solutions instead of a single one. This way, the decision about the best solution depends on several factors, i.e., budget availability, risk level, administrative regulations, etc. This is the reason why the multi-objective algorithm was selected. As an example, the solution for the case of unlimited budget availability is presented. If there are no investment limitations, the solution will look for a scenario with no flooding. This scenario corresponds to the horizontal axis of Figure 5. The solution for zero flooding needs an investment of either 1,213,453€ in pipe renewal or 719,366€ in storage units. In accordance with the methodology presented in this paper, the problem can be solved with a joint investment of 517,559€. Solutions in terms of the volume of detention tanks and the diameters of pipes are shown in Figure 6. Pipes marked with dashed lines represent changed pipes with respect to the original network.

The solution to Scenario 3 shows that some pipes connected to downwards tanks have a reduced diameter, meaning that those pipes act as a hydraulic control device. This confirms what was stated by Starzec et al. [4], and opens up the possibility of including hydraulic controls in the optimization process.

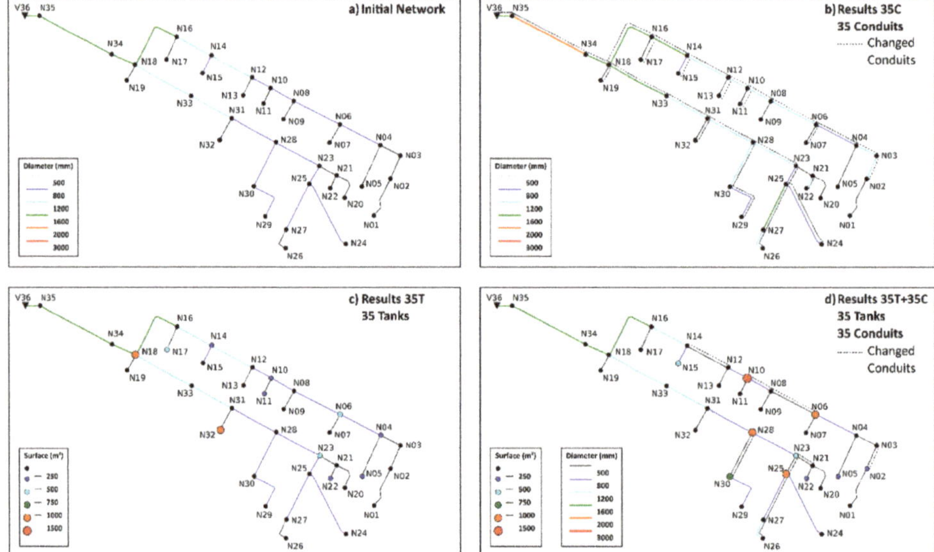

Figure 6. Original network (**a**) and solutions for no flooding for (**b**) Scenarios 1, (**c**) 2 and (**d**) 3. Changed pipes are represented as dashed lines.

In this figure, all scenarios provide the same damage costs for no investment. This is due to a slight modification in the algorithm. The initial population was not entirely random, since two additional individuals were added: the zero-investment and the maximum investment (which causes zero flood damage cost) solutions. The zero-investment solution cannot be improved, and hence, is present in every solution. In contrast, the maximum investment cost was simulated assuming that all the pipes were renewed using the maximum available diameter (2000 mm for this case) and that the tanks would have the maximum area. This solution might be improved during the simulation. As an example, Figure 7 shows the evolution of this zero-damage solution as the simulation progresses under two different simulations. It can be observed that the solution for zero flooding quickly become unchanged (after 20 to 50 generations). A stopping criterion based on the evolution of this parameter was tested, but the results suggested that this stopping criterion could be discarded.

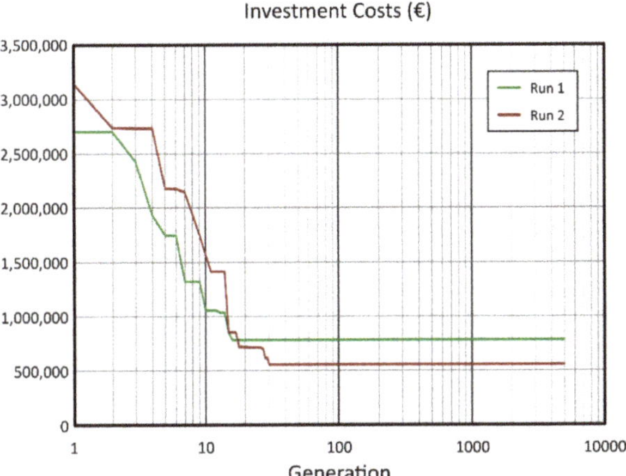

Figure 7. Evolution of the minimal investment costs for a no-flooding situation as a function of the generation. The scenario used corresponds to combining tanks and pipes (35T + 35C) under two different simulations.

Another indicator used to evaluate the solution was the standard deviation of the crowding distance in the first Pareto front. The dependence of this parameter with the population size was also studied. Figure 8 shows that there is a strong dependency between both parameters. In fact, this figure shows that this relation might be represented by a power function with an extremely high correlation. This relation makes it possible to determine the minimum population for a desired distribution of individuals in the Pareto front.

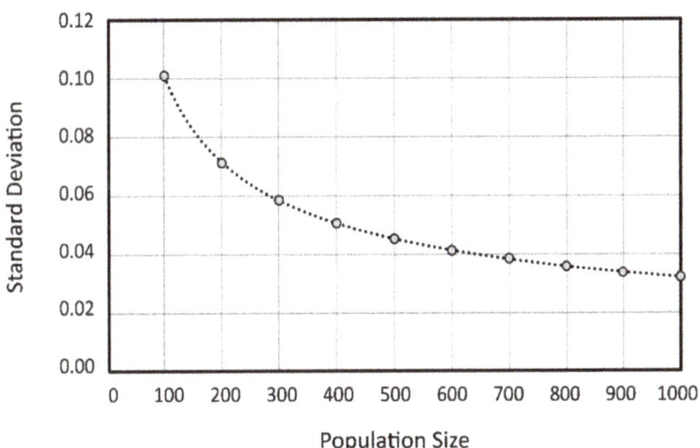

Figure 8. Standard deviation of first Pareto front as a function of population size in the final solution for Scenario 3 (35T + 35C).

Finally, some other relationships were studied. As expected, the evolution of the minimal investment costs to eliminate floods with a given population size concludes that the results are better when the number of generations increases. This conclusion was also reached by other researchers [37].

5. Conclusions

A multi-objective optimization algorithm based on NSGA-II for drainage network rehabilitation has been presented as a tool to help managers of drainage networks, or any other decision makers, to establish rehabilitation plans of their networks. Budget limitations are considered, and a set of solutions is provided, as opposed to just one, allowing decisions to be made based upon the objectives to be achieved and the available budget. The tool was tested on the network of E-Chico (Colombia) and the results obtained are in accordance with those of some previous studies. An EPA-SWMM hydraulic engine was used as the solver for the hydraulic conditions of the network.

The first conclusion obtained after this work was that the combined use of pipe rehabilitation and detention tanks improved the performance of the network in comparison to any strategy used separately.

The use of damage functions to evaluate the effect of flooding makes it possible to compare investment and damages, even though the first comprises expenses that depend on the probability of the occurrence of rainfall.

The capacity of the optimization tool to provide better results was also tested, and it stands out that, depending on the rehabilitation configuration, the population size must be large (greater than 500 individuals). Even though this conclusion was expected, the use of a medium population size was enough to achieve acceptable solutions in some cases. The computational effort needed to solve the case study indicates that some additional studies must be done which seek to reduce the size of the problem.

Different stopping criteria were used. The total number of generations or an objective value in the standard deviation of the crowding distance in the first Pareto front showed better performance than the evolution of the investment cost needed to avoid flooding.

The main conclusion of this paper is that the combination of the renewal of pipes and the installation of storage units can reduce the risk for sewer network utilities vis-à-vis eventual rainfall increases due to climate change. The results for the combined solution show that some of the conduits are substituted for smaller ones. This fact implies some type of flow control based on hydraulic principles.

The results obtained in the Pareto front consist of a set of feasible solutions that will help decision makers select the best option. However, there is no optimal solution; any model must be chosen in accordance with some criteria or rules, i.e., budget availability, risk evaluation, design criteria, etc.

The results of this paper open up a wide field for future investigation, including working with three different scenarios (separating investment costs on tanks from investments on pipe rehabilitation). On the one hand, the combination of tanks and pipes which conform with a hydraulic control device for flooding is a topic that should be studied. On the other, flood analysis requires great computational effort. In this sense, additional research must be done to reduce the size of the problem.

Supplementary Materials: The following are available online at http://www.mdpi.com/2073-4441/11/5/935/s1, Figure S1: Representation of E-Chicó drainage network, Figure S2: Design storm based on the Alternating Blocks Method, Table S1: Data for nodes and subcatchments in the network used as a case study, Table S2: Data for conduits in the network used as a case study, Table S3: Time series for the design storm used in the case study, Table S4: Series of suitable diameters and their associated for the case study, Table S5: Results for scenario 1, all 35 conduits are suitable to change their diameters, Table S6: Results for scenario 2, all 35 nodes are suitable locations for detention tanks, Table S7: Results for scenario 3, all 35 conduits are suitable to change their diameters and all 35 nodes are suitable locations for detention tanks.

Author Contributions: All authors contributed extensively to the work presented in this paper. F.J.M.-S. and P.L.I.-R. contributed to the subject of research, the development, the writing of the paper and the preparation of algorithms. U.A.N.-N. adjusted the parameters and performed the simulations and contributed to writing of the paper. D.M.-M. provided the basics for the optimization algorithm, contributed to the writing of the paper and helped in the final revision.

Funding: This work was supported by the Program Fondecyt Regular (Project 1180660) of the Comision Nacional de Investigación Científica y Tecnológica (Conicyt), Chile.

References

1. Consorcio de Compensación de Seguros. *Estadística Riesgos Extraordinarios*; Serie 1971-2017; Ministerio de Economía y Empresa: Madrid, Spain, 2016. (In Spanish)

2. Kordana, S. The identification of key factors determining the sustainability of stormwater systems. *E3S Web Conf.* **2018**, *45*, 00033. [CrossRef]

3. Yazdi, J.; Lee, E.H.; Kim, J.H. Stochastic multiobjective optimization model for urban drainage network rehabilitation. *J. Water Resour. Plan. Manag.* **2015**, *141*, 04014091. [CrossRef]

4. Starzec, M.; Dziopak, J.; Słyś, D.; Pochwat, K.; Kordana, S. Dimensioning of required volumes of interconnected detention tanks taking into account the direction and speed of rain movement. *Water* **2018**, *10*, 1826. [CrossRef]

5. Mailhot, A.; Duchesne, S. Design criteria of urban drainage infrastructures under climate change. *J. Water Resour. Plan. Manag.* **2010**, *136*, 201–208. [CrossRef]

6. Gulizia, C.; Camilloni, I. Comparative analysis of the ability of a set of CMIP3 and CMIP5 global climate models to represent precipitation in South America. *Int. J. Climatol.* **2015**, *35*, 583–595. [CrossRef]

7. Ma, M.; He, B.; Wan, J.; Jia, P.; Guo, X.; Gao, L.; Maguire, L.; Hong, Y. Characterizing the flash flooding risks from 2011 to 2016 over China. *Water* **2018**, *10*, 704. [CrossRef]

8. Kirshen, P.; Caputo, L.; Vogel, R.M.; Mathisen, P.; Rosner, A.; Renaud, T. Adapting urban infrastructure to climate change: A drainage case study. *J. Water Resour. Plan. Manag.* **2015**, *141*, 04014064. [CrossRef]

9. Moselhi, O.; Shehab-Eldeen, T. Classification of defects in sewer pipes using neural networks. *J. Infrastruct. Syst.* **2000**, *6*, 97–104. [CrossRef]

10. Driessen, P.; Hegger, D.; Kundzewicz, Z.; van Rijswick, H.; Crabbé, A.; Larrue, C.; Matczak, P.; Pettersson, M.; Priest, S.; Suykens, C.; et al. Governance strategies for improving flood resilience in the face of climate chang. *Water* **2018**, *10*, 1595. [CrossRef]

11. Reyna, S.M.; Vanegas, J.A.; Khan, A.H. Construction Technologies for Sewer Rehabilitation. *J. Constr. Eng. Manag.* **1994**, *120*, 467–487. [CrossRef]

12. Abraham, D.M.; Wirahadikusumah, R.; Short, T.J.; Shahbahrami, S. Optimization modeling for sewer network management. *J. Constr. Eng. Manag.* **1998**, *124*, 402–410. [CrossRef]

13. Sebti, A.; Fuamba, M.; Bennis, S. Optimization model for BMP selection and placement in a combined sewer. *J. Water Resour. Plan. Manag.* **2015**, *142*, 04015068. [CrossRef]

14. Zahmatkesh, Z.; Burian, S.J.; Karamouz, M.; Tavakol-Davani, H.; Goharian, E. Low-impact development practices to mitigate climate change effects on urban stormwater runoff: Case study of New York City. *J. Irrig. Drain. Eng.* **2015**, *141*, 04014043. [CrossRef]

15. Rossman, L.A. *Storm Water Management Model User's Manual*; National Risk Management Research Laboratory: Cincinnati, OH, USA, 2015.

16. Mora-Melià, D.; López-Aburto, C.; Ballesteros-Pérez, P.; Muñoz-Velasco, P. Viability of green roofs as a flood mitigation element in the central region of Chile. *Sustainability* **2018**, *10*, 1130. [CrossRef]

17. Ugarelli, R.; Di Federico, V. Optimal scheduling of replacement and rehabilitation in wastewater pipeline networks. *J. Water Resour. Plan. Manag.* **2010**, *136*, 348–356. [CrossRef]

18. Ngamalieu-Nengoue, U.; Iglesias-Rey, P.; Martínez-Solano, F.; Mora-Melià, D.; Saldarriaga Valderrama, J. Urban drainage network rehabilitation considering storm tank installation and pipe substitution. *Water* **2019**, *11*, 515. [CrossRef]

19. Lee, E.; Kim, J. Development of resilience index based on flooding damage in urban areas. *Water* **2017**, *9*, 428. [CrossRef]

20. Iglesias-Rey, P.L.; Martínez-Solano, F.J.; Saldarriaga, J.G.; Navarro-Planas, V.R. Pseudo-genetic model optimization for rehabilitation of urban storm-water drainage networks. *Procedia Eng.* **2017**, *186*, 617–625. [CrossRef]

21. Fadel, A.W.; Marques, G.F.; Goldenfum, J.A.; Medellín-Azuara, J.; Tilmant, A. Full flood cost: Insights from a risk analysis perspective. *J. Environ. Eng.* **2018**, *144*, 04018071. [CrossRef]

22. Duan, H.-F.; Li, F.; Yan, H. multi-objective optimal design of detention tanks in the urban stormwater drainage system: LID implementation and analysis. *Water Resour. Manag.* **2016**, *30*, 4635–4648. [CrossRef]

23. Starzec, M. A critical evaluation of the methods for the determination of required volumes for detention tank. *E3S Web Conf.* **2018**, *45*, 00088. [CrossRef]

24. Pochwat, K.B.; Słyś, D. Application of artificial neural networks in the dimensioning of retention reservoirs. *Ecol. Chem. Eng. S* **2018**, *25*, 605–617. [CrossRef]

25. Cunha, M.C.; Zeferino, J.A.; Simões, N.E.; Saldarriaga, J.G. Optimal location and sizing of storage units in a drainage system. *Environ. Model. Softw.* **2016**, *83*, 155–166. [CrossRef]

26. De Martino, G.; De Paola, F.; Fontana, N.; Marini, G.; Ranucci, A. Pollution reduction in receivers: Storm-water tanks. *J. Urban Plan. Dev.* **2011**, *137*, 29–38. [CrossRef]

27. Andrés-Doménech, I.; Montanari, A.; Marco, J.B. Efficiency of storm detention tanks for urban drainage systems under climate variability. *J. Water Resour. Plan. Manag.* **2012**, *138*, 36–46. [CrossRef]

28. Wang, M.; Sun, Y.; Sweetapple, C. Optimization of storage tank locations in an urban stormwater drainage system using a two-stage approach. *J. Environ. Manage.* **2017**, *204*, 31–38. [CrossRef] [PubMed]

29. Cunha, M.C.; Zeferino, J.A.; Simões, N.E.; Santos, G.L.; Saldarriaga, J.G. A decision support model for the optimal siting and sizing of storage units in stormwater drainage systems. *Int. J. Sustain. Dev. Plan.* **2017**, *12*, 122–132. [CrossRef]

30. Dziopak, J. A wastewater retention canal as a sewage network and accumulation reservoir. *E3S Web Conf.* **2018**, *45*, 00016. [CrossRef]

31. Słyś, D. An innovative retention canal—A case study. *E3S Web Conf.* **2018**, *45*, 00084. [CrossRef]

32. Deb, K.; Pratap, A.; Agarwal, S.; Meyarivan, T. A fast and elitist multiobjective genetic algorithm: NSGA-II. *IEEE Trans. Evol. Comput.* **2002**, *6*, 182–197. [CrossRef]

33. Martínez-Solano, F.J.; Iglesias-Rey, P.L.; Saldarriaga, J.G.; Vallejo, D. Creation of an SWMM toolkit for its application in urban drainage networks optimization. *Water* **2016**, *8*, 259. [CrossRef]

34. Wang, Q.; Zhou, Q.; Lei, X.; Savić, D.A. Comparison of multiobjective optimization methods applied to urban drainage adaptation problems. *J. Water Resour. Plan. Manag.* **2018**, *144*, 04018070. [CrossRef]

35. Raquel, C.R.; Naval, P.C. An effective use of crowding distance in multiobjective particle swarm optimization. In Proceedings of the 2005 conference on Genetic and evolutionary computation—GECCO, Washignotn, DC, USA, 25–29 June 2005; p. 257.

36. Mora-Melia, D.; Iglesias-Rey, P.L.; Martinez-Solano, F.J.; Ballesteros-Pérez, P. Efficiency of evolutionary algorithms in water network pipe sizing. *Water Resour. Manag.* **2015**, *29*, 4817–4831. [CrossRef]

37. Mora-Melià, D.; Javier Martínez-Solano, F.; Iglesias-Rey, P.L.; Gutiérrez-Bahamondes, J.H. Population size influence on the efficiency of evolutionary algorithms to design water networks. *Procedia Eng.* **2016**, *186*, 341–348. [CrossRef]

*water*

Article

Suppress Numerical Oscillations in Transient Mixed Flow Simulations with a Modified HLL Solver

Zhonghao Mao, Guanghua Guan and Zhonghua Yang *

State Key Laboratory of Water Resources and Hydropower Engineering Science, Wuhan University, Wuhan 430072, China; 2011301580373@whu.edu.cn (Z.M.); ggh@whu.edu.cn (G.G.)
* Correspondence: yzh@whu.edu.cn

Received: 21 February 2020; Accepted: 21 April 2020; Published: 27 April 2020

Abstract: Transition between free-surface and pressurized flows is a crucial phenomenon in many hydraulic systems. During simulation of such phenomenon, severe numerical oscillations may appear behind filling-bores, causing unphysical pressure variations and computation failure. This paper reviews existing oscillation-suppressing methods, while only one of them can obtain a stable result under a realistic acoustic wave speed. We derive a new oscillation-suppressing method with first-order accuracy. This simple method contains two parameters, P_a and P_b, and their values can be determined easily. It can sufficiently suppress numerical oscillations under an acoustic wave speed of 1000 ms^{-1}. Good agreement is found between simulation results and analytical results or experimental data. This paper can help readers to choose an appropriate oscillation-suppressing method for numerical simulations of flow regime transition under a realistic acoustic wave speed.

Keywords: flow regime transition; finite volume methods; numerical oscillations; numerical viscosity; Preissmann slot model

1. Introduction

In water conveyance systems, water flows under free-surface flow condition or pressurized flow condition. Under certain circumstances, transition between the two flow regimes may occur (i.e., the flow regime transition phenomenon). Following the transition, force exerting on structures changes violently and causes structural damage [1–3]. Numerical simulation of flow regime transition can provide substantial information for the design and management of river-crossing bridges, tunnels, conducts and culverts [4–10].

The complexity of flow regime transition lies in the presence of free-surface and pressurized flows, which are governed by different equations. This problem can be avoided by adopting one set of governing equations for two flow regimes. Based on this idea, the Preissmann slot model (PSM) is proposed [11]; it was adopted by many researchers and commercial software packages [12–18]. The strong gradient in piezometric head at the interface between two flow regimes forms a discontinuity in flow. Finite volume methods can capture the discontinuity in the flow implicitly, which makes them popular in simulating flow regime transition [19].

Despite of all the fine properties that finite volume methods have, the numerical oscillations in a flow regime transition simulation have troubled many hydraulic engineers [12]. These numerical oscillations have the same origin with "post-shock oscillations" in gas dynamics [20–23]. In analytical results, the thickness of the filling-bore is infinitely small, and the flow states at the two sides of a filling-bore satisfy the Rankine–Hugoniot condition. In numerical simulations, a filling-bore spreads over several computational cells, and the flow states at the two adjacent cells do not satisfy the Rankine–Hugoniot condition. This causes trivial discrepancies in the mass and momentum fluxes, which are amplified in simulation results due to the large acoustic wave speed. High-order finite volume methods cause more numerical oscillations because of low dissipation away from the shocks [21].

First-order upwind finite volume methods fail to prevent numerical oscillations without compromising the representation of the filling-bore [23–25].

A lot of effort has been spent to obtain a stable and accurate result of flow regime transition [12,23,24]. In this paper, some existing oscillation-suppressing methods are tested on a benchmark model. Considering the lack of an efficient and simple method, the authors derive a new method, which can suppress numerical oscillations and capture the filling-bore nicely under a high acoustic wave speed. The structure of this paper is as follows: Section 2 introduces the governing equations and the discretization method. Section 3 reviews the existing oscillation-suppressing methods, and their effects are evaluated on the benchmark model that was adopted by Malekpour and Karney [24]. Section 4 proposes a new and simple modified HLL solver to suppress numerical oscillations. Its accuracy and robustness are tested against the analytical results and experiment data in Section 5. Conclusions are drawn in the last section.

2. Governing Equations and Discretization Method

The PSM places an infinitely high narrow slot on the top of the conduct, so that it becomes an open-channel with a composite cross-section. The water depth in the slot represents the piezometric head of the pressurized flow inside the original conduct, as shown in Figure 1. The slot width needs to be very small so that the gravity wave speed inside it is identical to the acoustic wave speed.

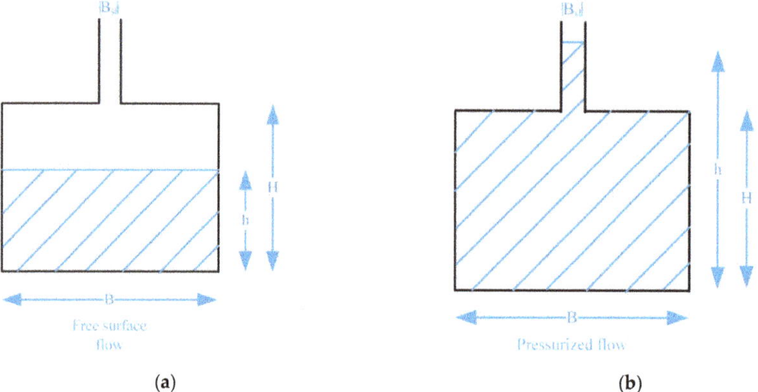

Figure 1. A rectangular conduct with a slot on its top: (**a**) free surface flow; (**b**) pressurized flow.

Under the framework of PSM, the governing equations of one-dimensional flow regime transition with uniform cross-sections can be written as [26]:

$$\frac{\partial \mathbf{U}}{\partial t} + \frac{\partial \mathbf{F}(\mathbf{U})}{\partial x} = \mathrm{S}(\mathbf{U})$$

$$\mathbf{U} = \begin{pmatrix} A \\ Q \end{pmatrix}, \mathbf{F}(\mathbf{U}) = \begin{pmatrix} Q \\ \frac{Q^2}{A} + gI(h) \end{pmatrix}, \mathrm{S}(\mathbf{U}) = \begin{pmatrix} 0 \\ gA(S_b - S_f) \end{pmatrix} \tag{1}$$

$$I(h) = \int_0^h (h - \xi)l(x, \xi)d\xi$$

where Q is the volume flow rate, A is the wetted area, g is the acceleration of gravity, h is the water depth, I is the moment of inertia, l is the cross-sectional width and $l(x, h) = b(x)$ the water surface width.

The external force acting on the flow is accounted in the source term, where S_b is the bed slope and S_f is the friction slope, which can be computed using the Manning relation:

$$S_f = \frac{n^2 u |u|}{R^{4/3}} \tag{2}$$

where u is the flow velocity, n is the Manning coefficient and R is the hydraulic radius. For a rectangular cross-section with a slot on its top, the parameters b, A, I and wave speed c can be expressed as functions of h:

$$b(h) = \begin{cases} B & h \le H \\ B_{sl} & h > H \end{cases} \tag{3}$$

$$A(h) = \begin{cases} Bh & h \le H \\ BH + B_{sl}(h - H) & h > H \end{cases} \tag{4}$$

$$I(h) = \begin{cases} 0.5Bh^2 & h \le H \\ BH(h - 0.5H) + 0.5B_{sl}(h - H)^2 & h > H \end{cases} \tag{5}$$

$$c(h) = \begin{cases} \sqrt{gh} & h \le H \\ \sqrt{g\frac{A}{B_{sl}}} & h > H \end{cases} \tag{6}$$

where H and B are the cross-sectional height and width, and B_{sl} is slot width. In order to make the gravity wave speed inside the slot equal to the acoustic wave speed a, the slot width $B_{sl} = gA_f a^{-2}$, where A_f is the full cross-sectional area of the conduct [27]. Using the Godunov-type finite volume methods with first-order accuracy and assuming a piecewise constant data construction, the governing equations are discretized as

$$\mathbf{U}_i^* = \mathbf{U}_i^n - \frac{\Delta t_i}{\Delta x_i}(\mathbf{F}_{i+1/2} - \mathbf{F}_{i-1/2}) \tag{7}$$

$$\mathbf{U}_i^{n+1} = \mathbf{U}_i^* + \Delta t_i \mathbf{S}(\mathbf{U}_i^*) \tag{8}$$

3. Review of Current Oscillation-Suppressing Methods

The benchmark model was proposed by Malekpour and Karney [24], it consists of a conduct with square-unit cross-sections that are connected to a reservoir at the upstream end. The acoustic wave speed is 1000 ms^{-1} and the slot width is 9.8×10^{-6} m. Under the initial condition, 0.6 m-deep stagnant water is in the conduct while the water level inside the reservoir is constantly 4 m, as shown in Figure 2.

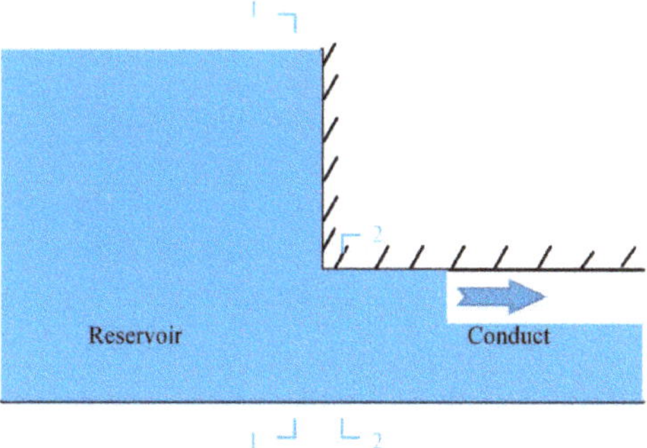

Figure 2. Front view of the benchmark model.

Under initial condition, flow states $\mathbf{U}_L$ in the reservoir and $\mathbf{U}_R$ in the conduct are discontinuous:

$$\begin{bmatrix} h_L \\ u_L \end{bmatrix} = \begin{bmatrix} 4\,\text{m} \\ 0\,\text{ms}^{-1} \end{bmatrix}, \begin{bmatrix} h_R \\ u_R \end{bmatrix} = \begin{bmatrix} 0.6\,\text{m} \\ 0\,\text{ms}^{-1} \end{bmatrix} \tag{9}$$

Since h_L is larger than h_R, a shock wave (filling-bore) belonging to the second characteristic field is formed at the conduct inlet and propagates downstream. Consider 1–1 as a cross-section in the reservoir and 2–2 as a cross-section at conduct inlet: flow sates at 1–1 and 2–2 are $\mathbf{U}_1$ and $\mathbf{U}_2$, respectively. Then $\mathbf{U}_2$ can be obtained by solving the following equations iteratively [24]:

$$h_1 = h_2 + \frac{u_2^2}{2g} \tag{10}$$

$$u_2 = u_R + \sqrt{\frac{[gI(A_R) - gI(A_2)](A_R - A_2)}{A_R A_2}} \tag{11}$$

where $\mathbf{U}_1 = \mathbf{U}_L$ by assuming that flow velocity inside the reservoir is negligible. In this case, h_2 and u_2 are 3.167 m and 4.0334 ms^{-1}, a ghost cell is set at the upstream boundary adopting h_2 and u_2. Since $\mathbf{U}_2$ is connected to $\mathbf{U}_R$ through a right shock, the flow states inside the conduct ultimately will take place by $\mathbf{U}_2$. The propagation speed of the filling-bore is given by the Rankine–Hugoniot condition, which is 10.067 ms^{-1}. Then we can construct the analytical result in the benchmark model at t_0:

$$h(x) = \begin{cases} 3.167\,\text{m} & x < 10.067 t_0 \\ 0.6\,\text{m} & x > 10.067 t_0 \end{cases}, u(x) = \begin{cases} 4.0334\,\text{ms}^{-1} & x < 10.067 t_0 \\ 0\,\text{ms}^{-1} & x > 10.067 t_0 \end{cases} \tag{12}$$

As customary, x denotes the distance to the conduct inlet. In a numerical simulation, the size of each computational cell is 1 m, the time step is 0.008 s and the Courant number is 0.8. When the acoustic wave speed adopted in the simulation exceeds 100 ms^{-1}, the magnitude of the numerical oscillations become so large that the simulated piezometric head become negative, and the simulation will not proceed [24]. In the remaining part of this section, the readers will see that only one method can get a satisfactory result under a high acoustic wave speed, while its performance rely on two parameters which must be well tuned. This shows the importance of devising an alternative method, which is stable and convenient.

3.1. Numerical Filtering Method

In this method, the exact Riemann solver is adopted to solve the Riemann problem at each cell boundary. Flow states $\mathbf{U}_{i+1/2}$ at $x_{i+1/2}$ satisfy the following equations:

$$u_{i+1/2} = \begin{cases} u_i - \sqrt{g\frac{[I(A_i)-I(A_{i+1/2})](A_i-A_{i+1/2})}{A_i A_{i+1/2}}} & A_{i+1/2} > A_i \\ u_i + \int_0^{A_L} \sqrt{\frac{g}{ab}}\,d\alpha - \int_0^{A_{i+1/2}} \sqrt{\frac{g}{ab}}\,d\alpha & A_{i+1/2} < A_i \end{cases} \tag{13}$$

$$u_{i+1/2} = \begin{cases} u_{i+1} + \sqrt{g\frac{[I(A_{i+1})-I(A_{i+1/2})](A_{i+1}-A_{i+1/2})}{A_{i+1} A_{i+1/2}}} & A_{i+1/2} > A_{i+1} \\ u_{i+1} - \int_0^{A_{i+1}} \sqrt{\frac{g}{ab}}\,d\alpha + \int_0^{A_{i+1/2}} \sqrt{\frac{g}{ab}}\,d\alpha & A_{i+1/2} < A_{i+1} \end{cases} \tag{14}$$

Equations (13) and (14) can be solved iteratively, then the wave structure in the Riemann problem can be determined and utilized to compute the flux; see Kerger et al. [26] for detail. Although the exact solver can obtain the complete wave structure, serious numerical oscillations appear in the simulation result. Vasconcelos et al. [23] proposed to suppress the numerical oscillations by averaging the flow states among the three conjunct cells at each time step:

$$\mathbf{U}_i^{n+1} = (1 - 2\varepsilon)\mathbf{U}_i^{n+1} + \varepsilon\left(\mathbf{U}_{i-1}^{n+1} + \mathbf{U}_{i+1}^{n+1}\right) \tag{15}$$

The authors suggest ε to be between 0.025 and 0.050. This method will increase the spreading length of the filling-bore front and remove any physical oscillations that appear in the solution. The simulation result of this method using $\varepsilon = 0.04$ is drawn in Figure 3.

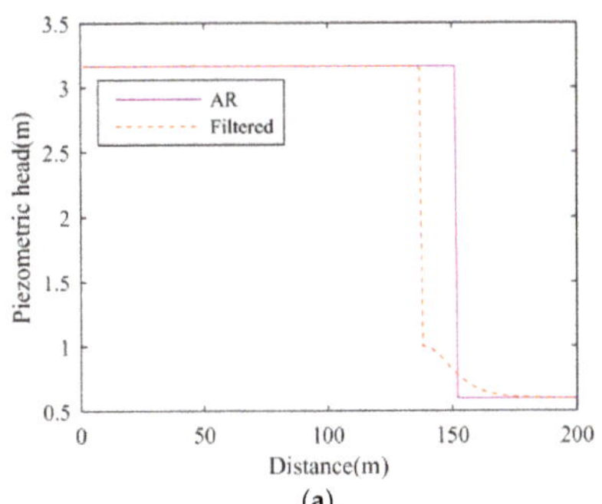

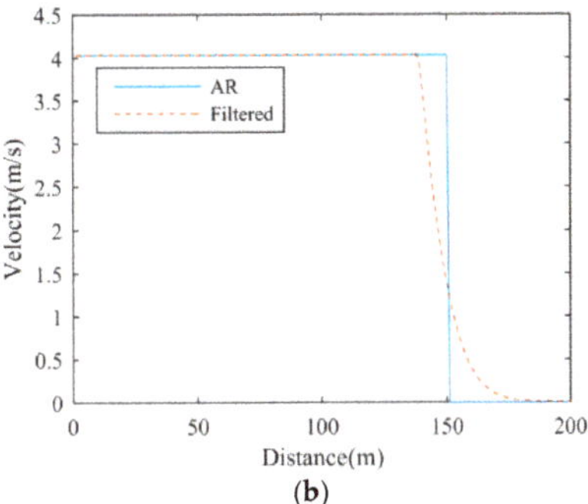

(a)

(b)

Figure 3. Comparison of results simulated by the numerical filtering method (filtered) and the analytical result (AR): (**a**) piezometric head; (**b**) flow velocity.

3.2. Hybrid Flux Method

The hybrid flux method uses two types of numerical fluxes alternatively. The first one is based on the Roe solver [28] and the LxF scheme [23]. Define $\lambda_{i+1/2}^j$ as eigenvalues of the linearized Jacobian matrix, $\mathbf{R}_{i+1/2}^j$ as the corresponding right eigenvectors and $\alpha_{i+1/2}^j$ as the wave strengths across the cell boundary:

$$\lambda_{i+1/2}^1 = \frac{\widetilde{Q}_{i+1/2}}{\widetilde{A}_{i+1/2}} - \widetilde{c}_{i+1/2}, \lambda_{i+1/2}^2 = \frac{\widetilde{Q}_{i+1/2}}{\widetilde{A}_{i+1/2}} + \widetilde{c}_{i+1/2} \tag{16}$$

$$\mathbf{R}_{i+1/2}^1 = \left[1, \lambda_{i+1/2}^1\right]^T, \mathbf{R}_{i+1/2}^2 = \left[1, \lambda_{i+1/2}^2\right]^T \tag{17}$$

$$\alpha_{i+1/2}^1 = \frac{\lambda_{i+1/2}^2(A_{i+1} - A_i) - (Q_{i+1} - Q_i)}{\lambda_{i+1/2}^2 - \lambda_{i+1/2}^1}, \alpha_{i+1/2}^2 = \frac{(Q_{i+1} - Q_i) - \lambda_{i+1/2}^1(A_{i+1} - A_i)}{\lambda_{i+1/2}^2 - \lambda_{i+1/2}^1} \tag{18}$$

where $\widetilde{Q}_{i+1/2}, \widetilde{A}_{i+1/2}$ and $\widetilde{c}_{i+1/2}$ are Roe averages:

$$\widetilde{A}_{i+1/2} = (A_i A_{i+1})^{1/2} \tag{19}$$

$$\widetilde{Q}_{i+1/2} = \frac{Q_{i+1}(A_i)^{1/2} + Q_i(A_{i+1})^{1/2}}{(A_i)^{1/2} + (A_{i+1})^{1/2}} \tag{20}$$

$$\widetilde{c}_{i+1/2} = \begin{cases} \left[g\frac{I(A_{i+1}) - I(A_i)}{A_{i+1} - A_i}\right]^{1/2} & A_{i+1} \neq A_i \\ \left[g\frac{A_i + A_{i+1}}{b_i + b_{i+1}}\right]^{1/2} & A_{i+1} = A_i \end{cases} \tag{21}$$

Then the fluxes obtained by the Roe solver are written as

$$\mathbf{F}^{\text{Roe}}_{i+1/2} \;=\; \frac{1}{2}\Big[\mathbf{F}\big(\mathbf{U}^{n}_{i+1}\big)+\mathbf{F}\big(\mathbf{U}^{n}_{i}\big)\Big]-\frac{1}{2}\sum_{j=1}^{2}\Big|\lambda^{j}_{i+1/2}\Big|\alpha^{j}_{i+1/2}\mathbf{R}^{j}_{i+1/2} \tag{22}$$

The Roe solver is known to be vulnerable to numerical oscillations [21,29], while the LxF scheme is robust against numerical oscillations but causes too much diffusion of the filling-bore:

$$\mathbf{F}^{\text{LxF}}_{i+1/2} \;=\; \frac{1}{2}\Big[\mathbf{F}\big(\mathbf{U}^{n}_{i+1}\big)+\mathbf{F}\big(\mathbf{U}^{n}_{i}\big)\Big]-\frac{\Delta x}{2\Delta t}\big(\mathbf{U}_{i+1}-\mathbf{U}_{i}\big) \tag{23}$$

Compare Equations (22) with (23): The difference between the Roe solver and LxF scheme is the choice of eigenvalues. Vasconcelos et al. [23] proposed a new method to determine the eigenvalues:

$$\Big|\lambda^{1,2}_{i+1/2}\Big| \;=\; \min\left[\frac{\Delta x}{\Delta t},\;\left|\frac{\widetilde{Q}_{i+1/2}}{\widetilde{A}_{i+1/2}}\mp\widetilde{c}_{i+1/2}\right|+L(\Delta c_{i+1/2})^{L}\frac{\Delta x}{\Delta t}\right] \tag{24}$$

$$\Delta c_{i+1/2} \;=\; \frac{\big|\widetilde{c}_{i+1}-\widetilde{c}_{i}\big|}{\max\big(\big|\widetilde{c}_{i+1}-\widetilde{c}_{i}\big|\big)_{i\,=\,1...N-1}} \tag{25}$$

This method is referred to as the hybrid one from here on further. At the cell boundary between the free-surface and pressurized flows, $\Delta c_{i+1/2}=1$, as L changes from 0 to 1, the eigenvalues switches from those obtained by the Roe solver to those adopted in the LxF scheme. At the other cell boundaries, $\Delta c_{i+1/2}$ is approximately 0, thus the eigenvalues in Equation (24) remain close to those obtained by the Roe solver. In this way, numerical viscosity is added at the cell boundary where flow condition transition happens, and its amount increases with L. The simulation results of the hybrid one with $L=0.6$ are drawn in Figure 4. This method overestimates the spreading length of the filling-bore, and it fails to suppress the numerical oscillations under a high acoustic wave speed.

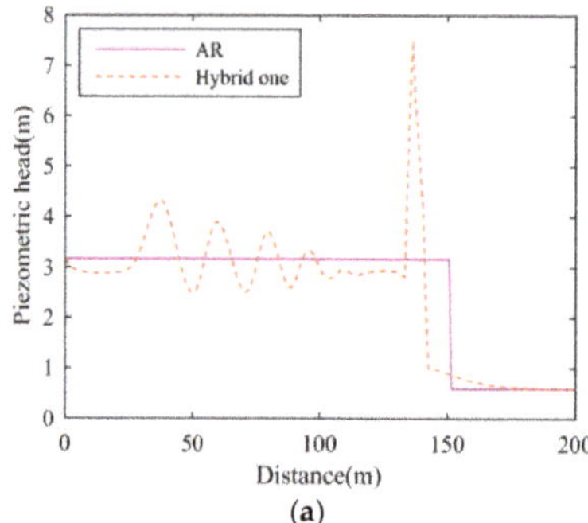

(a)

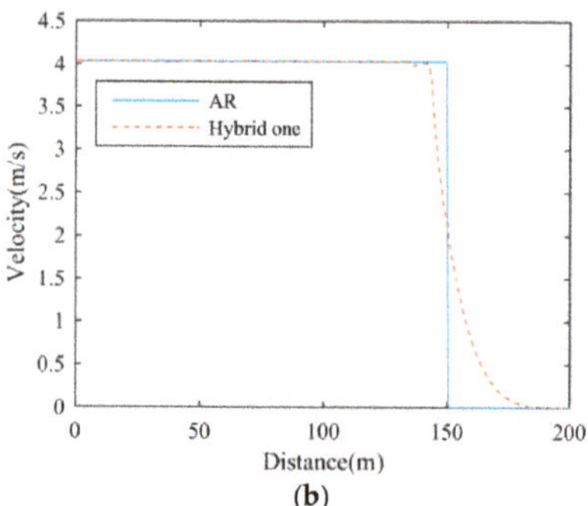

(b)

Figure 4. Comparison of the results simulated by hybrid one and the analytical result (AR): (a) piezometric head; (b) flow velocity.

Hyunuk et al. [12] chose different fluxes; they used the FORCE scheme [30] at the cell boundary between the free-surface and pressurized flows, and the HLL solver [30] elsewhere:

$$\mathbf{F}^{\text{hybrid}}_{i+1/2} \;=\; \begin{cases} \mathbf{F}^{\text{FORCE}}_{i+1/2} & (h_{L}-H_{L})(h_{R}-H_{R})<0 \\ \mathbf{F}^{\text{HLL}}_{i+1/2} & \text{otherwise} \end{cases} \tag{26}$$

This method is referred to as hybrid two from here on further. The HLL solver starts by estimating the largest wave speed S_{wR} and smallest wave speed S_{wL} in the Riemann problem. The fluxes are computed based on the signs of S_{wL} and S_{wR}:

$$\mathbf{F}^{HLL}_{i+1/2} = \begin{cases} F(\mathbf{U}_i) & S_{wL} > 0 \\ \frac{S_{wR}F(\mathbf{U}_i) - S_{wL}F(\mathbf{U}_{i+1}) + S_{wR}S_{wL}(\mathbf{U}_{i+1} - \mathbf{U}_i)}{S_{wR} - S_{wL}} & S_{wL} \leq 0 \quad \text{and} \quad S_{wR} \geq 0 \\ F(\mathbf{U}_{i+1}) & S_{wR} < 0 \end{cases} \tag{27}$$

The choices of S_{wL} and S_{wR} follow Toro [31]:

$$S_{wL} = u_i - \Omega_i, S_{wR} = u_{i+1} + \Omega_{i+1} \tag{28}$$

$$\Omega_{K(K = i, i+1)} = \begin{cases} \sqrt{g \dfrac{[I(A_{i+1/2}) - I(A_K)]A_{i+1/2}}{A_K(A_{i+1/2} - A_K)}} & A_{i+1/2} > A_K \\ c_K & A_{i+1/2} \leq A_K \end{cases} \tag{29}$$

where $A_{i+1/2}$ is an estimate of the wetted area at $x_{i+1/2}$; we adopt the one proposed by Leno et al. [32], which admits the minimum amount of numerical viscosity:

$$A_{i+1/2} = \frac{A_i + A_{i+1}}{2}\left(1 + \frac{u_i - u_{i+1}}{c_i + c_{i+1}}\right) \tag{30}$$

The FORCE flux can be written as the algebraic average of the LxF scheme and Lax–Wendorff scheme:

$$\mathbf{F}^{LW}_{i+1/2} = F(\mathbf{U}^{LW}_{i+1/2}), \mathbf{U}^{LW}_{i+1/2} = \frac{1}{2}(\mathbf{U}^n_i + \mathbf{U}^n_{i+1}) - \frac{\Delta x}{2\Delta t}\left[F(\mathbf{U}^n_{i+1}) - F(\mathbf{U}^n_i)\right] \tag{31}$$

$$\mathbf{F}^{FORCE}_{i+1/2} = \frac{1}{2}\left(\mathbf{F}^{LW}_{i+1/2} + \mathbf{F}^{LxF}_{i+1/2}\right) \tag{32}$$

The FORCE scheme is a centred scheme; thus, it is robust against numerical oscillations. It is less diffusive than the LxF scheme, which can reduce the over-smearing at strong gradients [33]. The simulation results of hybrid two are depicted in Figure 5.

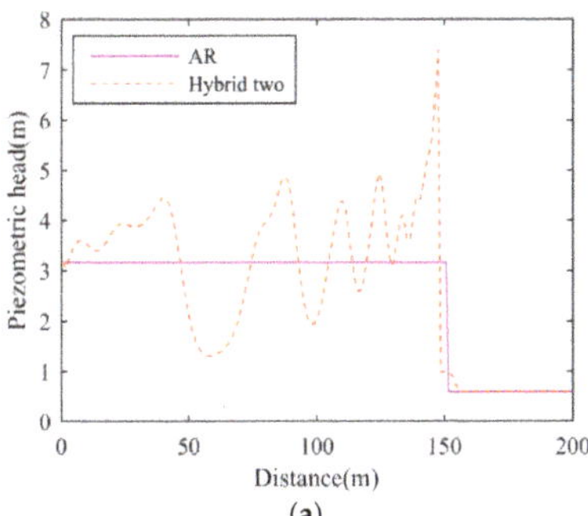
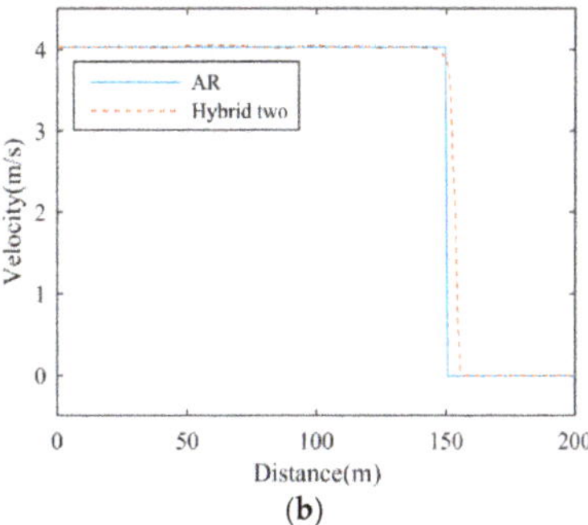

Figure 5. Comparison of the results simulated by hybrid two and the analytical result (AR): (**a**) piezometric head; (**b**) flow velocity.

The hybrid flux methods have added enough numerical viscosity at the cell boundary between the two flow regimes but, still, serious numerical oscillations appear in the simulation results. The reason is that the numerical oscillations appear as soon as the flow regime transition happens, and while the two methods start to add numerical viscosity one time-step falls behind of it. If one method can foresee the happening of the flow regime transition, and admits numerical viscosity ahead of it, it will achieve a more stable result.

3.3. Modified HLL Solver

Malekpour and Karney [24] pointed out that the amount of numerical viscosity in the HLL fluxes increases with the magnitude of S_{wL} and S_{wR}. In fact, the HLL fluxes equal the LxF fluxes when $|S_{wL}|$ and $|S_{wR}|$ equal $\Delta x/\Delta t$. To increase the amount of numerical viscosity, they proposed a modified HLL solver (referred to as the M_HLL solver). In the M_HLL solver, $A_{i+1/2}$ in Equation (29) is computed according to a reference depth h_G:

$$A_{i+1/2} = A(h_G), h_G = K_a \max(h_{i-NS}, h_{i-NS+1}, \cdots, h_{i-1}, h_i, h_{i+1}, \cdots, h_{i+NS-1}, h_{i+NS}) \tag{33}$$

The depth h_G is defined as K_a, multiplying the highest piezometric head in the $2NS+1$ cells surrounding cell i, while $K_a > 1$ and $NS \geq 3$. The solution of Equation (33) produces a larger magnitude of S_{wL} and S_{wR}; thus, increasing the numerical viscosity before the flow regime transition happens. The simulation results of M_HLL with $K_a = 1.4$ and $NS = 5$ are drawn in Figure 6.

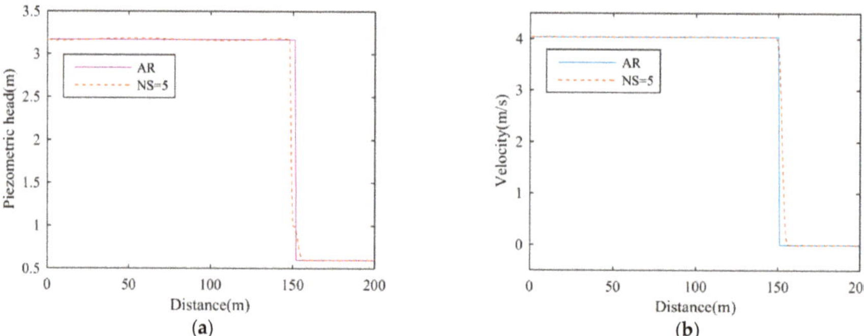

Figure 6. Comparison of the results simulated by M_HLL, with $K_a = 1.4$ and $NS = 5$, and the analytical result (AR): (**a**) piezometric head; (**b**) flow velocity.

The M_HLL solver can suppress numerical oscillations in the benchmark model, and it only slightly increases the spreading length at the filling-bore front. However, the values of K_a and NS can affect the diffusion and accuracy of the M_HLL solver to a great extent; see Figure 7.

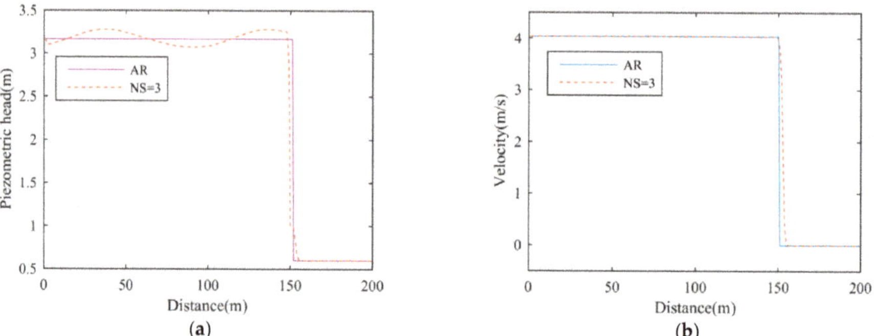

Figure 7. Comparison of the results simulated by M_HLL, with $K_a = 1.4$ and $NS = 3$, and the analytical result (AR): (**a**) piezometric head; (**b**) flow velocity.

Thus, the values of K_a and NS must be well-tuned. Meanwhile, the way to determine h_G makes the HLL solver hard to use in parallelized computation. In the next section, the authors present an alternative method, which is equally efficient as the M_HLL solver.

4. A New Modified HLL Solver

In this section, we present a new modified HLL solver (referred to as the P_HLL solver) to suppress numerical oscillations. In the P_HLL solver, the solution to $A_{i+1/2}$ depends on the water depths at cell i and $i + 1$. When h_i and h_{i+1} are below P_bH, $A_{i+1/2}$ is computed using Equation (30) to admit the minimum amount of numerical viscosity, otherwise $A_{i+1/2}$ is computed according to a constant depth P_aH:

$$A_{i+1/2} = A(P_aH), \quad h_i > P_bH \quad \text{or} \quad h_{i+1} > P_bH \tag{34}$$

P_b must be smaller than one, and a preferable value is between 0.6 and 0.8. P_aH must be larger than the piezometric head peak during the transition to admit enough numerical viscosity.

To illustrate the effect of P_a and P_b, we study the Riemann problem at $x_{i+1/2}$ in the benchmark model. Suppose h_i and h_{i+1} is 3.167 m and 0.6 m, respectively, and u_i and u_{i+1} is 4.0334 ms^{-1} and 0 ms^{-1}, respectively. The solution of Equation (30) lies between A_i and A_{i+1}, and after substituting it into Equation (28), one will get S_{wL} (noted as S_{wL1}) as the speed of the left rarefaction wave head, and S_{wR} (noted as S_{wR1}) as the speed of right shock wave:

$$S_{wL1} = u_i - c_i, \; S_{wR1} = u_{i+1} + \sqrt{g\frac{[I(A_{i+1/2}) - I(A_{i+1})]A_{i+1/2}}{A_{i+1}(A_{i+1/2} - A_{i+1})}} \tag{35}$$

A sketch of two waves in the Riemann problem is shown in Figure 8.

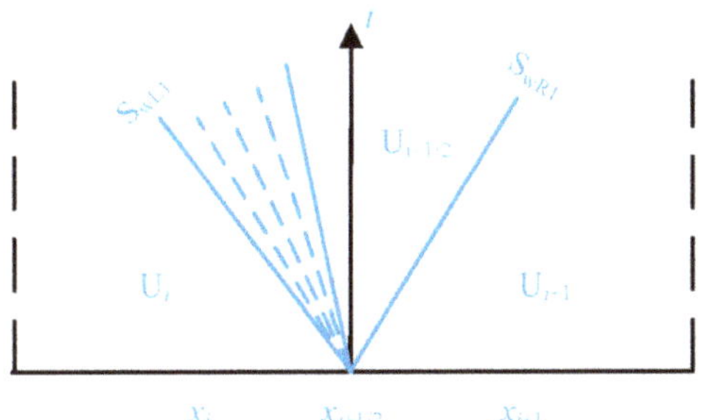

Figure 8. A left rarefaction wave with a right shock wave at $x_{i+1/2}$.

Because cell i is under a pressurized flow condition, it is easy to see that $|S_{wL1}|$ nearly equals the acoustic wave speed. The entropy condition of right shock wave requires

$$u_{i+1/2} + c_{i+1/2} > S_{wR1} > u_{i+1} + c_{i+1} \tag{36}$$

The magnitude of S_{wR1} equals the propagation speed of the filling-bore, which is 10.067 ms^{-1}, and it is much smaller than the acoustic wave speed.

The solution of Equation (34) is unconditionally larger than A_i and A_{i+1}, which produces two shock waves in the Riemann problem. Consequently, S_{wR2} is the speed of the right shock wave, and S_{wL2} is the speed of the left shock wave; see Figure 9.

The entropy condition of the left shock wave requires

$$u_{i+1/2} - c_{i+1/2} < S_{wL2} < u_i - c_i \tag{37}$$

Since cell i is under a pressurized flow condition, $u_i \ll c_i$; thus, $|S_{wL2}| > |S_{wL1}|$ and they are both close to the acoustic wave speed. The speed of the right shock wave increases with $A_{i+1/2}$; thus, $S_{wR2} > S_{wR1}$. This larger magnitude of S_{wR} admits more mass and momentum into cell $i + 1$ before it becomes pressurized. The loci of U_{i+1} simulated by HLL and P_HLL ($P_a = 5$, $P_b = 0.7$) are drawn in Figure 10.

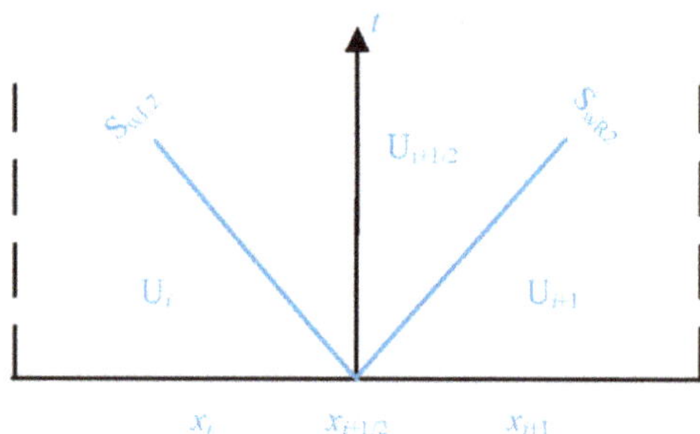

Figure 9. A left shock wave with a right shock wave at $x_{i+1/2}$.

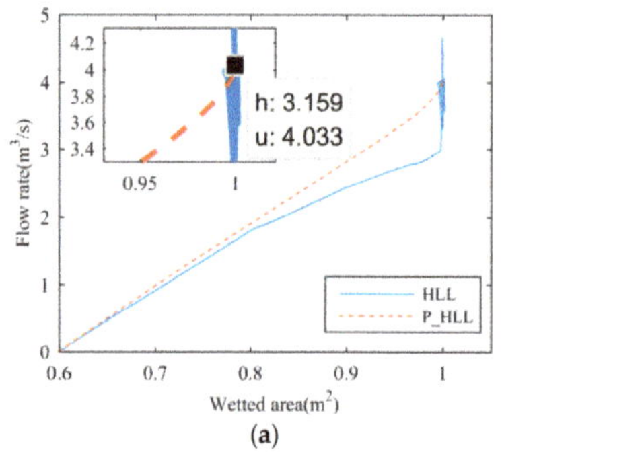

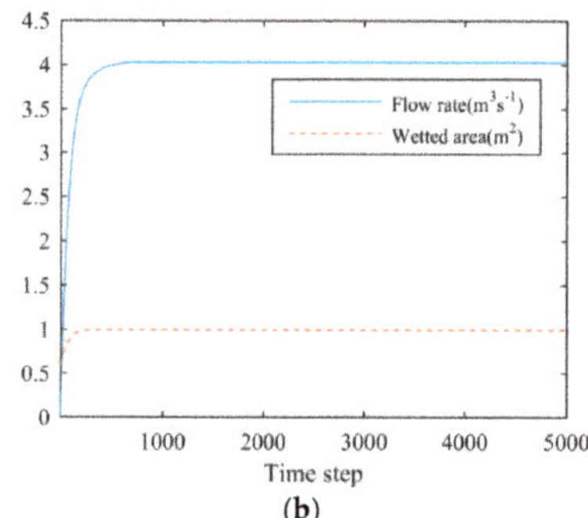

(a)

(b)

Figure 10. (a) Locus of the flow states in cell $i + 1$ simulated by HLL and P_HLL ($P_a = 5$, $P_b = 0.7$); (b) history of the flow states in cell $i + 1$ simulated by P_HLL ($P_a = 5$, $P_b = 0.7$).

Vertexes appear in the locus simulated by HLL after cell $i + 1$ becomes pressurized, which represent numerical oscillations in the simulation result. In contrast, P_HLL achieves a smooth and stable transition between the free-surface flow and pressurized flow. The locus simulated by P_HLL converges to a 3.159 m depth and 4.033 ms^{-1} velocity, which is very close to the flow states at the entrance of the conduct. The discrepancy in water depth is more pronounced due to the small value of the slot width. Therefore, P_HLL preserves the conservation in mass and momentum.

A larger magnitude of S_{wR2} admits more mass and momentum into cell $i + 1$ before it is pressurized; thus, it increases the diffusion of the filling-bore. The magnitude of S_{wL2} and S_{wR2} are related to the value of $A_{i+1/2}$, which consequently depends on the value of P_a; see Figures 11 and 12.

Figure 11 shows that S_{wR2} increases with P_a, while $|S_{wL2}|$ barely changes with P_a and stays close to the acoustic wave speed. However, S_{wR2} does not increase significantly when P_a changes from 1 to 10, which denotes that the diffusion of the filling-bore does not increase significantly when P_a changes from 1 to 10. The simulation results using the P_HLL solver with $P_b = 0.7$ and different values of P_a are drawn in Figure 13.

Although $P_a = 10$ produces a more diffusive filling-bore, the spreading length of the filling-bore does not increase significantly compared to that using $P_a = 5$. During a realistic transition phenomenon, the piezometric head peak seldom exceeds 10 times the cross-sectional depth. Therefore, one can always start by adopting a large P_a (for example 10) in the P_HLL solver and do not worry about significantly compromising the representation of the filling-bore.

Compared to P_a, the value of P_b has a more significant effect on the numerical oscillations, for it determines the threshold depth where numerical viscosity starts to increase. P_b must be smaller than one so that the numerical viscosity is added before the flow regime transition happens. A smaller P_b

leads to more stable result, but it may cause more diffusion. The simulation results using $P_a = 5$ and $P_b = 0.7$ or 0.9 are drawn in Figure 14.

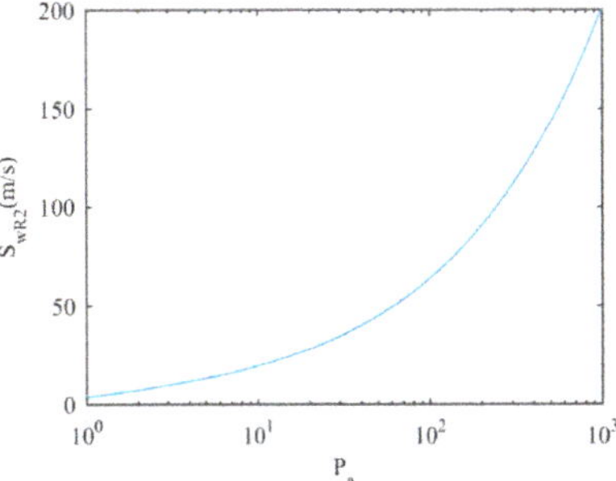

Figure 11. S_{wR2} under different values of P_a.

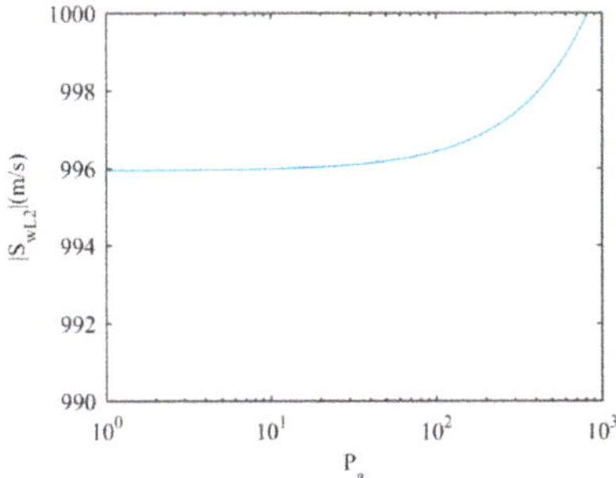

Figure 12. $|S_{wL2}|$ under different values of P_a.

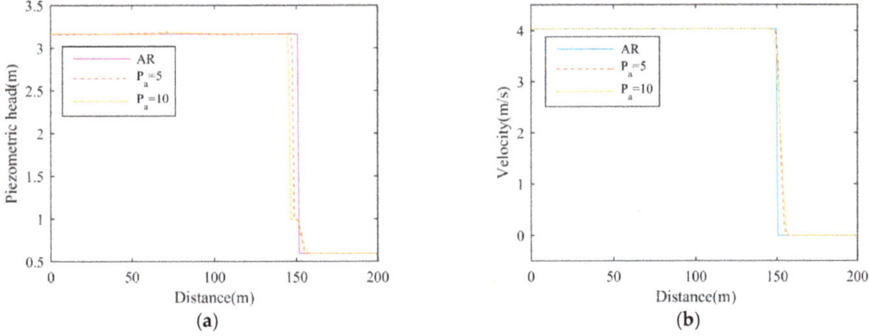

Figure 13. Comparison of the results simulated by P_HLL, with $P_b = 0.7$ and $P_a = 5$ or $P_a = 10$, and the analytical result (AR): (**a**) piezometric head; (**b**) flow velocity.

When $P_b = 0.9$, a smooth transition between the free-surface and pressurized flows cannot be guaranteed. Therefore, we suggest a P_b between 0.6 and 0.8 to avoid causing two much diffusion of the filling-bore. This is also supported by the numerical tests in the next section.

In conclusion, at a free-surface cell, P_HLL admits numerical viscosity once the water depth is higher than a threshold value $P_b H$. Thus, a smooth transition from the free-surface flow to pressurized flow can be obtained. Meanwhile, P_HLL causes diffusion of the filling-bore. In realistic applications, a P_a of 10 is large enough to suppress numerical oscillations without significantly increasing the spreading-length of the filling-bore. The value of P_b is suggested to be between 0.6 and 0.8.

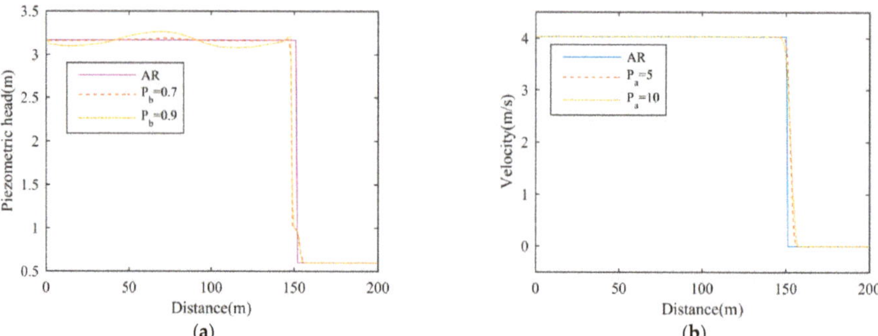

Figure 14. Comparison of the results simulated by P_HLL, with $P_a = 5$ and $P_b = 0.7$ or $P_b = 0.9$, and the analytical result (AR): (**a**) piezometric head; (**b**) flow velocity.

5. Numerical Tests

5.1. Two Filling-Bores

This test evaluates the accuracy of P_HLL under the presence of two filling-bores. It consists of a 200 m-long conduct with square-unit cross-sections and two reservoirs connected to it at the upstream end and downstream end; the acoustic wave speed is 1000 ms^{-1}. At initial conditions, water in the conduct is stationary with a depth of 0.6 m, while water depth at the upstream and downstream reservoir is 4 m and 3 m, respectively. The model set up is sketched in Figure 15.

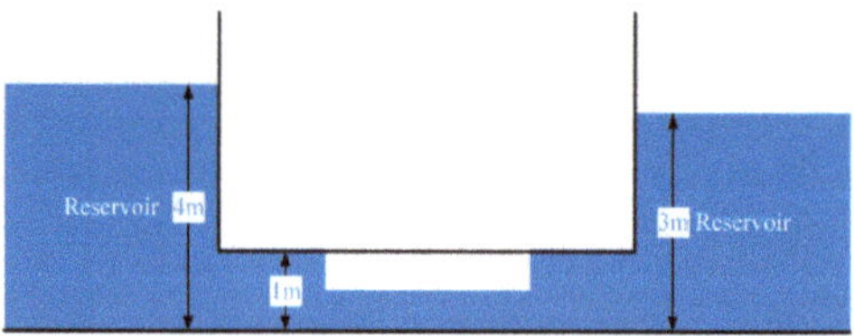

Figure 15. A sketch of the reservoir–conduct–reservoir model.

At $t = 0$ s, the conduct inlet and outlet are opened immediately, forming two filling-bores that propagate in the opposite direction. The upstream filling-bore is identical with that in the benchmark model. The downstream filling-bore can be derived analogously; its propagation speed is 8.429 ms^{-1}, and the flow states behind it are 2.42 m and -3.3717 ms^{-1}. Boundary conditions adopt ghost cells at the conduct inlet and outlet. Before the two filling-bores collide with each other, the analytical result at t_0 is

$$
\begin{aligned}
h(x) &= \begin{cases} 3.167\text{ m} & x < 10.067t_0 \\ 0.6\text{ m} & 10.067t_0 < x < 200 - 8.429t_0 \\ 2.42\text{ m} & 200 - 8.429t_0 < x \end{cases} \\
u(x) &= \begin{cases} 4.0334\text{ ms}^{-1} & x < 10.067t_0 \\ 0\text{ ms}^{-1} & 10.067t_0 < x < 200 - 8.429t_0 \\ -3.3717\text{ ms}^{-1} & 200 - 8.429t_0 < x \end{cases}
\end{aligned}
\tag{38}
$$

In P_HLL, we choose $P_a = 5$ and $P_b = 0.8$, optionally. In M_HLL, we choose $K_a = 1.4$ and $NS = 5$ as suggested by Malekpour and Karney. The computational cell is 1 m, time step is 0.0008 s and the Courant number is 0.8. The simulation results in the two tests at $t = 6$ s are drawn in Figure 16. In this paper, an error indicator based on the L_2-norm [34] is used to evaluate the accuracy of P_HLL and M_HLL. In the following equation, φ_i stands for the simulation result at cell i, while φ_{ref} stands for the analytical result.

$$L_2 = \left(\frac{1}{N} \sum_{i=1}^{N} (\varphi_i - \varphi_{\text{ref}})^2 \right)^{0.5} \tag{39}$$

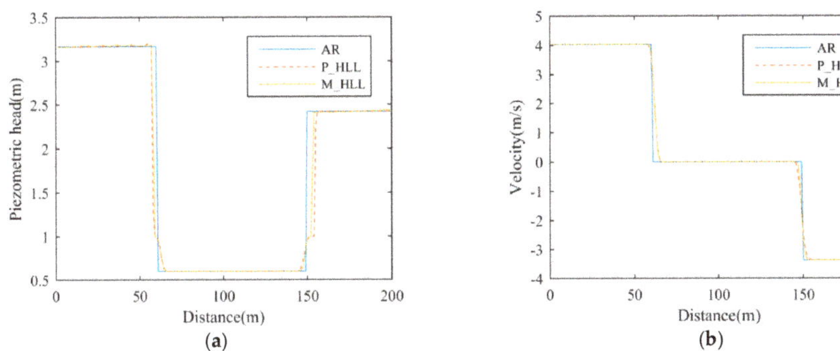

Figure 16. Comparison of the results simulated by P_HLL ($P_a = 5$ and $P_b = 0.8$), M_HLL ($K_a = 1.4$ and $NS = 5$) and the analytical result (AR): (**a**) piezometric head; (**b**) flow velocity.

The L_2 in the piezometric head of P_HLL and M_HLL are 0.2963 and 0.2913, respectively; the L_2 in the velocity of P_HLL and M_HLL are 0.2879 and 0.2873, respectively. In P_HLL, the spreading length of the right filling-bore is slightly longer than that in M_HLL. This denotes that P_HLL is more diffusive than M_HLL in there. At the same time, P_HLL has eliminated some minor numerical oscillations while M_HLL does not. Both solvers are very robust and stable.

5.2. Negative Pressure Flow

In this test, P_HLL and M_HLL are adopted to simulate a water hammer phenomenon in a 600 m-long circular pipe with a 0.5 m diameter and acoustic wave speed is 1200 ms^{-1}. The pipe is horizontal and frictionless; a steady flow rate of 0.477 m^{3}s^{-1} is initially in it. It connects to a reservoir at the downstream end, and water depth inside it is 45 m; see Figure 17.

At $t = 0$ s, the inflow rate is decreased to 0.4 m^{3}s^{-1}, which triggers a water hammer phenomenon; the water hammer pressure is 48.05 m according to Kerger et al. [26]. In P_HLL, the values of P_a and P_b are 100 and 0.8. In M_HLL, the values of K_a and NS are 1.2 and 12, as suggested by Malekpour and Karney. The computational cell is 1.2 m, the time-step is 0.0008 s and Courant number is 0.8. A ghost cell is set at the upstream boundary, and flow rate in it is constantly 0.4 m^{3}s^{-1}, while the piezometric head adopts the transmissive condition. Another ghost cell is set at the downstream boundary in which the piezometric head is constantly 45 m and the flow rate adopts the transmissive condition.

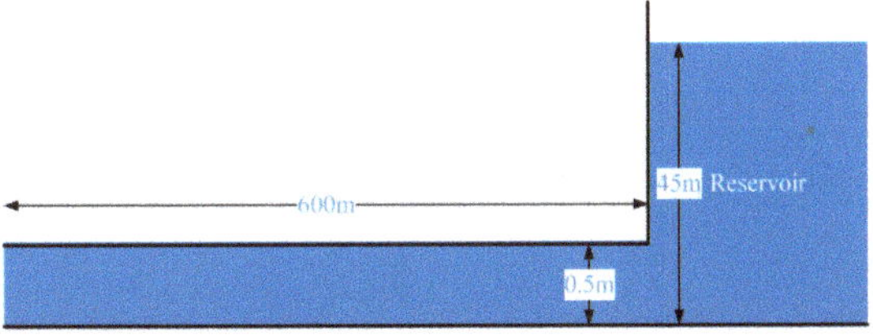

Figure 17. A sketch of the conduct–reservoir model.

The L_2 indicator is adopted to evaluate the accuracy of the two solvers; it is defined as

$$L_2 = \left(\frac{1}{N} \sum_{i=1}^{N} (\varphi_i - \varphi_{\text{ref}})^2 \right)^{0.5} \tag{40}$$

where N_t is the number of time step, φ_i is the simulation result at the midpoint of the pipe, and φ_{ref} is the analytical result, which is given as

$$h_i = \begin{cases} 45 \text{ m} & n < i\Delta t < n + 0.25 \\ -3.05 \text{ m} & n + 0.25 < i\Delta t < n + 0.75 \\ 45 \text{ m} & n + 0.75 < i\Delta t < n + 1.25 \quad , n = 0,1,2,3 \ldots \\ 93.05 \text{ m} & n + 1.25 < i\Delta t < n + 1.75 \\ 45 \text{ m} & n + 1.75 < i\Delta t < n + 2 \end{cases}$$

$$u_i = \begin{cases} 2.4293 \text{ ms}^{-1} & n < i\Delta t < n + 0.25 \\ 2.0377 \text{ ms}^{-1} & n + 0.25 < i\Delta t < n + 0.75 \\ 1.6461 \text{ ms}^{-1} & n + 0.75 < i\Delta t < n + 1.25 \quad , n = 0,1,2,3 \ldots \\ 2.0377 \text{ ms}^{-1} & n + 1.25 < i\Delta t < n + 1.75 \\ 2.4293 \text{ ms}^{-1} & n + 1.75 < i\Delta t < n + 2 \end{cases} \tag{41}$$

The history of the flow states at the pipe midpoint in the simulation and analytical results are drawn in Figure 18. Both solvers nicely capture the reflection of the water-hammer wave in the pipe. The L_2 in the piezometric head of P_HLL and M_HLL are 6.3965 and 6.3970, respectively, while L_2 in the velocity of P_HLL and M_HLL are 0.1332 and 0.1333, respectively.

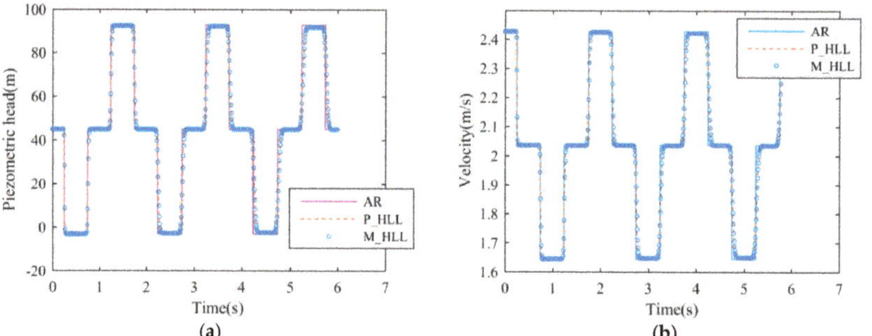

Figure 18. Locus of the flow states at the midpoint of the pipe simulated by P_HLL ($P_a = 100$ and $P_b = 0.8$), M_HLL ($K_a = 1.2$ and $NS = 12$) and the analytical result (AR): (**a**) piezometric head; (**b**) flow velocity.

Interestingly, in this test, the P_HLL solver can obtain almost the same result with an arbitrary value of P_a; for example, the simulation result using $P_a = 1$ is drawn in Figure 19.

In Section 4, we have proven that under the framework of PSM, any non-negative value of P_a will produce a wave speed that is close to the acoustic wave speed, provided that the cell is under pressurized flow condition; see Figure 12 for detail. In this test, all the computational cells are under a pressurized flow condition, which makes the simulation result of $P_a = 100$ and $P_a = 1$ almost the same. In the flow regime transition simulation, $P_a H$ must be larger than the highest piezometric head.

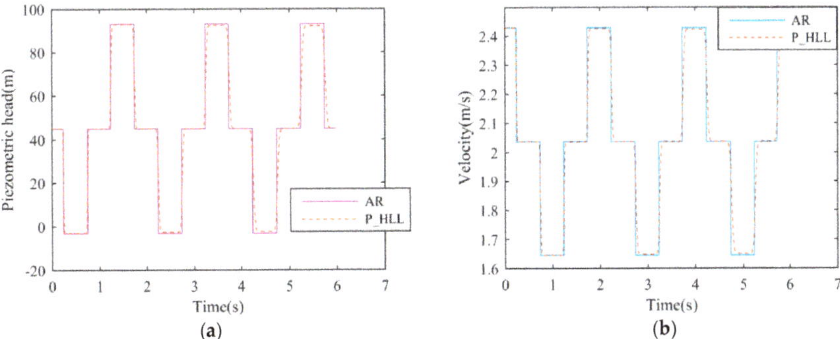

Figure 19. Locus of the flow states at the midpoint of pipe simulated by P_HLL ($P_a = 1$ and $P_b = 0.8$) and the analytical result (AR): (**a**) piezometric head; (**b**) flow velocity.

5.3. Vasconcelos's Experiment

This experiment was designed by Vasconcelos et al. [35] to study the filling process in a realistic stormwater storage tunnel. It is a 14.6 m-long horizontal tunnel with circular cross-sections of 9.4 cm in diameter; initially, stagnant water of 7.3 cm depth is established in the tunnel. A fill box with a 25 cm × 25 cm bottom and 31 cm height connects to the tunnel inlet. A surge tank with a constant diameter of 19 cm connects to the tunnel outlet. The experiment starts by constantly supplying 3.1 Ls^{-1} water into the fill box, and when water level inside the fill box reaches its top, water is simply spilled away. A gate is installed at the tunnel outlet; its opening is smaller than the initial water depth. When the filling bore collides with the gate, it triggers a water-hammer phenomenon. A ventilation tower is fixed upstream of the gate so that no air is trapped in the tunnel. The experiment setup is drawn in Figure 20.

Figure 20. A sketch of the experimental setup.

The manning coefficient is 0.016 m$^{1/6}$, acoustic wave speed is 300 ms^{-1} and head loss coefficient associated with the partially opened gate is 12, as suggested by Malekpour and Karney. In P_HLL, the values of P_a and P_b are 5 and 0.8. In M_HLL, the values of K_a and NS are 1.2 and 12. The computational cell is 0.1 m, and the time-step is set for a Courant number of 0.8. At the upstream end, the three unknowns are the discharge, the wetted area at the tunnel inlet and the water level in the fill box. At the downstream end, the three unknowns are discharge, the wetted area at tunnel outlet and the water level in the surge tank. The continuity, energy and characteristic equations are applied to obtain the three unknowns at each boundary [36]. The loci of flow states at $x = 9.9$ m in the simulation results of P_HLL and M_HLL are shown in Figures 21 and 22.

The simulation results are in good agreement with the experimental data, and the two solvers have correctly computed the arrival time of the filling-bore front at $x = 9.9$ m. P_HLL has computed a piezometric head peak slightly larger than M_HLL. Most importantly, P_HLL and M_HLL do not smear the physical pressure oscillations.

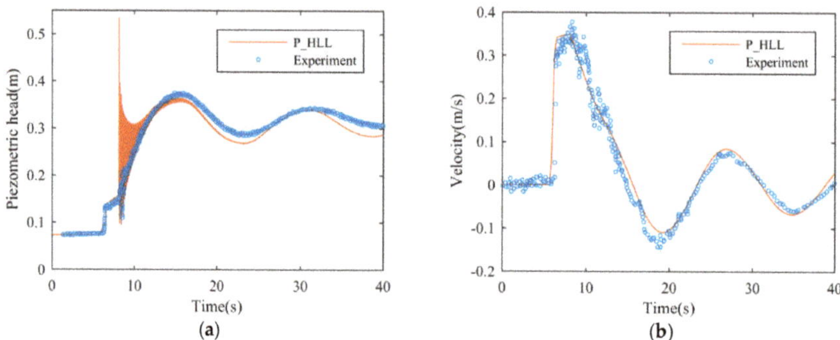

Figure 21. Locus of flow states at $x = 9.9$ m simulated by P_HLL ($P_a = 5$ and $P_b = 0.8$) and experimental data: (**a**) piezometric head; (**b**) flow velocity.

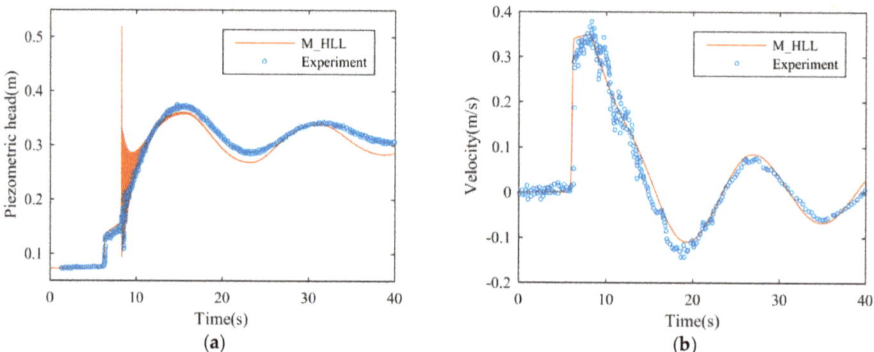

Figure 22. Locus of flow states at $x = 9.9$ m simulated by M_HLL ($K_a = 1.2$ and $NS = 12$) and experimental data: (**a**) piezometric head; (**b**) flow velocity.

5.4. Aureli's Experiment

This experiment is conducted on a 12.12 m-long pipe with a 19.2 cm-diameter and 4 mm-wall thickness [25]. We used x to present the longitudinal distance to the pipe inlet. There is a sharp turn at about $x = 7$ m; the downward part has a slope of about 8.4% and the upward part has a slope of -27.7%. Six piezometric transducers were installed in the bottom at $x = 1$ m, 3 m, 4.5 m, 6.8 m, 7.06 m and 8.52 m. A sluice gate was installed at approximately $x = 5$ m; it is closed at the initial conditions. Stagnant water with a 22.5 cm piezometric head at transducer 1 was established behind the sluice gate; the rest of the pipe was empty. As the experiment began, the gate was lifted within 0.2 s, setting flush into the pipe. The pipe inlet was blocked so that no water flows through it, while the outlet was totally open. The experimental setup is shown in Figure 23.

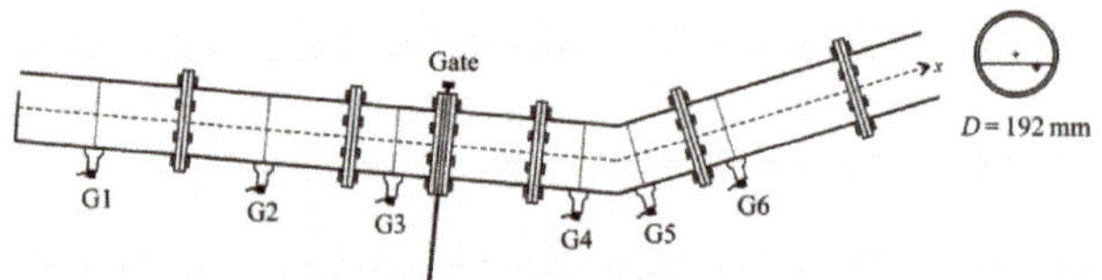

Figure 23. A sketch of the experimental setup, drawn by Aureli et al. [25].

In P_HLL, the values of P_a and P_b are 4 and 0.7; in M_HLL, the values of K_a and NS are 1.4 and 5. The computational cell is 0.04 m, acoustic wave speed is 200 ms^{-1} and time-step is set for a Courant number of 0.8. A reflective boundary condition was set at the upstream end, while a transmissive boundary condition was set at the downstream end. At the wet/dry interface, the estimates of wave speed followed Leon et al. [27]. The loci of the flow states at $x = 6.8$ m simulated by P_HLL and M_HLL are drawn in Figures 24 and 25.

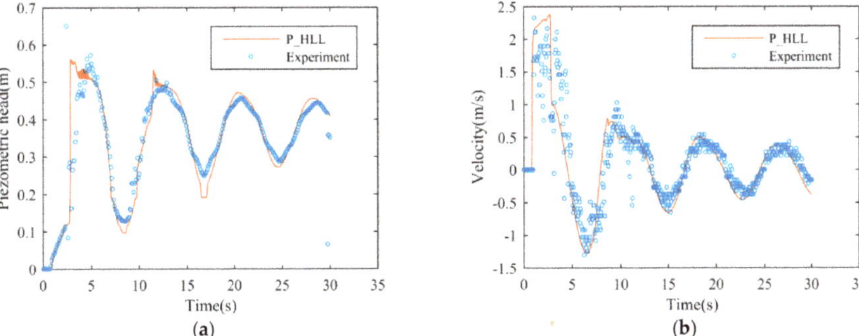

Figure 24. Locus of flow states at $x = 6.8$ m simulated by P_HLL ($P_a = 4$ and $P_b = 0.7$) and experimental data: (**a**) piezometric head; (**b**) flow velocity.

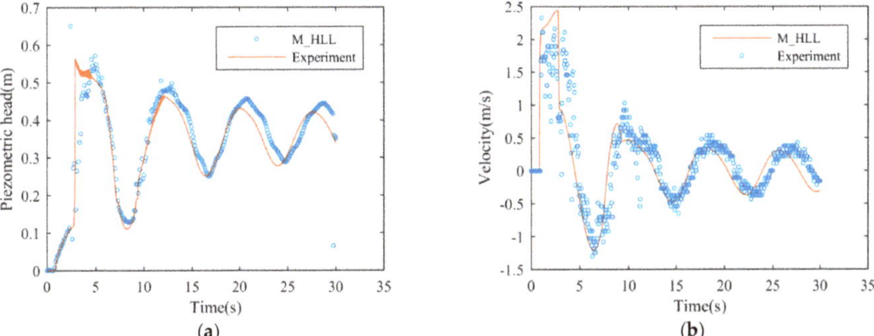

Figure 25. Locus of flow states at $x = 6.8$ m simulated by M_HLL ($K_a = 1.4$ and $NS = 5$) and experimental data: (**a**) piezometric head; (**b**) flow velocity.

The simulation results of P_HLL and M_HLL show certain discrepancies: M_HLL overestimates the wave speed. It is because P_HLL adds numerical viscosity only if the depth is higher than the threshold value of $P_b H$, while M_HLL adds numerical viscosity at any free-surface cells. Nonetheless, the simulation results of the two solvers are in good agreement with the experimental data.

6. Conclusions

Numerical oscillation is a critical problem in transient mixed flow simulations. These numerical oscillations arise from the ambiguity about the location of the filling-bore within one computational cell, which cannot be captured even with high-order finite volume methods. First order finite volume methods have failed to suppress them while capturing the filling-bore front.

Four oscillation-suppressing methods were tested on a benchmark model, with three of them failing to get a satisfactory result under a high acoustic wave speed. The key is to admit numerical viscosity before the flow regime transition occurs. Numerical viscosity can be added by artificially increasing the magnitude of the wave speed in the HLL solver. Following this idea, this paper presents a P_HLL solver that has two parameters: P_a and P_b. P_a needs to be larger than the highest piezometric head while P_b needs to be between 0.7 and 0.9. This solver adds numerical viscosity when the water depth is above P_bH so that a smooth transition between the free-surface and pressurized flows can be achieved. The amount of numerical viscosity increases with P_a, while a large P_a does not smear the simulation result significantly. Therefore, one can always start by adopting a P_a that is large enough under realistic applications, for example 10.

The P_HLL solver is tested in several numerical tests, in which a good agreement between the simulation results and analytical results or experimental data is found. In the test where the wet/dry interface is presented, the P_HLL solver achieves a more accurate result than M_HLL. Meanwhile, P_HLL solver has sufficiently suppressed the numerical oscillations, and accurately captured the propagation of the filling-bore. Compared to the M_HLL solver proposed by Malekpour and Karney, the P_HLL is more convenient to use as the parameters in this solver are easier to determine.

The result in this paper can provide useful information for readers to design an oscillation-suppressing method. It provides an alternative method to the M_HLL solver, which may be used in the parallelization of computation.

This method is proposed under the framework of PSM; it is limited to the simulation of flow regime transition. Besides, since the air pressure is not counted, this method cannot be applied to simulate flow regime transition where air pockets are present, which is out of the scope of this paper.

Author Contributions: Conceptualization, Z.M., G.G., Z.Y.; resources, Z.M., G.G., Z.Y.; writing—original draft, Z.M.; writing—review and editing, Z.M. All authors have read and agreed to the published version of the manuscript.

Funding: The authors acknowledge the support of the NSFC grant 51979202 and NSFC grant 51879199.

Conflicts of Interest: The authors declare no conflict of interest.

References

1. Cong, J.; Chan, S.N.; Lee, J.H.W. Geyser formation by release of entrapped air from horizontal pipe into vertical shaft. *J. Hydraul. Eng.* **2017**, *143*, 04017039. [CrossRef]
2. Liu, L.J.; Shao, W.Y.; Zhu, D.Z. Experimental study on stormwater geyser in vertical shaft above junction chamber. *J. Hydraul. Eng.* **2020**, *146*. [CrossRef]
3. Chegini, T.; Leon, A.S. Numerical investigation of field-scale geysers in a vertical shaft. *J. Hydraul. Res.* **2019**. [CrossRef]
4. Cataño-Lopera, Y.A.; Tokyay, T.E.; Martin, J.E.; Schmidt, A.R.; Lanyon, R.; Fitzpatrick, K.; Scalise, C.F.; García, M.H. Modeling of a transient event in the tunnel and reservoir plan system in Chicago, Illinois. *J. Hydraul. Eng.* **2014**, *140*, 05014005. [CrossRef]
5. Fuamba, M.; Bouaanani, N.; Marche, C. Modeling of dam break wave propagation in a partially ice-covered channel. *Adv. Water Resour.* **2007**, *30*, 2499–2510. [CrossRef]
6. Holly, F.M.; Merkley, G.P. Unique problems in modeling irrigation canals. *J. Irrig. Drain. Eng.* **1993**, *119*, 656–662. [CrossRef]
7. Guo, Q.; Song, C.C.S. Surging in urban storm drainage systems. *J. Hydraul. Eng.* **1990**, *116*, 1523–1537. [CrossRef]
8. Liu, G.Q.; Guan, G.H.; Wang, C.D. Transition mode of long distance water delivery project before freezing in winter. *J. Hydroinformatics* **2013**, *15*, 306–320. [CrossRef]
9. Mao, Z.H.; Guan, G.H.; Yang, Z.H.; Zhong, K. Linear model of water movements for large-scale inverted siphon in water distribution system. *J. Hydroinformatics* **2019**, *21*, 1048–1063. [CrossRef]
10. Guan, G.; Clemmens, A.J.; Kacerek, T.F.; Wahlin, B.T. Applying water-level difference control to central arizona project. *J. Irrig. Drain. Eng.* **2011**, *137*, 747–753. [CrossRef]
11. Cunge, J.A.; Wegner, M. Intégration numérique des équations d'écoulement de barré de Saint-Venant par un schéma implicite de différences finies. *Houille Blanche Rev. Int. De L Eau* **1964**, *1*, 33–39. [CrossRef]
12. An, H.; Lee, S.; Noh, S.J.; Kim, Y.; Noh, J. Hybrid numerical scheme of preissmann slot model for transient mixed flows. *Water* **2018**, *10*, 899. [CrossRef]
13. Dazzi, S.; Maranzoni, A.; Mignosa, P. Local time stepping applied to mixed flow modelling. *J. Hydraul. Res.* **2016**, *54*, 145–157. [CrossRef]

14. Fernandez-Pato, J.; Garcia-Navarro, P. A pipe network simulation model with dynamic transition between free surface and pressurized flow. In *12th International Conference on Computing and Control for the Water Industry, Ccwi2013, Perugia, Italy, 2–4 September 2013*; Brunone, B., Giustolisi, O., Ferrante, M., Laucelli, D., Meniconi, S., Berardi, L., Campisano, A., Eds.; Elsevier Science Bv: Amsterdam, The Netherlands, 2014; Volume 70, pp. 641–650.

15. Kerger, F.; Erpicum, S.; Dewals, B.J.; Archambeau, P.; Pirotton, M. 1D unified mathematical model for environmental flow applied to steady aerated mixed flows. *Adv. Eng. Softw.* **2011**, *42*, 660–670. [CrossRef]

16. Ferreri, G.B.; Freni, G.; Tomaselli, P. Ability of preissmann slot scheme to simulate smooth pressurisation transient in sewers. *Water Sci. Technol.* **2010**, *62*, 1848–1858. [CrossRef] [PubMed]

17. Maranzoni, A.; Mignosa, P. Numerical treatment of a discontinuous top surface in 2D shallow water mixed flow modeling. *Int. J. Numer. Methods Fluids* **2018**, *86*, 290–311. [CrossRef]

18. Maranzoni, A.; Dazzi, S.; Aureli, F.; Mignosa, P. Extension and application of the Preissmann slot model to 2D transient mixed flows. *Adv. Water Resour.* **2015**, *82*, 70–82. [CrossRef]

19. Toro, E.F.; Garcia-Navarro, P. Godunov-type methods for free-surface shallow flows: A review. *J. Hydraul. Res.* **2007**, *45*, 736–751. [CrossRef]

20. Kitamura, K.; Shima, E. Numerical experiments on anomalies from stationary, slowly moving, and fast-moving shocks. *AIAA J.* **2019**, *57*, 1763–1772. [CrossRef]

21. Johnsen, E. Analysis of numerical errors generated by slowly moving shock waves. *AIAA J.* **2013**, *51*, 1269–1274. [CrossRef]

22. Zaide, D.W.; Roe, P.L. Flux functions for reducing numerical shockwave anomalies. In Proceedings of the Seventh International Conference on Computational Fluid Dynamics, Big Island, Hawaii, 9–13 July 2012.

23. Vasconcelos, J.G.; Wright, S.J.; Roe, P.L. Numerical oscillations in pipe-filling bore predictions by shock-capturing models. *J. Hydraul. Eng.* **2009**, *135*, 296–305. [CrossRef]

24. Malekpour, A.; Karney, B.W. Spurious numerical oscillations in the preissmann slot method: Origin and suppression. *J. Hydraul. Eng.* **2016**, *142*, 04015060. [CrossRef]

25. Aureli, F.; Dazzi, S.; Maranzoni, A.; Mignosa, P. Validation of single- and two-equation models for transient mixed flows: A laboratory test case. *J. Hydraul. Res.* **2015**, *53*, 440–451. [CrossRef]

26. Kerger, F.; Archambeau, P.; Erpicum, S.; Dewals, B.J.; Pirotton, M. An exact Riemann solver and a Godunov scheme for simulating highly transient mixed flows. *J. Comput. Appl. Math.* **2011**, *235*, 2030–2040. [CrossRef]

27. Leon, A.S.; Ghidaoui, M.S.; Schmidt, A.R.; Garcia, M.H. Application of Godunov-type schemes to transient mixed flows. *J. Hydraul. Res.* **2009**, *47*, 147–156. [CrossRef]

28. Vasconcelos, J.G.; Wright, S.J.; Roe, P.L. Improved simulation of flow regime transition in sewers: Two-component pressure approach. *J. Hydraul. Eng.* **2006**, *132*, 553–562. [CrossRef]

29. Kim, S.S.; Kim, C.; Rho, O.H.; Hong, S.K. Cures for the shock instability: Development of a shock-stable Roe scheme. *J. Comput. Phys.* **2003**, *185*, 342–374. [CrossRef]

30. Toro, E.F.; Billett, S.J. Centred TVD schemes for hyperbolic conservation laws. *IMA J. Numer. Anal.* **2000**, *20*, 47–79. [CrossRef]

31. Toro, E. *Shock Capturing Methods for Free Surface Shallow Water Flows*; John Wiley: New York, NY, USA, 2001.

32. Leon, A.S.; Ghidaoui, M.S.; Schmidt, A.R.; Garcia, M.H. Godunov-type solutions for transient flows in sewers. *J. Hydraul. Eng.-Asce* **2006**, *132*, 800–813. [CrossRef]

33. Toro, E.F.; Hidalgo, A.; Dumbser, M. FORCE schemes on unstructured meshes I: Conservative hyperbolic systems. *J. Comput. Phys.* **2009**, *228*, 3368–3389. [CrossRef]

34. Bertaglia, G.; Ioriatti, M.; Valiani, A.; Dumbser, M.; Caleffi, V. Numerical methods for hydraulic transients in visco-elastic pipes. *J. Fluids Struct.* **2018**, *81*, 230–254. [CrossRef]

35. Vasconcelos, J.; Wright, S. Experimental Investigation of Surges in a Stormwater Storage Tunnel. *J. Hydraul. Eng.* **2005**, *131*, 853–861. [CrossRef]

36. Vasconcelos, J.; Wright, S. Numerical Modeling of the Transition between Free Surface and Pressurized Flow in Storm Sewers. *J. Water Manag. Model.* **2004**. [CrossRef]

Article

A Stochastic Model to Predict Flow, Nutrient and Temperature Changes in a Sewer under Water Conservation Scenarios

Olivia Bailey [1], Ljiljana Zlatanovic [2,3], Jan Peter van der Hoek [2,3,4], Zoran Kapelan [2,5], Mirjam Blokker [6], Tom Arnot [1] and Jan Hofman [1,6,*]

[1] Water Innovation & Research Centre, Department of Chemical Engineering, University of Bath, Claverton Down, Bath BA2 7AY, UK; ob219@bath.ac.uk (O.B.); t.c.arnot@bath.ac.uk (T.A.)

[2] Department of Water Management, Faculty of Civil Engineering and Geosciences, Delft University of Technology, P.O. Box 5, 2600 AA Delft, The Netherlands; l.zlatanovic-1@tudelft.nl (L.Z.); j.p.vanderhoek@tudelft.nl (J.P.v.d.H.); z.kapelan@tudelft.nl (Z.K.)

[3] Amsterdam Institute for Advanced Metropolitan Solutions, Kattenburgerstraat 5, 1018 JA Amsterdam, The Netherlands

[4] Waternet, P.O. Box 94370, 1090 GJ Amsterdam, The Netherlands

[5] College of Engineering, Mathematics & Physical Sciences, University of Exeter, Harrison Building, North Park Road, Exeter EX4 4QF, UK

[6] KWR Water Research Institute, P.O. Box 1072, 3430 BB Nieuwegein, The Netherlands; Mirjam.Blokker@kwrwater.nl

* Correspondence: j.a.h.hofman@bath.ac.uk

Received: 24 March 2020; Accepted: 15 April 2020; Published: 21 April 2020

Abstract: Reducing water use could impact existing sewer systems but this is not currently well understood. This work describes a new flow and wastewater quality model developed to investigate this impact. SIMDEUM WW® was used to generate stochastic appliance-specific discharge profiles for wastewater flow and concentration, which were fed into InfoWorks® ICM to quantify the impacts within the sewer network. The model was validated using measured field data from a sewer system in Amsterdam serving 418 households. Wastewater concentrations of total suspended solids (TSS), chemical oxygen demand (COD), total Kjeldahl nitrogen (TKN) and total phosphorus (TPH) were sampled on an hourly basis, for one week. The results obtained showed that the InfoWorks® model predicted the mass flow of pollutants well (R-values 0.69, 0.72 and 0.75 for COD, TKN and TPH respectively) but, due to the current lack of a time-varying solids transport model within InfoWorks®, the prediction for wastewater concentration parameters was less reliable. Still, the model was deemed capable of analysing the effects of three water conservation strategies (greywater reuse, rainwater harvesting and water-saving appliances) on flow, nutrient concentrations, and temperature in sewer networks. Results show through a 62% reduction in sewer flow, COD, TKN and TPH concentrations increased by up to 111%, 84% and 75% respectively, offering more favourable conditions for nutrient recovery.

Keywords: sewer design; stochastic sewer modelling; wastewater quality; household discharge; reduced water consumption

---

1. Introduction

Contemporary water cycle infrastructure has typically been developed to promote public health and safety by supplying wholesome drinking water and by transporting wastewater and stormwater out of urban areas as quickly as possible. This has led to linear water use (take, use, throwaway) that is sub-optimal on grounds of sustainability. With growing environmental awareness, the idea of a circular economy has emerged, and a paradigm shift is required to close the water cycle and re-classify wastes as resources to recover and reuse. Resource recovery from wastewater is more effective at high concentrations. This can be achieved through dewatering processes at treatment plants [1–3] but another option is to limit wastewater dilution in the collection process [4]. Limiting wastewater dilution can be achieved by reducing domestic drinking water use, separation of storm/wastewater systems and preventing groundwater inflow by repairing/replacing broken pipes. This reduces nutrient loss from the cycle whilst reduced drinking water demand and wastewater transportation volume could save cost by reducing demands on existing infrastructure. Transporting more concentrated flow with a smaller pipe/equipment size requirement is also facilitated. Urban water cycles could enable resource recovery if considered from this new value proposition. This philosophy has prompted the development of a water cycle model to investigate the effects of future water use behaviours on the urban water system, and ultimately highlight how these systems could deliver enhanced resource recovery. This paper describes the development of a stochastic wastewater quality model and the comparison of this model to monitored field data. The sewer model forms part of a wider aim to develop an integrated water cycle model using a combination of SIMDEUM® and InfoWorks® WS/ICM packages. The integrated model will predict flow and wastewater quality changes in both drinking water and wastewater infrastructures, to evaluate the consequences of future water use scenarios.

Water demand and water quality models can be developed as deterministic or stochastic models. In a deterministic model, the results are fully based on pre-set parameter values and initial conditions. Stochastic models will include randomness and each time the model is used it will produce a different output. The advantage of deterministic models is the relative ease of use, whilst stochastic models will provide better insight in the system's dynamics. Because water use at the household level is extremely dynamic and follows random patterns, we have chosen to use a stochastic approach for this project as it gives a better reflection of reality.

A number of models have been developed to predict the impacts of various water conservation measures on the sewer system. These models have been largely deterministic [5–7] and have tested specific impacts of rainwater harvesting (RWH) and greywater reuse (GWR) on wastewater quality. Penn et al. [7] reported pollutant concentration increases of 6–42% COD, 7–73% TSS, 9–57% NH_4-N and 7–52% PO_4-P for flow decreases of 8–41%. However, these deterministic approaches model domestic wastewater production as a continuous discharge based on averaged data, assuming an identical water use pattern for all residents. In reality, individual household wastewater profiles are a discontinuous series of discrete points, and hence a stochastic model is needed to model household discharges which are more representative of this reality. Penn, et al. [8] published a stochastic wastewater generator that does not require a great amount of input data, but which is based on empirical sampling, and assumes that the observed flow data (from 15 households) represents the flow of the target population. The flow generator was used as an input to a network model that assessed ability of flow to move gross solids (GS) in the sewer. GS movement was assessed through calculating critical flow required to move solids, but this does not link solids/pollutant generation to the discharges themselves. If we are to model water use changes that have not yet been observed, a model based on deterministic methods or empirical sampling is insufficient. There is therefore need for a stochastic sewer model that is independent of observed data for predicting impacts of changing water use. To our knowledge there is currently no sewer model that links unique appliance-discharge patterns to the specific water quality attributes produced by household appliances. Developing a model with this capability will offer a better understanding of how and when pollutants/nutrients build up in sewers, and how various water use changes could affect this in future.

This paper utilises a more complex stochastic generator than that developed by Penn et al. [8]. This tool, SIMDEUM® [9], generates appliance-specific flow patterns based on probability parameters linked to appliance usage, household composition, and consumer water use behaviour [10]. Patterns produced by SIMDEUM® are specific to each appliance (e.g., toilet, sink and washing machine) which makes it possible to investigate explicit water use changes without assuming typical water usage patterns based on historical data. SIMDEUM WW® extends from SIMDEUM® to convert demand patterns into wastewater discharges, including thermal and nutrient loads [11]. This conversion is achieved through correcting the flow rate or delaying the time of discharge, e.g., toilets can take minutes to fill but seconds to discharge. Thermal and nutrient loads from each appliance are incorporated into the discharge profile by assigning typical (per use) load to each appliance.

Bailey et al. [12] developed a stochastic flow model to assess the impact of water conservation on the sewer. This model utilised stochastic household discharge patterns (generated with SIMDEUM WW®) as input to a sewer network model based in InfoWorks® ICM. The flow model was validated using data from an English catchment, provided by Wessex Water (UK-based water utility). The flow model was extended to include wastewater pollutant concentrations by linking typical wastewater quality data to appliance-specific discharges within SIMDEUM WW® and utilising the InfoWorks® ICM wastewater quality model [13]. The flow/quality model was used to simulate and compare a series of future water use scenarios. The wastewater quality aspect of this model, however, has not previously been compared to field data to assess its validity. This paper details a wastewater quality monitoring campaign conducted in a small housing estate in Amsterdam with that objective.

The paper is organised as follows: firstly, we describe the model development and the methodology behind the wastewater quality monitoring campaign. Followed by the framing of six future water use scenarios that were tested using the model. Then, a description of the Amsterdam-based catchment used to analyse the model precedes the model predictions and a comparison of modelled parameters with the measured data. Finally, we make key conclusions.

2. Methodology

A model was developed to simulate the effects of future water use scenarios in sewers. The Infoworks® ICM (Sewer Edition; Innovyze Ltd., Oxfordshire, UK) hydraulic and wastewater quality model was used to simulate the sewage system. This model was integrated with stochastic discharge patterns generated using SIMDEUM® and SIMDEUM WW® [10,14]. The MATLAB® codes behind SIMDEUM® were edited to make its outputs compatible with InfoWorks® ICM. Six future water use scenarios were framed and simulated using the validated model, allowing flow and concentration effects to be evaluated.

Infoworks® ICM Sewer Edition is an industry standard for 1-dimensional sewer network modelling. The software offers accurate analysis of hydraulics and water quality in sewer and stormwater networks. The model uses a network of nodes and conduits and solves the flow and mass balances for the network, based on water quantity and quality input, fed into the model via the nodes. The geometry of the network and the shape of the conduits is defined by geographical input and data from the real network.

2.1. Household Discharge Modelling

2.1.1. Hydraulic Discharge Model

The SIMDEUM® software tool was developed in the Netherlands for accurate water demand modelling. It can generate household water demand patterns based on statistical and probabilistic information about inhabitants and their appliance usage [10]. The SIMDEUM® pattern generator was calibrated for use in the studied catchment (Prinseneiland), which is described in Section 2.4.1, details of the studied catchment are shown in Section 3.

2.1.2. Wastewater Quality Loading

SIMDEUM WW® was used to link each wastewater discharge with an appliance-specific wastewater quality profile. SIMDEUM WW® originally included very little detail on pollutant discharges, having been used simply to demonstrate the possibility of nutrient discharge modelling [11,15]. Therefore, a review of relevant literature [5,15–19] was conducted to find appropriate input values for nutrient simulation.

These input parameters describe pollutant mass per discharge for each household appliance (see Table 1), and the derivation of these parameters is described in Bailey et al. [13]. The nutrient discharge aspect of SIMDEUM WW® has never been validated. Through comparison of the wastewater quality model with measured data from this work, the phosphorus (TPH) parameters reported in literature were found to be too high. This is due to recent changes in EU legislation reducing phosphorus use in detergents [20]. The phosphorus parameters were corrected to align with this legislation and are highlighted in bold in Table 1. The phosphorus associated with the kitchen tap was approximated as in Comber et al. [21] where it was found to be 0.03 g person^{-1} day^{-1}. It was assumed that this much phosphorus enters the sewer through the disposal of food scraps. The other value shown in Table 1, i.e., 0.03 g use^{-1}, which depicts quality profile for each discharge, was found by calibration based on observed wastewater data and above assumed phosphorus value. The phosphorus from toilet use was updated in accordance with Comber et al. [21], and assuming, on average, six toilet uses per person, per day.

Quality of non-potable water sources was quantified using data from Penn et al. [6] (greywater) also Ward et al. [22] and Farreny et al. [23] (rainwater)—see Supplementary Information. This was combined with appliance pollutant quantities, shown in Table 1.

Table 1. Appliance-specific pollutant concentrations for improved SIMDEUM WW® (adapted from Bailey et al. [13]). Bold values were defined in this work using observed wastewater data.

Appliance	Temperature (°C)	Sewage Quality (g use^{-1})				Ref.
		COD	TKN	TPH	TSS	
Bath	36	25.90	0.85	**0.00**	8.88	[5,16]
Shower	35	12.60	0.49	**0.00**	4.32	[5,16]
Bathroom tap	40	1.48	0.04	**0.00**	0.56	[5,16]
Kitchen tap	40	7.48	0.35	**0.03**	4.68	[5,16,21]
Dish washer	35	30	1.35	**0.00**	13.20	[5,16]
Washing machine		65.25	0.638	**0.00**	17.10	[5,16]
- With GWR	(35, 35, 35, 45)	69.40	0.78	**0.00**	17.88	[6]
- With RWH		66.29	0.86	**0.00**	17.72	[22]
Toilet		11.22	1.99	**0.22**	3.04	[15,21]
- With GWR	23	11.48	2.00	**0.22**	3.09	[6]
- With RWH		11.28	2.00	**0.22**	3.08	[22]

2.2. Stochastic Sewer Model

Wastewater flow and quality were simulated through a sewer network using InfoWorks® ICM (Sewer Edition; Innovyze Ltd., Oxfordshire, UK). Stochastic household discharge patterns, described in Section 2.1, were imported into InfoWorks® ICM to produce time-varying domestic wastewater event. Each property has a unique flow and associated wastewater concentration profile as input to the sewer; discharges were input with one-minute intervals.

InfoWorks® ICM incorporates both hydraulic and wastewater quality modelling components. The hydraulic component was validated by Bailey et al. [12] using measured flow, depth and velocity data. Saint-Venant equations govern hydraulics in InfoWorks® ICM. The wastewater quality model runs parallel to the hydraulic model, as described in Bailey et al. [13], but was not validated. The concentration of dissolved pollutants and suspended sediment at every node in the sewer network is calculated for every time step using the InfoWorks® *Network Model*. The governing equation at a node is given by conservation of mass, Equation (1). Pollutant inflows arrive from incoming conduits and any external sources, in this case, wastewater events (household discharges). It is assumed that nodes are well-mixed and there is no deposition or accumulation.

$$\frac{dM_J}{dt} = \sum_i Q_i c_i + \frac{dM_{sJ}}{dt} - \sum_o Q_o c_o \tag{1}$$

where:

M_J = mass of suspended sediment or dissolved pollutant in node J (kg)
Q_i = flow into node J from link i (m^3 s^{-1})
c_i = concentration in the flow into node J from link i (kg m^{-3})
M_{sJ} = additional mass entering node J from external sources (kg)
Q_o = flow from node J to link o (m^3 s^{-1})
c_o = concentration in the flow from node J to link o (kg m^{-3})

The InfoWorks® *Conduit Model* then calculates the concentration of dissolved pollutants and suspended sediment in each conduit. A conduit is a conceptual link of defined length between two nodes. One-dimensional flow is assumed in the conduit, as are well-mixed concentrations across each section of the conduit. Pollutants are assumed move through the conduit with the local mean flow velocity, and dispersion along the conduit is negligible. Wastewater determinants were all treated as dissolved pollutants because InfoWorks® ICM software fails to recognise time-varying suspended solid input. The authors have been advised that this shortfall will be corrected in a future software update. Therefore, wastewater determinants in the model are transported through advection, with no erosion, deposition, or accumulation of sediments. The advective mass flow between each element is shown in Equation (2).

$$F_a = F_m \times c_{upwind} \tag{2}$$

where:

F_a = mass flow through the face due to advection (kg s^{-1})
F_m = volumetric flow through the face (m^3 s^{-1})
c_{upwind} = c_l if volumetric flow goes from left to right element, c_r otherwise (kg m^{-3}); c_l, c_r = determinant concentration in respectively the left and right element

Adjusting to Allow for Mixing in the Sampling Tank

The sampling campaign, described in Section 2.3.2, generated data on wastewater in the pump feed tank rather than wastewater flowing in the sewer system (see Figure 1). As the sewage flows into the tank it mixes with the held-up water and thus the samples will reflect a dampened wastewater concentration compared to model predictions. The sewer model output was adjusted to allow for this mixing to allow comparison of model predictions with sampled concentration data. Equation (3) is the derived expression for concentration in the tank (C_A), assuming the volume remains approximately constant (average volume of 1.6 m^3, midway between high and low levels). It also assumes that no reactions occur in the tank and the wastewater has a constant density.

$$C_A(t) = (C_{A,in}(t) - C_{A,o})\left(1 - e^{\left(-\frac{Q(t)}{V}t\right)}\right) \tag{3}$$

where:

C_A = Concentration of pollutant A in the tank (kg m^{-3})
$C_{A,\,in}$ = Concentration of pollutant A into the tank (kg m^{-3})
$C_{A,o}$ = Initial concentration of pollutant A (kg m^{-3})
Q = Flowrate into tank (m^3 s^{-1})
V = Tank volume (m^3)

t = Time (s)

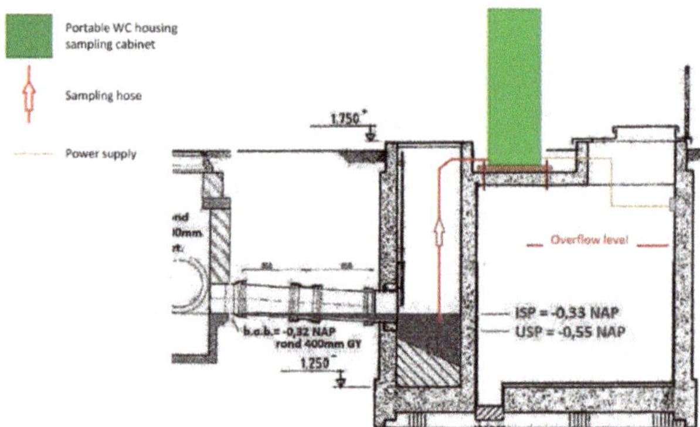

Figure 1. Wastewater sampling campaign equipment set up. Portable toilet housing the sampling cabinet that draws wastewater from the wet well of the pumping station in Prinseneiland. ISP is the level at which the pump switches on, USP is the level where the pump switches off. The tank area is 2 m^2.

2.3. Methodology for Field Testing

2.3.1. Data Availability for Validating the Hydraulic Discharge Model

The Prinseneiland catchment (See Section 3.1) has three sources of hydraulic water network data. Two water mains supply drinking water to the island; a flow meter was present in each, providing live data recording of water demand. Fifty-eight percent of catchment households have a water meter recording specific water use, but this is mainly for billing purposes as data is summed over the period between physical meter readings. The final data source was provided by pump flow and tank level readings, recorded at the wastewater pumping station. Readings are recorded every 2–5 min dependant on changes recorded by the level controller. A pump switches on when the tank level reaches the programmed high level (above the inlet pipe) and off when the level reaches the programmed low level (above the pump). The volumetric flowrate through the pump was measured using an ECOFLUX electromagnetic flowmeter (www.krohne.com) (accuracy ± 0.5% of the measured value at velocities ≥ 0.4 m s^{-1} and ± 0.002 m s^{-1} if velocity is below 0.4 m s^{-1}). The tank level was measured using two VEGABAR 52 (www.vega.com) sensors, where the deviation is reported to be less than 0.075%. By performing a mass balance on the flow through the pump and the changing level in the tank (Equation (4)), it was possible to convert these readings into a sewer flow profile (Equation (5)).

$$V_{[\tau_n,\tau_{n+1}]} = PC_{(n)}\left(\tau_{(n+1)} - \tau_{(n)}\right) + \frac{A(LS_{n+1} - LS_n)_{S1} + A(LS_{n+1} - LS_n)_{S2}}{2} \tag{4}$$

$$Q_t = \frac{\sum_{t}^{n=0} V_{[\tau_n,\tau_{n+1}]}}{t} \tag{5}$$

where:

$V_{[\tau_n,\,\tau_{n+1}]}$ = Volume entering the tank between level sensor readings (m^3)
PC = Pumping capacity (m^3 s^{-1})
LS = Tank level (m)
A = Tank area (m^2)
$S1, S2$ = Level sensors
τ = Sample time (s)
Q_t = Wastewater flowrate into the tank (m^3 s^{-1})

t = Time (s)

At the end of August 2019, a wastewater quality campaign was carried out on Prinseneiland to collect data necessary for validating the wastewater quality component of the stochastic sewer model. The campaign was conducted continuously over 7 days, under dry weather conditions. Wastewater was sampled from the pump wet well at the end of the catchment. All Water Services (www.aws-water.nl) carried out the fieldwork and the wastewater samples were analysed by Eurofins Omegam. A vacuum sampling device was used (photographs in the Supplementary Information). The sampling cabinet was placed within a portable toilet at street level to comply with space constraints and protect apparatus from damage. The sampling hose was secured at the sewer inlet to the wet well in such a way that the end of the hose was approximately 3 cm below the cut-off level of the pump. This ensured that the wastewater was as "fresh" as possible when sampled from the tank, and thus most representative of the sewer flow. This method meant it was always possible to draw samples from the chamber, but during the night where wastewater flow is low, there is the possibility that stagnant wastewater is sampled. The sampling cabinet contained 24 1 L bottles into which a 50 mL sub-sample was drawn every 3 min, i.e., 20 sub-samples per hour make up the 1 L sample for that hour. The sample collection vessels were held at 1–5 °C. Sampling was carried out according to Dutch standard 'NEN 6600-1 (NL) Water—Sampling—Part 1: Waste water from March 2009. Every 24 h the completed samples were removed from the cabinet and decanted into three separate packages for separate analysis (see Table 2), and nitrogen and phosphorus were analysed from the same package. Samples were preserved on site according to Dutch standard 'NEN-EN-ISO 5667-3 (s) Water—Sampling—Part 3: Conservation and treatment of water samples' and were delivered daily to the analysis laboratory under cooling.

Table 2. Wastewater quality parameters analysed and specific methodology associated with each parameter.

Parameter Sampled	Parameter Description	Method (Eurofins Omegam)	Limit of Determination (mg l^{-1})	Required Sample Volume (ml Sample^{-1})	Measurement Uncertainty (+/−)
COD (mg l^{-1})	Chemical oxygen demand	Conforms to NEN 6633	5.00	100	15%
TKN (mg l^{-1})	Total Nitrogen-Kjeldahl	Conforms to NEN-ISO 5663	1.00	100	13%
TPH (mg l^{-1})	Total Phosphorus	Own method based on NEN-EN-ISO 15681_2	0.05	50	12%
TSS (mg l^{-1})	Total suspended solids	Conforms to NEN-EN 872 and NEN 6499	1.00	750	16%

2.3.2. Quality of Sampling and Analysis Work

AWS are accredited according to the requirements as laid down in NEN-EN-ISO/IEC 17025: 2005 and Dutch Accreditation Council (RvA) regulations under number L599. Eurofins Omegam laboratory in Amsterdam (who carried out the sample analysis) is also accredited by RvA.

2.3.3. Wastewater Quality Parameters

The parameters analysed and the procedures followed by the laboratory are shown in Table 2.

2.4. Model Validation

2.4.1. Procedure for Model Calibration

The SIMDEUM® model was calibrated by adjusting input variables describing household occupancy, home–presence, and specific details of household water use in the area. Households are characterised as either a single, dual, or family occupancy. Average occupancy and family size are also defined. The household data was derived from census data from the local government of the study area. Home presence data is culture and area-specific, and details typical times that people rise, go to work and go to bed. These data were obtained from the Netherlands Institute for Social Research (SCP) that conducts a five-year time-budget survey. Comparison of the model output with monitored catchment data showed a local deviation from the national survey data on wake-up time, so this was adjusted on a case-specific basis. Household water use data is available from local water companies and should be input to the model to describe typical water use for each household appliance. The specific model adaptions made for the studied catchment are detailed in Section 3.2.

2.4.2. Procedure for Model Validation

Validation of the model was conducted by assessing the model performance over an average week. Dry weather flow data was selected at various points of the year (2 weeks from each season) to produce an average water use pattern of the catchment in order to compare with the model. The goodness of fit of model output was evaluated by computation of the Nash–Sutcliffe efficiency (*NSE*) and the root mean squared error (*RMSE*). The similarity of the flow patterns was evaluated with the correlation coefficient (*R*). The equations for *NSE*, *RMSE* and *R* are found below in Equations (6–8).

$$NSE = 1 - \frac{\sum_{i=1}^{n}(x_{obs} - x_{sim})^2}{\sum_{i=1}^{n}(x_{obs} - \bar{x})^2} \tag{6}$$

$$RMSE = \sqrt{\frac{1}{n-1}\sum_{i=1}^{n}(x_{obs} - x_{sim})^2} \tag{7}$$

$$R\ (X,\ Y) = \frac{\sum(x - \bar{x})(y - \bar{y})}{\sqrt{\sum(x - \bar{x})^2 \sum(y - \bar{y})^2}} \tag{8}$$

where:

x_{obs} = Observed parameter
x_{sim} = Simulated parameter
$\bar{x}, \bar{y}$ = Sample mean of parameters x, y

2.5. Impact Assessment for Water Conservation Technologies

The development and validation of the sewer model allow it to be used to predict the effect of future scenarios. Table 3 describes the future scenarios that were developed for testing in the Prinseneiland catchment. These scenarios were based on total area reform (100% implementation). Water use scenarios include "Eco", which involves an upgrade of household appliances (such as 1 L flush toilets and water-saving showers) and 'GWR'/'RWH', which utilise greywater or rainwater feed for toilet flushing and washing machines. Greywater and rainwater feed quality data are found in the Supplementary Material. Each scenario has been presented using future population statistics supplied by the Municipality of Amsterdam (Gemeente Amsterdam), as outlined in Table 4. The '(a)' scenarios are the maximum bound for occupation in the catchment, and the "(b)" scenarios explore the effect of a continued rise in single occupancy households, thus provides a minimum occupancy bound.

Table 3. Future scenario description.

Scenario	Demand (L cap^{-1} d^{-1})	Description
1—Baseline	112	Present-day scenario—validated hydraulic model
2a—Eco, max. occupancy	42	Water-saving appliances such as 1 L flush toilets and water-saving showers (as presented by Agudelo and Blokker [24])
2b—Eco, min. occupancy	44	
3a—GWR, max. occupancy	67	Greywater reuse utilised for toilet flushing and washing machines
3b—GWR, min. occupancy	68	
4a—RWH, max. occupancy	67	Rainwater harvesting utilised for toilet flushing and washing machines
4b—RWH, min. occupancy	68	

Table 4. Population statistics for present and future scenarios (based data and projections obtained from Gemeente Amsterdam).

	Single	Dual	Family	Family Size	Occupancy
Baseline	58%	23%	19%	3.4	1.7
(a) Max.	55%	21%	24%	3.5	1.8
(b) Min.	91%	4%	5%	3.1	1.1

(a) Amsterdam projected population statistics, (b) Reduction in average occupancy to 1.1.

SIMDEUM® generates household discharge patterns based on the specific usage and discharge characteristics of household appliances. Figure 2 shows how these household micro-components vary between the scenarios. Differences in drinking water demand and discharge occur through the use of non-potable water sources (not included in water demand) or outdoor use (does not enter the sewer). In the case of greywater reuse and rainwater harvesting, household appliances were held at baseline water consumption. Water was only redirected to appliances, i.e., no internal mass balance for water movement was incorporated into the model. It is assumed that there will always be sufficient water in a storage tank to allow these appliance discharges.

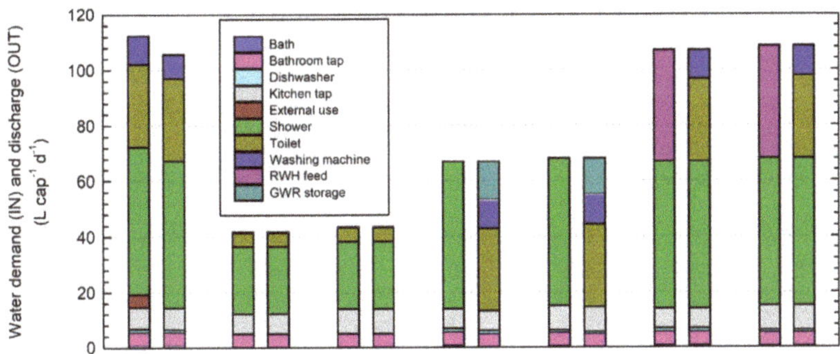

Figure 2. Outline of appliance demand and discharge for each of the future scenarios.

3. Catchment Used for Model Analysis

3.1. Description of the Modelled Catchment

Prinseneiland is a small housing estate located in Amsterdam, which is the capital and most populous municipality of the Netherlands. A map of Prinseneiland is found in Figure 3. There are 418 domestic households and 55 other premises (offices, ateliers, storage buildings) located in the housing estate.

The sewer system is a looped and combined network (i.e., stormwater and wastewater). Concrete sewer pipes, measuring 684 m (400–600 mm diameter and 1:1961 to 1:133 slope, the average slope was 1:615), lead to a pumping station where wastewater is pumped away from the housing estate for treatment. Flow and level monitors at the pumping station provide data for model validation every 2–5 min.

Thirty-second time steps were used in calculations and simulations were conducted for 5 days. Wastewater quality modelling parameters remained as the default with the exception of the temperature model parameters in which the heat transfer coefficient was 4×10^{-5} m s^{-1}, and the equilibrium water temperature was 23 °C, to align with the warm weather at the time of sampling.

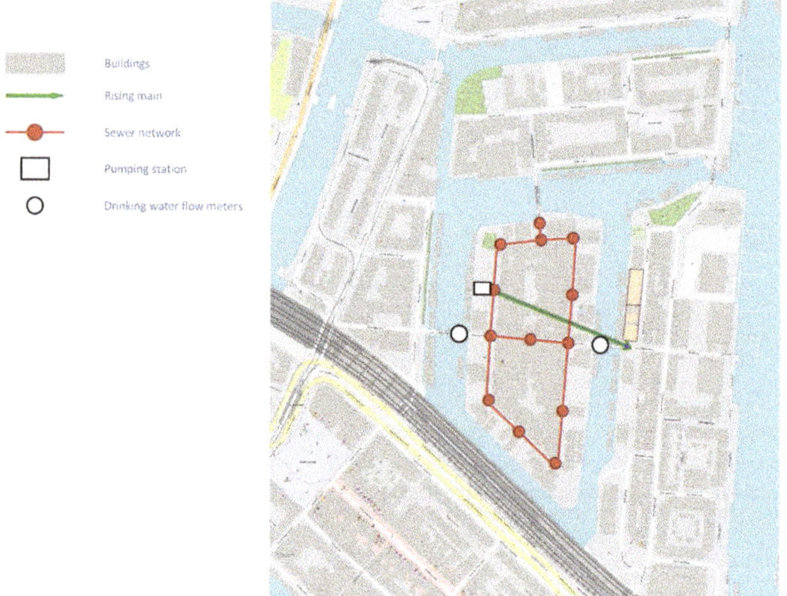

Figure 3. Map of modelled catchment—Prinseneiland, NL (Waternet, Amsterdam).

3.2. Model Calibration Details

The SIMDEUM® model was calibrated by changing input variables describing household occupancy, home–presence data and specific details of household water use in the area. The average household size in Prinseneiland is 1.7 people household^{-1}, where single, dual occupancy and family households are divided 58%, 23% and 19% respectively (see Table 4). This information was put into SIMDEUM® along with the data shown in Figure 4, which details the typical distribution of water use between household appliances (micro-components) on Prinseneiland. The split of water use between appliances was determined by applying a scale factor to the micro-component statistics for the whole of Amsterdam [25], as in Figure 4. Water and wastewater flow into and away from the island were monitored by the local water company, Waternet. The model output was compared with measured demand data from the island, and it was found that inhabitants seemed to rise an hour later than the Dutch average. The home presence schedules were therefore updated to give an average wake up time of 8 am (9 am for stay-at-home adults and seniors).

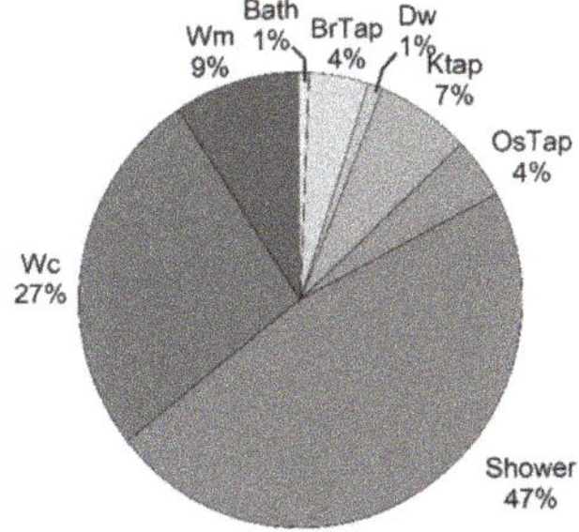

Appliance	Average water use (L cap⁻¹ day⁻¹)	
	Amsterdam	Prinseneiland
Bath	1.0	0.8
Bathroom tap (BrTap)	5.8	4.9
Dishwasher (Dw)	1.4	1.2
Kitchen Tap (Ktap)	9.6	8.1
Outside tap (OsTap)	5.7	4.8
Shower	62.7	52.9
Toilet (Wc)	35.3	29.7
Washing machine (Wm)	12.3	10.3
TOTAL DAILY USAGE	**133.8**	**112.7**

Figure 4. Appliance-specific water use in Amsterdam, Netherlands [25] and the derived appliance usage of Prinseneiland assuming the Amsterdam average micro-component trend.

4. Results and Discussion

4.1. Calibration and Validation of the Stochastic Sewer Flow Model

Figure 5 shows the drinking water flow measured on entrance to the modelled catchment, demonstrating about a one-hour delay between clean water entering the catchment and the sewer flow leaving the catchment. This is due to a combination of time in flow and hold up time of water used in household appliances before discharge. Figure 5 also shows that in the early hours of the morning this delay extends to almost two hours, which is likely due to the longer hold-up derived from increased use of dishwashers and washing machines. The water balance between drinking water and wastewater data in Prinseneiland revealed an average excess of 1.3 m³ day⁻¹ in the wastewater. This excess is likely due to infiltration to the sewer and runoff from the street and represents approximately 2% of the dry-weather flow. This external inflow to the system could also explain some of the difference between drinking water and wastewater flows, particularly at night when flow is low.

Once SIMDEUM® had been calibrated as described in Section 2.4.1, the model represented the sewer system described in Section 3 reasonably well. Comparison of the model output with the sewer flow data can be seen in Figure 6 along with the model evaluation statistics (correlation coefficient, Nash-Sutcliff coefficient and the root mean squared efficiency, *RMSE*).

The model under-predicts the sewer flow during working hours, this is due to the assumption that the housing estate is purely domestic. There is an average discrepancy of 10 m³ between the hours of 10:00 and 18:00, which can be explained by the metered usage of the business premises. Nine percent of the registered properties on Prinseneiland are business addresses and these vary in function from warehouses to offices. These businesses were not modelled as they are not easy to describe well, and this study primarily investigates the impacts of varying water use on domestic wastewater.

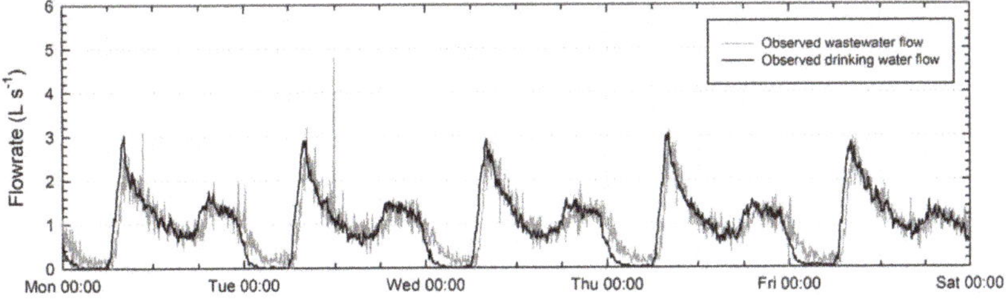

Figure 5. Comparison of the mean drinking water and wastewater flow in the studied catchment.

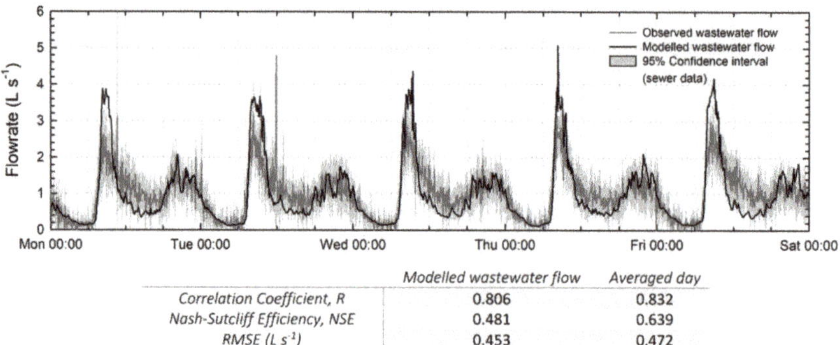

	Modelled wastewater flow	Averaged day
Correlation Coefficient, R	0.806	0.832
Nash-Sutcliff Efficiency, NSE	0.481	0.639
RMSE (L s⁻¹)	0.453	0.472

Figure 6. Performance of stochastic sewer model when compared to measured sewer flow data.

4.2. Sampling Wastewater for Quality Analysis

To confirm that the wastewater quality model provides a good representation of real life, a week-long wastewater sampling campaign was carried out, described in Section 2.4.2. The sampling campaign began on a Thursday at 11 am and ran through until the following Thursday at 11 am. These results have been reordered to represent a Monday–Friday profile for ease of analysis—but it should be noted that the Thursday and Friday measurements were taken the week before the Monday—Wednesday measurements. The weekends have not been modelled due to the limited capacity of SIMDEUM® to describe weekend water use. Weekend water use is less strongly linked to a daily routine and SIMDEUM® has yet to be developed to incorporate this difference. The results of the sampling campaign are shown in Figures 7–9. Figure 7 shows how the measured wastewater flow over the sampling week compared to the measured wastewater flow used to validate the hydraulic model, see Section 4.1. There was heavy rainfall from 20:35 until 21:05 on the Tuesday evening of the sampling campaign; this explains the flow peak shown in Figure 7 (indicated by an arrow) and its deviation from the calibration flow. Figure 8; Figure 9 show the hourly measurements of wastewater concentration that were taken for total suspended solids (TSS), chemical oxygen demand (COD), total Kjeldahl nitrogen (TKN) and total phosphorus (TPH).

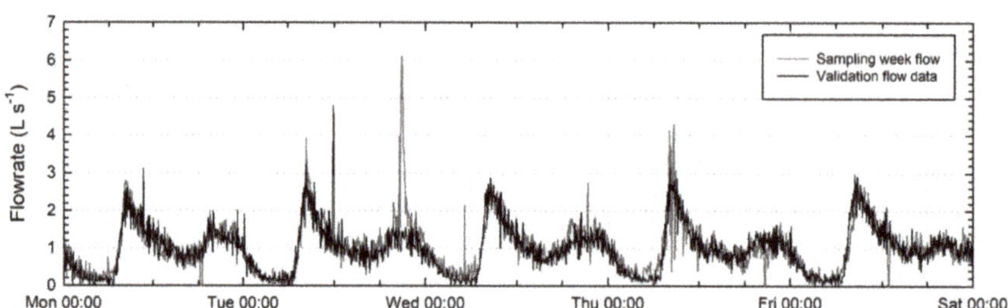

Figure 7. Wastewater flow over sampling week compared to flow data used for model validation.

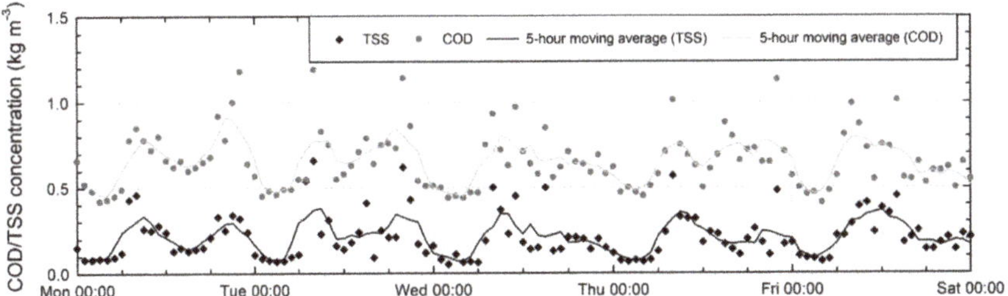

Figure 8. Hourly concentration of suspended solids and chemical oxygen demand (COD) in wastewater over sampling week.

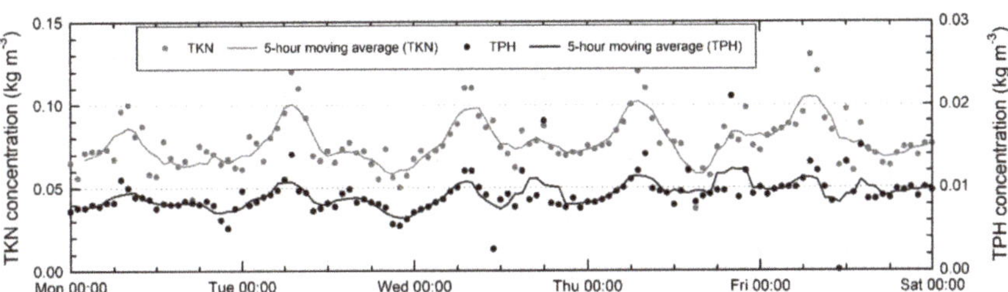

Figure 9. Hourly concentration of total Kjeldahl nitrogen (TKN) and total phosphorus (TPH) in wastewater over sampling week.

There was a good correlation between TSS and COD (R = 0.82) and a reasonable correlation between TKN and TPH (R = 0.55) but the correlation with suspended solids is weak (R = 0.38 for TKN and R = 0.20 for TPH). This indicates that the bulk of the COD is combined within the suspended solids but the TKN and TPH are present in a more dilute form. It is also notable that there is a reasonable correlation between the flowrate and the concentration of TSS and COD (R = 0.78 and R = 0.73 respectively). This seems to indicate that higher pollutant concentrations are produced at peak flow, but it is more likely that accumulated solids are washed through the system during high flow. This could be a consequence of sampling the wastewater downstream, where the highest concentration of COD/suspended solids occurs in the morning peak flow and the evening peak flow, but this is not necessarily the case upstream. This is discussed further in Section 4.3. TKN concentration also peaks with the morning surge in flow but then drops early afternoon, before steadily increasing throughout the evening until the next morning. TPH follows a very similar pattern to TKN but has a second evening peak in concentration. This is likely due to phosphorus sources now being restricted for the toilet and kitchen sink discharges, whereas the nitrogen is discharged more often.

4.3. Model Comparison with Sewer Quality Data

Figure 10 shows a comparison of the modelled mass flow compared to the observed data (calculated as the product of the measured concentration and the measured wastewater flowrate). The shaded areas represent the sampling error associated with each parameter, highlighted in Table 2. As indicated in Section 4.2, there was heavy rainfall from 20:35 to 21:05 on the Tuesday evening of the sampling campaign, and this is reflected in the concentration peak on the second evening of the plots in Figure 10 (indicated by an arrow). Apart from this, the model represents the observed mass flow reasonably well, as the timing and magnitude of the mass flow profiles are in alignment with the measured values. The predicted mass flow overnight is, on average, higher than the observed mass flow, and the observed morning peak is higher than predicted. This confirms the hypothesis, in Section 4.2, that these flow peaks likely include accumulation of solids rather than higher concentration discharges from households. This build-up of suspended solids has not been accounted for in this version of the model as time-varying solid generation is not available in InfoWorks® (see Section 2.2).

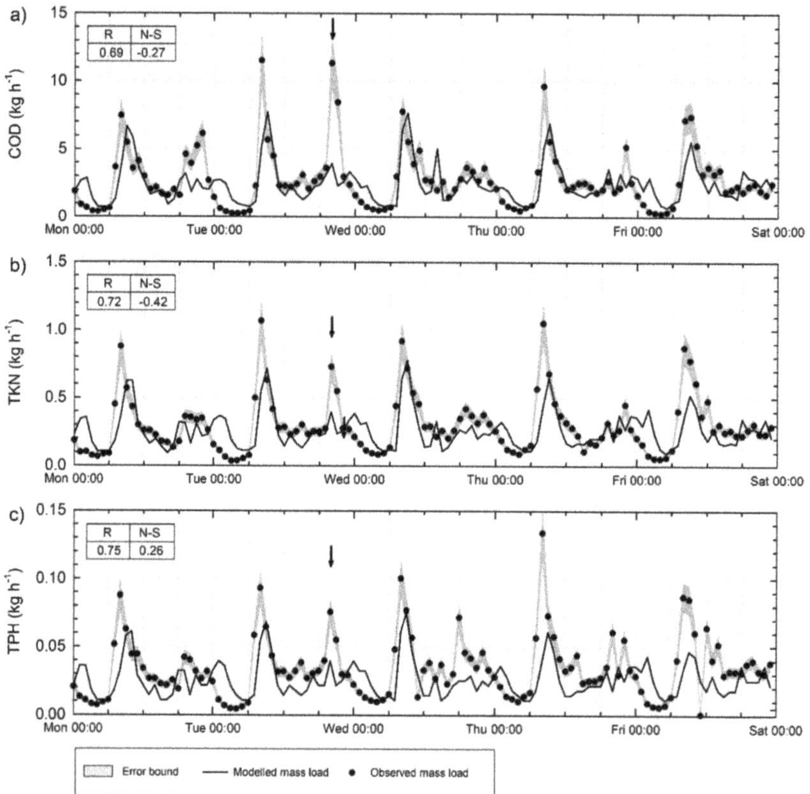

Figure 10. Mass flow of COD (**a**), TKN (**b**) and TPH (**c**) predicted by the model compared to the mass calculated from measured concentration and measured flow rate at the wastewater pumping station. The correlation coefficient (CC) and Nash–Sutcliff coefficient (N–S) are given for each plot.

Figure 11 shows the comparison of the predicted and measured nutrient concentration. The modelled tank concentrations were calculated according to Equation (3). This also supports the conclusion that the discrepancy between the modelled wastewater concentration and the observed is due to the lack of differential solids transport modelling in the network. The model predicts concentration to be highest during the night as most water use at night is from toilets, but this cannot be confirmed by the measured data. Following the design of the sampling campaign, the high concentration wastewater produced at night would only be accounted for during the first few 3-min sub-samples of the peak flow the following morning. The subsequent sub-samples are likely to be diluted substantially, leading to a morning peak in a lower concentration than the more concentrated night flows. SIMDEUM WW® appears to be performing well as a wastewater generator, but as the solids transport has not been adequately modelled within the sewer system (InfoWorks® ICM), the concentration cannot be aligned with the measured data. The modelled TKN and TPH follow the measured concentration data better than the COD, this is likely due to their lower correlation with suspended solids, and hence, dilute modelling is more appropriate here.

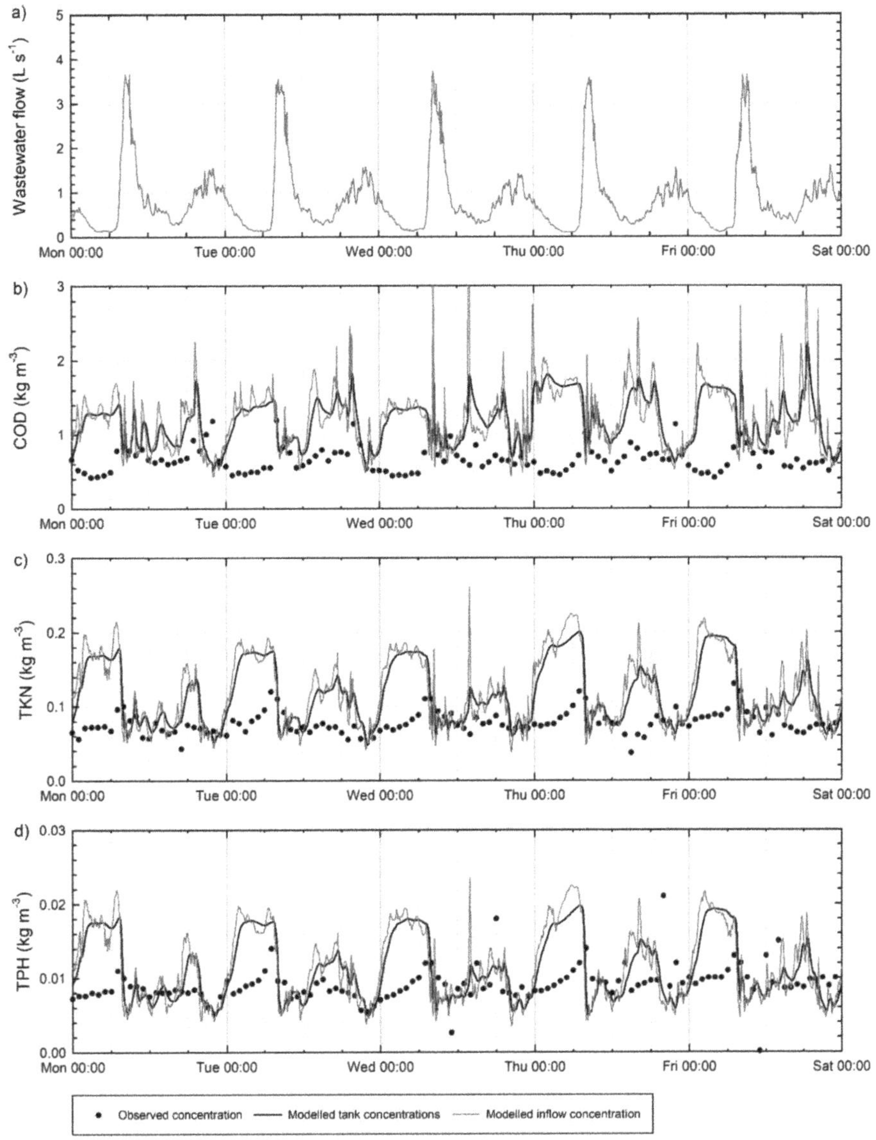

Figure 11. Wastewater flow (**a**) and modelled COD (**b**), TKN (**c**) and TPH (**d**) concentration in comparison with the measured concentration.

4.4. Variability of the Model

To address the variability of the stochastic model, each weekday was evaluated on factors of flow and nutrient mass—see Figure 12, where each day is compared to the first simulated day. The sample point for comparison was the final pipe of the network, before the pumping station. The stochastic model results are relatively consistent as the gradient of the line of best fit, m, for each day is close to 1. Correlation between Day 1 of the simulation and the subsequent days is very high for flowrate but the correlation is less strong for the nutrient mass flow. COD showed the smallest variability followed by TKN and then TPH. This is thought to be due to TKN and TPH being linked more strongly to appliances that follow a less strict daily usage pattern, e.g., kitchen taps, dishwashers and washing machines. Whereas the toilet and shower use (more strongly linked to COD generation) happen at similar times of day. Elias-Maxil [26] assessed the variability in SIMDEUM® with over 200 simulations and concluded that the pattern generator reaches a steady state after 75 simulations, i.e., the variability approaches zero. As the studied catchment includes 418 households, this confirms that the variability at the outfall is low.

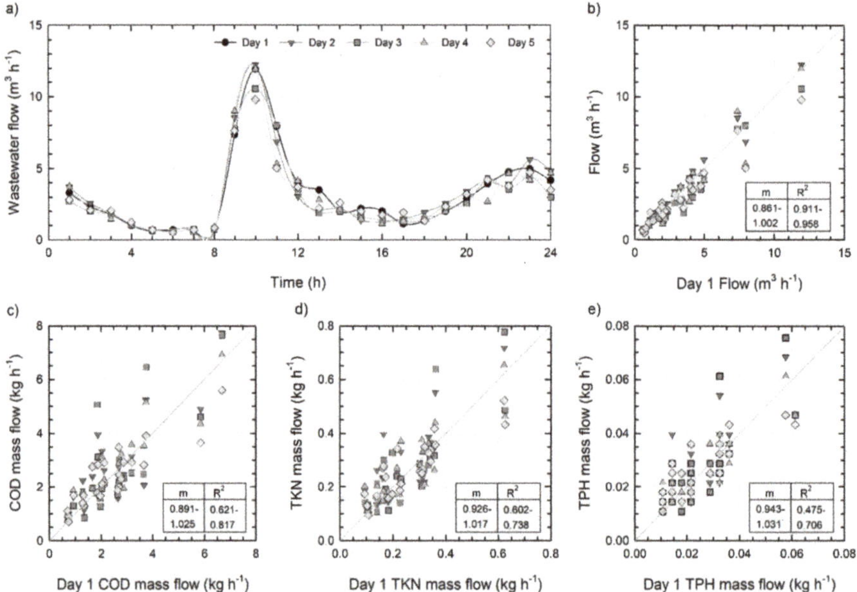

Figure 12. (**a**) Variation in stochastic modelled flow over 5 days, (**b**) Flow variation over 5 days compared to Day 1, (**c**) COD mass flow variation over 5 days compared to Day 1, (**d**) TKN mass flow variation over 5 days compared to Day 1, (**e**) TPH mass flow variation over 5 days compared to Day 1.

4.5. Future Scenario Testing

Six future scenarios (Section 2.5) were tested using the stochastic flow and wastewater quality model to observe the effects of different water conservation technologies on flow and wastewater concentration. Increased wastewater concentration can offer benefits for resource recovery, whilst reducing household water use is beneficial for water security and sustainability reasons.

Figure 13 shows the results from this simulation, analysed over a 5-day period (Monday–Friday). It can be seen in Figure 13a, that the effect of Eco (2a/2b) and GWR (3a/3b) scenarios is the dramatic reduction in the morning peak. The sewer system experiences a much narrower range of flowrates in these scenarios, which warrants smaller pipe diameters. Penn, et al. [27] stated that for a 1–6 mm diameter solid, the critical shear is 0.867–1.42 Pa, respectively, so without reducing pipe diameters, these water use scenarios may struggle to transport larger solids (see Figure 13b).

Figure 13c–f shows the consequence on wastewater quality parameters, and there is little impact of population changes between the scenarios (a and b scenarios). RWH produces a very similar situation to the baseline as it is simply replacing potable sources with a non-potable alternative. The impact of this scenario is better addressed by evaluating the impact on the drinking water system, as it will likely increase water residence time in the distribution network, which may compromise water quality. The Eco scenario produces the highest concentration of wastewater, although the range of concentrations is similar to the baseline/RWH scenarios. GWR produces wastewater at concentrations between the other two scenarios but in a much narrower range. This scenario could, therefore, be preferable for resource recovery as there is a narrower operating range for treatment units. However, GWR is the poorest performing water use scenario in terms of wastewater temperature, as shower and bath water do not directly enter the sewer, hence sewer temperature reduces. This model has been demonstrated as a useful tool for analysis of various resource recovery options for future urban water planning.

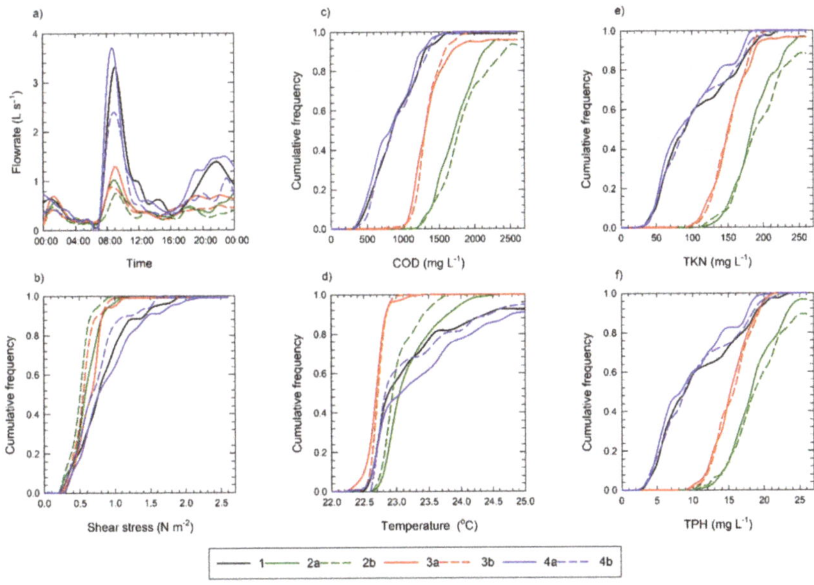

Figure 13. (**a**) Effect of scenarios on the flowrate at the catchment outfall, (**b–f**) Cumulative frequency of the shear stress achieved, COD, Temperature, TKN and TPH concentration in wastewater at the catchment outfall over 5 day (respectively).

Bailey et al. [13] concluded that this model over-predicts phosphorus concentrations, but with the results from the sampling campaign, and the changes made in the estimated wastewater composition due to the removal of phosphorus in detergents (Section 2.1.2), the model now predicts in line with reality. Daily pollutant load produced per capita in these scenarios ranged from 86–122 g COD, 8–12 g TKN and 0.8–1.2 g TPH—these values align with independently published values [21,28–30].

5. Conclusions

A new stochastic wastewater flow and quality model has been developed to address the impacts of water use changes on wastewater flow concentration. The hydraulic model was tested and validated in previous work. This paper presents the validation of the wastewater quality model using measured data. The model was used to investigate the impact of three water-saving strategies (greywater recycling, rainwater harvesting and installation of smart water appliances) on water quantity and quality in the sewer network.

The results obtained lead to the following key findings:

1. Stochastic sewer model wastewater quality validation: The predicted mass flows of COD, TKN and TPH compared well with the corresponding observed data values. The same, however, cannot be said for the COD, TKN and TPH concentrations. These concentrations were treated as dilute pollutants as InfoWorks® does not currently incorporate differential solids transport, leading to the misalignment of the predicted and measured concentration data. High concentration flows are produced by the stochastic generator during the night but only washed through the system in the morning. As the concentrations were measured at a downstream point in the network, there was a lag time in transporting suspended solids which was not accounted for in the network model.

2. Implications for three water-saving strategies on the quantity and quality of flow in the receiving sewer network: It was found that wastewater flow can be reduced by up to 62% with concentrations of COD, TKN and TPH increasing by up to 111%, 84% and 75% respectively with the installation of water-saving appliances. In addition, it was found that the use of water-saving appliances and greywater recycling dramatically reduced the peak flows, whereas rainwater harvesting produced similar flow and concentration results in the baseline case. The greywater recycling case produced the most consistent wastewater concentrations and the lowest wastewater temperature.

3. Proposals for future work: This will involve incorporation of the time-varying component for suspended solids entry to the sewer system, and differential solids transport in the sewer. This advancement will be combined with a drinking water simulation to create a comprehensive urban water model for observing effects of future water use scenarios on the entire system. This project will ultimately highlight a future vision for the urban water cycle and support recommendations for optimal resource recovery within drinking and wastewater systems.

Supplementary Materials: The following are available online at http://www.mdpi.com/2073-4441/12/4/1187/s1

Author Contributions: Conceptualization, O.B., L.Z. and J.H.; Data curation, O.B.; Formal analysis, O.B. and L.Z.; Funding acquisition, J.P.v.d.H. and T.A.; Investigation, O.B.; Methodology, O.B. and L.Z.; Software, O.B., L.Z. and M.B.; Supervision, T.A. and J.H.; Validation, O.B. and L.Z.; Visualization, O.B. and J.H.; Writing—original draft, O.B.; Writing—review and editing, O.B., L.Z., J.P.v.d.H., Z.K., M.B., T.A. and J.H. All authors have read and agreed to the published version of the manuscript.

Funding: This study was conducted as part of the Water Informatics Science and Engineering (WISE) Centre for Doctoral Training (CDT), funded by the UK Engineering and Physical Sciences Research Council, Grant No. EP/L016214/1. Olivia Bailey is supported by a research studentship from this CDT. Funding was also obtained from the Topsector Water & Maritime TKI Watertechnology Program of the Dutch Ministry of Economic Affairs and Climate Change (No. 2016TUD003, project New Urban Water Transport Systems), water utility Waternet, drinking water companies Brabant Water, Limburg and Evides, water authority De Dommel, Amsterdam Institute for Advanced Metropolitan Solutions and Royal Haskoning DHV Consultancy.

Acknowledgments: The authors thank Waternet, the water utility of the city of Amsterdam, for providing expertise, data, and access to the sewer networks for field tests.

Conflicts of Interest: The authors declare no conflict of interest.

References

1. Diamantis, V.; Verstraete, W.; Eftaxias, A.; Bundervoet, B.; Vlaeminck, S.E.; Melidis, P.; Aivasidis, A. Sewage pre-concentration for maximum recovery and reuse at decentralized level. *Water Sci. Technol.* **2013**, *67*, 1188–1193. [CrossRef] [PubMed]

2. Mezohegyi, G.; Bilad, M.R.; Vankelecom, I.F.J. Direct sewage up-concentration by submerged aerated and vibrated membranes. *Bioresour. Technol.* **2012**, *118*, 1–7. [CrossRef] [PubMed]

3. Bianchini, A.; Bonfiglioli, L.; Pellegrini, M.; Saccani, C. Sewage sludge drying process integration with a waste-to-energy power plant. *Waste Manag.* **2015**, *42*, 159–165. [CrossRef] [PubMed]

4. Verstraete, W.; Vlaeminck, S.E. ZeroWasteWater: Short-cycling of wastewater resources for sustainable cities of the future. *Int. J. Sustain. Dev. World Ecol.* **2011**, *18*, 253–264. [CrossRef]

5. Parkinson, J.; Schütze, M.; Butler, D. Modelling the impacts of domestic water conservation on the sustainability of the urban sewerage system. *J. Chart. Inst. Water Environ. Manag.* **2005**, *19*, 49–56. [CrossRef]

6. Penn, R.; Hadari, M.; Friedler, E. Evaluation of the effects of greywater reuse on domestic wastewater quality and quantity. *Urban Water J.* **2012**, *9*, 137–148. [CrossRef]

7. Penn, R.; Schütze, M.; Friedler, E. Modelling the effects of on-site greywater reuse and low flush toilets on municipal sewer systems. *J. Environ. Manag.* **2013**, *114*, 72–83. [CrossRef] [PubMed]

8. Penn, R.; Schütze, M.; Gorfine, M.; Friedler, E. Simulation method for stochastic generation of domestic wastewater discharges and the effect of greywater reuse on gross solid transport. *Urban Water J.* **2017**, *14*, 846–852. [CrossRef]

9. Watershare. Available online: https://www.watershare.eu/tool/water-use-info/ (accessed on 1 March 2020).

10. Blokker, E.J.M.; Vreeburg, J.H.G.; van Dijk, J.C. Simulating Residential Water Demand with a Stochastic End-Use Model. *J. Water Resour. Plan. Manag.* **2010**, *136*, 19–26. [CrossRef]

11. Pieterse-Quirijns, E.J.; Agudelo-Vera, C.M.; Blokker, E.J.M. Modelling sustainability in water supply and drainage with SIMDEUM®. In Proceedings of the CIBW062 Symposium, Melbourne, Australia, 8–10 September 2019.

12. Bailey, O.; Arnot, T.C.; Blokker, E.J.M.; Kapelan, Z.; Vreeburg, J.; Hofman, J.A.M.H. Developing a stochastic sewer model to support sewer design under water conservation measures. *J. Hydrol.* **2019**, *573*, 908–917. [CrossRef]

13. Bailey, O.; Arnot, T.C.; Blokker, E.J.M.; Kapelan, Z.; Hofman, J.A.M.H. Predicting impacts of water conservation with a stochastic sewer model. *Water Sci. Technol.* **2019**, *80*, 2148–2157. [CrossRef] [PubMed]

14. Blokker, E.J.M. Stochastic Water Demand Modelling. In *Hydraulics in Water Distribution Networks*; IWA publishing: London, UK, 2011.

15. Blokker, E.J.M.; Agudelo-Vera, C.A. *Doorontwikkeling Simdeum: Waterverbruik over de dag, Energie voor Warmwater en Volume, Temperatuur en Nutriënten in Afvalwater*; Report BTO 2015.210(s); KWR Water Research Institute: Nieuwegein, The Netherlands, 2015.

16. Parkinson, J.N. *Modelling Strategies for Sustainable Domestic Wastewater Management in a Residential Catchment*; Imperial College for Science, Technology and Medicine: London, UK, 1999.

17. Butler, D.; Friedler, E.; Gatt, K. Characterising the quantity and quality of domestic wastewater inflows. *Water Sci. Technol.* **1995**, *31*, 13. [CrossRef]

18. Siegrist, R.; Witt, M.; Boyle, W. Characterisation of rural household wastewater. *J. Environ. Eng. ASCE* **1976**, *102*, 533–548.

19. Surendran, S. Grey-water reclamation for non-potable re-use. *J. CIWEM* **1998**, *12*, 406–413. [CrossRef]

20. The European Parliament; The Council of the European Union. Regulation (EU) No 259/2012 on the use of phosphates and other phosphorus compounds in consumer laundry detergents and consumer automatic dishwasher detergents. 2012. Available online: https://eur-lex.europa.eu/legal-content/EN/TXT/PDF/?uri=CELEX:32012R0259 (accessed on 1 March 2019).

21. Comber, S.; Gardner, M.; Georges, K.; Blackwood, D.; Gilmour, D. Domestic source of phosphorus to sewage treatment works. *Environ. Technol.* **2013**, *34*, 1349–1358. [CrossRef] [PubMed]

22. Ward, S.; Memon, F.A.; Butler, D. Harvested rainwater quality: The importance of building design. *Water Sci. Technol. J. Int. Assoc. Water Pollut. Res.* **2010**, *61*, 1707–1714. [CrossRef] [PubMed]

23. Farreny, R.; Morales-Pinzón, T.; Guisasola, A.; Tayà, C.; Rieradevall, J.; Gabarrell, X. Roof selection for rainwater harvesting: Quantity and quality assessments in Spain. *Water Res.* **2011**, *45*, 3245–3254. [CrossRef] [PubMed]

24. Agudelo, C.; Blokker, E.J.M. *How Future Proof Is Our Drinking Water Infrastructure*; Report BTO 2014.011; KWR Water Research Institute: Nieuwegein, The Netherlands, 2014.

25. Waternet, Average Water Use. Available online: https://www.waternet.nl/en/our-water/our-tap-water/average-water-use/ (accessed on 1 March 2020).

26. Elias-Maxil, J.A. *Heat Modelling of Wastewater in Sewer Networks: Determination of Thermal Energy Content from Sewage with Modeling Tools*; Technische Universiteit Delft: Delft, The Netherlands, 2015.

27. Penn, R.; Schütze, M.; Friedler, E. Assessment of the effects of greywater reuse on gross solids movement in sewer systems. *Water Sci. Technol.* **2014**, *69*, 99–105. [CrossRef] [PubMed]

28. Henze, M.; Loosdrecht, M.C.M.v.; Ekama, G.A.; Brdjanovic, D. 3 Wastewater Characterization. In *Biological Wastewater Treatment—Principles, Modelling and Design*; IWA Publishing: London, UK, 2008.

29. Tchobanoglous, G.; Burton, F.L.; Stensel, H.D. *Wastewater Engineering: Treatment, Disposal, and Reuse*, 3rd ed.; Tchobanoglous, G., Burton, F.L., Eds.; McGraw-Hill: London, UK, 1991.

30. Arildsen, A.L.; Vezzaro, L. *Revurdering af Person Ækvivalent for Fosfor—Opgørelse af Fosforindholdet i Dansk Husholdningsspildevand i Årene fra 1990 til 2017. Kgs*; Danmarks Tekniske Universitet (DTU): Lyngby, Denmark, 2019.

Article

Sediment Transport in Sewage Pressure Pipes, Part I: Continuous Determination of Settling and Erosion Characteristics by In-Situ TSS Monitoring Inside a Pressure Pipe in Northern Germany

Martin Rinas *, Jens Tränckner and Thilo Koegst

Department of Water Management, University of Rostock, Satower Straße 48, 18059 Rostock, Germany;
jens.traenckner@uni-rostock.de (J.T.); t.koegst@stalums.mv-regierung.de (T.K.)
* Correspondence: martin.rinas@uni-rostock.de

Received: 4 September 2019; Accepted: 9 October 2019; Published: 13 October 2019

Abstract: Continuous measurement systems are widely spread in sewers, especially in non-pressure systems. Due to its relatively low costs, turbidity sensors are often used as a surrogate for other indicators (solids, heavy metals, organic compounds). However, little effort is spent to turbidity sensors in pressurized systems so far. This work presents the results of one year in-situ turbidity/total suspended solids (TSS) monitoring inside a pressure pipe (600 mm diameter) in an urban region in northern Germany. The high-resolution sensor data (5 s interval) are used for the determination of solids sedimentation (within pump pauses) and erosion behavior (within pump sequences). In-situ results from sensor measurements are similar to laboratory results presented in previous studies. TSS is decreasing exponentially in pump pauses under dry weather inflow with an average of 0.23 mg/(L s). During pump sequences, solids eroded completely at a bed shear stress of 0.5 N/m^2. Sedimentation and erosion behavior changes with the inflow rate. Solids settle faster with increasing inflow: at storm water inflow with an average of 0.9 mg/(L s) and at diurnal inflow variation up to 0.6 mg/(L s) at 12:00 a.m. The results are used as calibration data for a sediment transport simulation in Part II.

Keywords: total suspended solids; in-situ; erosion; sedimentation; pressure pipe; sewage

---

1. Introduction

The physical characterization of sewage is indispensable for optimization efforts in all areas of wastewater management. Pumping processes are usually necessary in sewage and storm water transport. Because of their frequent use, all related processes offer high optimization potential, to name only a few: sediment transport, energy consumption, and storm water management.

The main key to process understanding and optimization lies in data quality and quantity. Advanced optimization tools (e.g., numerical simulations) are especially data greedy. Quality and quantity varies with the data collection method: either ex-situ or in-situ. Ex-situ methods are primary laboratory experiments. Experiments simulate real world conditions as accurately as possible and subsequently transfer the results into a representative model region. For example, most stream tests try to simulate more-or-less real-life conditions.

The advantage of in-situ methods is the proximity to real life. Thus, an imitation is not required. To measure undisturbed processes, impacts on the system should be kept by a minimum.

Erosion and sedimentation of particulate matter in sewage are the dominating physical effects regarding the above-mentioned themes (sediment transport, energy consumption, storm water management). In the past, settling and erosion behavior have been determined by ex-situ experiments [1–7]. In case of [1,2], several laboratory experiments were conducted to describe erosion and settling behavior of the raw sewage inflow to a pumping station (PS) in an urban drainage system in Rostock (northern Germany). Consequently, the derived results are only temporal snapshots of continuous highly dynamic sewer processes. A continuous description based on permanent (in-situ) measurement allows, by far, more statements about raw sewage transport behavior. In this study, the transport behavior inside a pressure pipe is of interest. Therefore, the measurement is located directly inside the pressure pipe. Thus, the pipe itself serves as the reaction chamber for the experiments.

This work aims at an in-depth characterization of the erosion and settling behavior of raw sewage by in-situ total suspended solids (TSS) online measurement for the period of one year. Providing a large amount of data helps to increase the accuracy of transport simulations and improves the efficiency of the sewer system. The following three objectives are defined:

- Determine applicability and quality of an in-situ TSS-online measurement system inside a pressure pipe
- Characterize raw sewage erosion and sedimentation behavior under dry weather inflow continuously by TSS-online monitoring
- Identify mechanisms changing the transport behavior and characterize modified erosion and sedimentation

Literature Review

In-situ measurements in the field of urban drainage concentrate usually on non-pressure systems (open channel flow), often in the context of combined sewer overflows (CSO) and pollutant loads in combined or storm sewers (e.g., [8–13]). Continuous monitoring systems are almost exclusively used for the calculation of loads or fluxes. Further data analysis, regarding solids transport behavior, is often not been conducted. An exception is provided by [12–14], all calculating mass curves from online data. An in-situ monitoring study, dealing with continuous TSS measurements within pressurized systems, have not been published as known to the authors.

The characterization of sediments by continuous measurements is mostly applied within ex-situ laboratory experiments: [6] performed ex-situ tests with wastewater to determine the sediment accumulation in a pilot flume (d = 300 mm open channel flow, average discharge = 4 L/s) using the same TSS sensor as used in this study (Hach Lange Solitax). Another example is [7], where sediments were collected for flushing experiments in the laboratory, equipped with a continuous turbidity measuring system. Similarly to [1], a continuous turbidity measurement was used to determine the erosion characteristics inside an ex-situ laboratory device.

The same sensor (Hach Lange Solitax) was also used by [9] inside a combined sewer (in-situ) to assess the dynamics of erosion and sedimentation events (load calculation).

Hybrids between ex-situ and in-situ are provided by [11,12], where the monitoring sensors were mounted in external tanks or flumes supplied by a pump from a sewer.

The applicability of online sensor data for urban drainage problems and related uncertainty was investigated with large effort by: [10–12] and [14–16]. The majority of the data processing methods in this study based on these publications.

One of the main differences to previous studies lies in the measurement interval (here 5 s). Sometimes, daily measurements were used as in [6], but with regard to systems dynamics, most studies used short intervals as in [11] or [14] with a 2 min time step, [7] with 20 s, or [9] with 15 s.

2. Materials and Methods

2.1. Study Side

One of the main PS in the city of Rostock ($\approx$200000 inhabitants) is PS Rostock-Schmarl, conveying raw sewage from approx. 40000 inhabitants. A special technical feature of the upstream, usually separating sewer system is the connection to main roads storm water runoff. The storm water system itself collects runoff from roof discharge and secondary roads. Whatever the inflow condition to PS Rostock-Schmarl is, the incoming sewage is filtered at first by a rake with a wide space bar opening (20 mm) before it is transported directly to the central wastewater treatment plant (wwtp) by four pumps (each of 55 kW) in two cast iron pipelines (diameter = 600 mm), each over 4500 m length. A schematic view of the catchment area and the PS in Figure 1 illustrates the study setup.

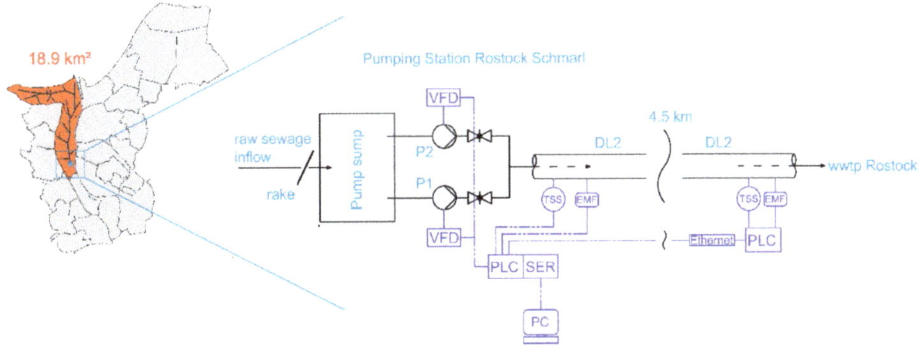

Figure 1. Visualization of the catchment area (18.9 km²) in Rostock (Germany) including a schematic view of the control and monitoring system during the study. The raw sewage inflow passes the rake and is collected inside the pump sump. Pumps P1 and P2 then conveying the raw sewage directly to the central wwtp in pressure pipe 2 (DL2). P1 and P2 are controlled over a variable-frequency drive (VFD) from a PC (connected over a serial port (SER) to the programmable logic controller (PLC)). The VFD adjusts pumps motor speed according to the control strategy [1,2,17]. All values from TSS sensors (TSS) and electromagnetic flowmeters (EMF) are stored in the PC.

2.2. In-Situ TSS Monitoring

For a studied period of one year, pumps P1 and P2 were controlled by a PC to perform a rule based, energy saving control strategy [1,2,17]. The sediment flux was monitored by online TSS measurements at the in- and outflow side of pressure pipe DL2. Table 1 shows the TSS sensors technical data. The sensors itself are shown in Figure 2.

Table 1. Technical data of TSS sensors

Sensor	Controller	Parameter	Measuring Range	Installed and Measured Duration	Interval	Service	Num. of Calibration Processes	Wiper Self-Cleaning Interval
Hach Lange Solitax inline Sc	Hach Sc 200 & Sc 1000	Turbidity, TSS	0.001–4000 FNU, 0.00–150.000 mg/L	343 days installed; 292 days measured	5 s	1 per month	5 processes with 73 samples	15 min

Furthermore, the following parameters were monitored every 5 s over 1 year: pump sump level (m), inflow to the PS (L/s), pumps power input (kW), frequency of the VFD (F) (Hz), engine speed (min⁻¹), pressure in DL2 directly after the pump (bar), flow in DL2 (Q_{pipe}) (L/s).

Figure 2. TSS sensors for monitoring sediment flux in pressure pipes. (**a**) TSS sensor in PS Rostock-Schmarl at the pressure side of the pump. (**b**) TSS sensor in the central wwtp Rostock at the outflow side of the pressure pipe.

2.3. Sample-Specific Sensor Calibration

The turbidity sensors measuring principle is an infrared due scattered light technique [18]. The TSS eventually calculated from the turbidity by an internal factory calibrated formula. Commonly, diatomaceous earth is used for the internal calibration process.

To adjust the TSS sensor values to the local raw sewage composition of PS Rostock-Schmarl, a sample-specific calibration based on a correlation method was repeated 5 times, with 73 separate raw sewage samples in total. The samples are collected with a ladle from the inflow channel, just before the rake. Afterwards, the samples are filtered, according to the rakes space bar opening of 20 mm, and subsequently separated into cylinders of a small volume (2.5 L). In the second step, the TSS concentration is artificially modified to obtain different cylinders with different TSS values. Therefore, the TSS concentration is decreased by mixing several dilutions using clear water, while settling increases the TSS values. This procedure provides a wide range of TSS values for calibration process. Thus, the resulting calibration function is applicable for a broad spectrum of TSS values without extrapolation. After that, a sample is filled into the calibration cylinder and continuously mixed by a magnetic stirrer. Next, the sensors are demounted from the pressure pipe and placed into the calibration cylinder. Subsequently, three sensor TSS values are noted from the controller board. Finally, each sample is analyzed for TSS by three-fold determination in the laboratory (analysis according to [19] by filtration and weight loss).

2.4. Fit Calibration Function and Analysis of Sensor Data

Sensor data must be validated before further processing into erosion and sedimentation behavior. Various literature deals with error assessment and validation of sensor data to find a correlation function and at least a true area of measured values with respect to uncertainties (i.e., [12] or [15]). According to these publications, the determination of uncertainties was processed after the commonly used "Guide to the Expression of Uncertainty in Measurement" (GUM) [20]. Therefore, the following data analysis scheme was applied:

1. Fit calibration function (TSS to TSS) with errors in y and x direction using the total least-squares regression;
2. Calculate function parameters uncertainties by Monte-Carlo simulation for 95% confidence level;
3. Transform original sensor data TSS_{sens} by the calibration function into calibrated sensor data TSS_{cal};

4. Remove TSS_{cal} values > 1.000 mg/L, based on local operators' expertise;
5. Further error assessment by Walsh's outlier test.

First, the calibration function is performed by the total least-squares regression. By this, errors in both directions, resulting from the TSS determination inside the laboratory (y) and the TSS measurement by the sensor (x), were accounted for the optimization problem. The resulting regression function is a first order polynomial function with the slope b (-) and intercept a (mg/L). The function calculates the calibrated TSS values TSS_{cal} (mg/L) from the original sensor data TSS_{sens} (mg/L), see Equation (1).

$$f(TSS_{sens}) = b \cdot TSS_{sens} + a \tag{1}$$

Second, the function parameters uncertainties are calculated for 95% confidence level by Monte-Carlo simulation in MATLAB (see also [12]). The calculation of a combined uncertainty resulting from the sensor measurement itself and the field influences (i.e., installation site) has been omitted. It is assumed, that field influences are already included in the sensor output. It is furthermore assumed that the field influence occurring during the in-pipe measurement is equal to the field influence occurring during the calibration process outside the pressure pipe.

Third, the original sensor output TSS_{sens} is transformed into calibrated sensor data TSS_{cal} to obtain the estimated TSS values by Equation (1). Furthermore, the 95% confidence interval is calculated based on the function parameters uncertainties.

Fourth, all TSS values > 1.000 mg/L are removed from the calibrated data set. The criterion based on the local operator's expertise.

Fifth, measurement errors are cleaned by the Walsh outlier test. The test requires no specific distribution, can easily be coded, enables fast computing, and detects outliers greater and less than the remaining values.

2.5. Determination of Settling- and Erosion Data

After calibration and error assessment, the final data processing followed, before determination of settling and erosion data is conducted. The transport characteristics are analyzed based on TSS sensor values in PS Rostock-Schmarl. The TSS sensor values from the central wwtp are used in Part II as reference for the sediment transport model.

The data processing scheme is visualized in Figure 3 and is described below. To maintain the data for determination, the complete data set (Figure 3a) must be split up in two parts: (i) the erosion data-part (Figure 3b), containing data while one of the two pumps is working; (ii) the sedimentation data-part (Figure 3c), containing data logged while pumps are shut off. Each data-part (erosion-part and sedimentation-part) is further split up into separate erosion (Figure 3d) and sedimentation events (Figure 3e). This is necessary, because the characterization is at least a mathematical approximation of a single erosion and sedimentation event. These single events are now the basis the mathematical description.

The sedimentation of solid fractions inside a fluid can be described by a settling velocity distribution (e.g., see [1] or [4]). However, since the turbidity sensors are not able to detect single particle fractions, the following approximation is applied. The settling events are described as a decay process, modeled by a differential Equation (2).

$$\frac{dC}{dt} = -\alpha \cdot C \tag{2}$$

Its solution is an exponential decay, called settling rate $C(t)$ (mg/L), see Equation (3). With t (s), the settling duration in each pump pause, C_0 (mg/L), a fixed value relating to the first TSS concentration in each single sedimentation event, C_{rest} (mg/L), the final solids concentration at the end of each single setting event and the exponential decay rate α (1/s), which is the key parameter to describe the settling behavior.

$$C(t) = C_0 \cdot e^{-\alpha \cdot t} + C_{rest} \tag{3}$$

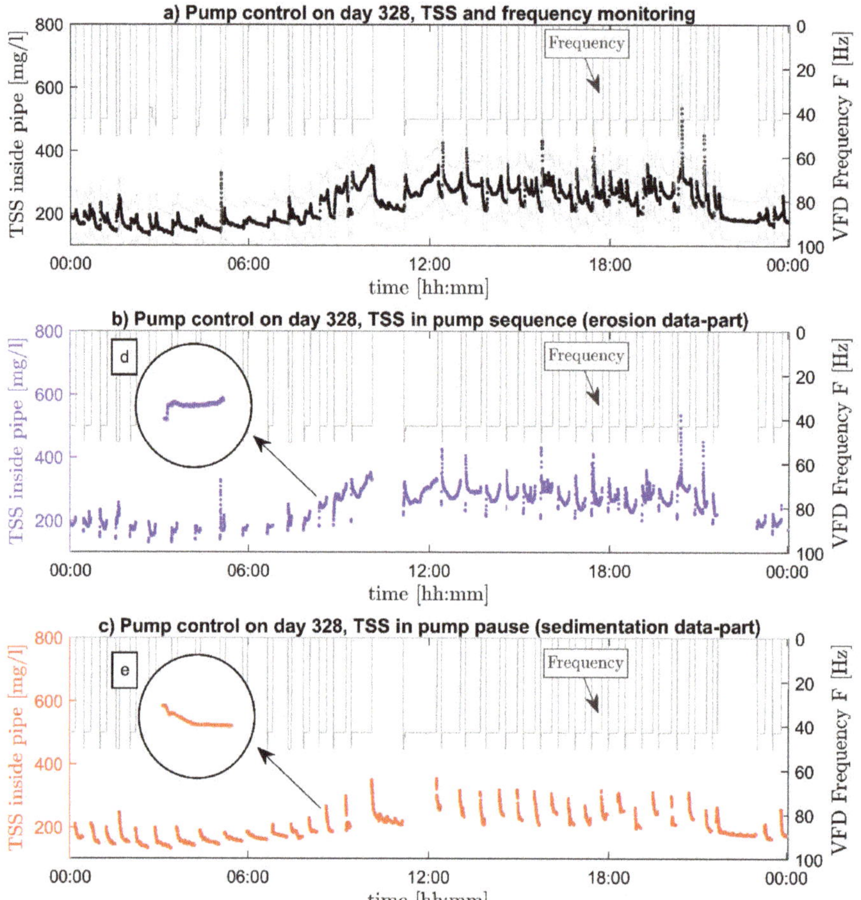

Figure 3. Data separation scheme: monitored TSS data (**a**) is split into an erosion- (**b**) and a sedimentation part (**c**). Frequency data from both VFD (for P1 and P2) is used as decision criterion for data separation (if VFD1 and VFD2 = 0, then settling sequence, else erosion sequence). The separation into single erosion (**d**) and sedimentation events (**e**) is based on a time difference between each value. If the time difference is larger the logging interval of 5 s (see Table 1), a single event is detected and separated.

Within each time step t, a proportion of the initial TSS concentration C_0 settles to pipes invert, according to the decay rate α, which is received by solving the optimization problem in Equation (4). With n, the number of values in each settling sequence and TSS_{cal} (mg/L), the measured and calibrated TSS concentration inside the pressure pipe.

$$\min_{\alpha} \sum_{i=1}^{n} \left(TSS_{cal,i} - C_i \right)^2 \tag{4}$$

The erosion events are described according to [1]. The measured erosion rate e_a (kg/(m s)) inside the pipe is calculated from the TSS concentration after pumps start (shown in Figure 3d), by Equation (5).

With $TSS_{cal,i} - TSS_{cal,i-1}$ (mg/L), the TSS difference between measurements, Δt (s), the time difference between measurements and A_s (m^2), the surface area of erosion (set to 1 m^2 for better comparability).

$$e_a = \frac{TSS_{cal,i} - TSS_{cal,i-1}}{\Delta t} \cdot A_s \tag{5}$$

The measured erosion rate e_a can be described as a function of the current bed shear stress, called erosion rate a (activation of sediments) (kg/(m s)), see Equation (6). With τ_{pipe} (N/m^2), the current bed shear stress, τ_{crit} (N/m^2), the critical bed shear stress where erosion starts and d (s), the erosion parameter, which describes the strength of the erosion (equal to slope of the first order polynomic function).

$$a\left(\tau_{pipe}\right) = \max\left(0, d \cdot \left(\tau_{pipe} - \tau_{crit}\right)\right) \tag{6}$$

τ_{pipe} is calculated by Equation (7), based on the fluid density $\rho = 1000$ (kg/m^3), the flow velocity v (m/s), and the friction factor λ (calculated after the Colebrook–White equation).

$$\tau_{pipe} = \rho \cdot \frac{v^2}{2} \cdot \frac{\lambda}{4} \tag{7}$$

The height of τ_{crit} depends on several parameters, as the formerly settling duration (higher τ_{crit} values for longer settling duration) and the composition of the sewage (organic components raises τ_{crit} due to biogenic changes). For a detailed description of τ_{crit} see [1]. The erosion rate a is adjusted to the measured erosion rate e_a by solving the optimization problem in Equation (8).

$$\min_{d,\,\tau_{crit}} \sum_{i=1}^{n} \left(e_{a,i} - a_i \cdot w_i\right)^2 \tag{8}$$

To consider real life conditions inside the pressure pipe, w (kg), the current particle mass on pipe bottom is multiplied with the erosion rate a. If the sediment bed is empty ($w = 0$), the erosion rate a becomes zero. By solving Equation (8), the function parameter d and additionally the critical bed shear stress τ_{crit} is received.

3. Results and Discussion

3.1. Sensor Calibration Results

The result of the calibration processes is shown in Figure 4. The laboratory TSS correlates to each sensor TSS by a first order polynomic function.

With the resulting calibration function, the sensor values are converted into the laboratory values, before further processing. The calibration functions are fitted with a R^2 value of 0.84 for TSS sensor PS Rostock-Schmarl and with a R^2 value of 0.85 for TSS sensor at the central wwtp Rostock. As found in literature, usual calibration functions used same functions with R^2 values between 0.83 and 0.92 [8] (calibration to turbidity) or, as already summarized in [15], between 0.80 and 0.95 [15] (calibration to turbidity). Reference [9] obtained a calibration function, for the same sensor used in this study, with a R^2 value of 0.94.

As one can see in the functions slope b, the measured values of both sensors are too high. Furthermore, we obtained two different calibration functions, although both sensors are identically and the same TSS samples used for calibration. A reason might be found in the differences to the internal calibration process. The used material for internal calibration differs from the raw sewage, as well as the calibration cylinder geometry. Another reason might be in the different controller devices used for the sensors. Differences in the internal signal processing may cause different values.

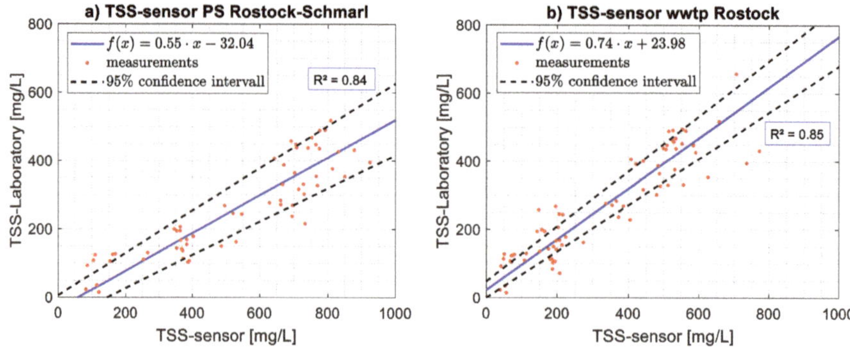

Figure 4. Calibration functions for calculating laboratory TSS values from in-situ measured sensor TSS values, including goodness of fit (R^2) and 95% confidence interval. (**a**) For TSS sensor in PS Rostock-Schmarl. (**b**) For TSS sensor at the central wwtp Rostock.

3.2. Evaluation of the Erosion and Settling Approximation

The evaluation of the raw sewage erosion and settling characteristics is based on an enormous amount of data. In sum, the TSS sensor in PS Rostock-Schmarl recorded 4238121 values (5 s interval over 1 year). 2203618 values of them account for erosion sequences and 2034503 values for settling sequences. The total number of single erosion sequences amount to 6653, while 6733 single settling events are recorded. This leads on average to ≈24 erosion and ≈24 settling events per day. Hence, the pumps are working every half hour for 30 min.

For each single erosion and settling event, a mathematical function is adjusted to the measured TSS values, automatically by a MATLAB code. This enables a fast, uncomplicated, and reproducible processing. The function itself is given by the settling rate $C(t)$, Equation (3), and by the erosion rate, written as $a(\tau,w)$ (following Equation (8)).

First, we will evaluate the approximations of the erosion and settling processes. Figure 5 evaluates the fit results graphically. In Figure 5a, all measured erosion rates e_a are plotted versus all fitted erosion rates $a(\tau,w)$. A perfect fit is given by $f(x) = x$ or $a(\tau,w) = e_a$. For the majority of erosion values, $a(\tau,w)$ follows the perfect fit course with deviations above and below. The fitting results are moderate. R^2 value of >=0.9 having 7.3% of the total approximations (n = 481 single events), 33.5% (n = 2100) were fitted with R^2 values of >=0.75, while R^2 values of >=0.5 having 54% (n = 3603). So the mathematical approximation by the erosion rate $a(\tau,w)$ is suitable to describe the real process of erosion. Because of the similar up- and downward deviation, a balance is assumed.

Figure 5b shows all measured TSS values in the pump pauses versus all approximations by $C(t)$. Here, the majority of the values are located just below the perfect fit line. Accordingly, the settling process is slightly overestimated. In contrast to erosion, a better fit is achieved for settling. R^2 values of >=0.9 having 31% of the total approximations (n = 2084 single events), 58.4% (n = 3934) were fitted with R^2 values of >=0.75 and R^2 values of >=0.5 having 76% (n = 5161).

Both models are able to describe real world conditions appropriately. The deficits in the erosion approximation may result from its dynamic process. If the TSS sensor measures a value immediately when the pump starts, it takes 5 s of pumping until the next value is recorded. Within these 5 s, some sediments are already eroded. This means that a shorter measuring interval is recommended for the erosion process in later studies. Furthermore, the limited flexibility of the erosion rate $a(\tau,w)$ itself, as it is based on a first order polynomic function (see Equation (6)), contributes to its moderate results. A transformation into a power function (e.g., $a\left(\tau_{pipe}\right) = d \cdot \left(\tau_{pipe} - \tau_{crit}\right)^{p}$) leads to slightly better results but with larger computation effort, e.g., for a sediment transport simulation.

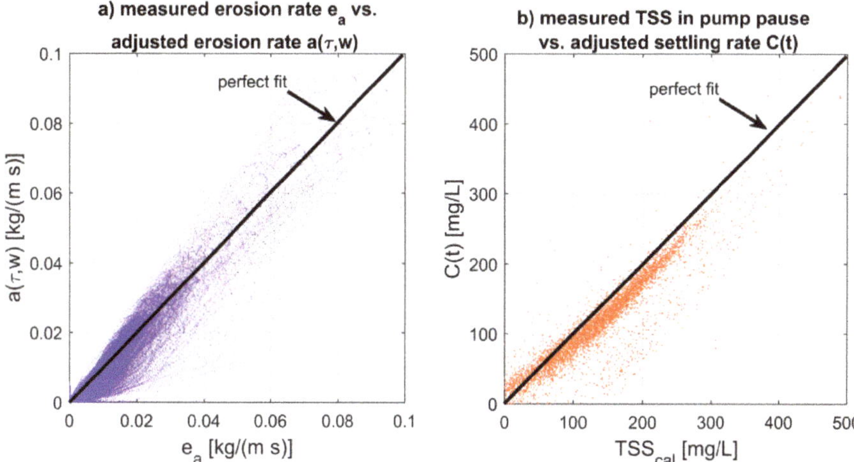

Figure 5. Evaluation of the erosion and settling approximation. (**a**) All measured erosion rates e_a vs. all mathematical approximations by $a(\tau,w)$. (**b**) All measured settling events inside the pipe vs. all mathematical approximations by $C(t)$.

3.3. Settling and Erosion Characteristics Inside the Pressure Pipe Under Dry Weather Inflow

Transport characteristics are essentially dealing with numerical simulation of sediment transport in open channel flow or dimensioning of facilities and treatment plants or solids transport inside pressure pipes. Hence, the in-pipe measurement helps to improve accuracy of the transport characterization and widen the spectrum of results substantially. Furthermore, it serves as a comparison to laboratory (ex-situ) results, obtained in [1,2].

The example in Figure 6a shows a typical situation in PS Rostock-Schmarl, comparable to usual urban drainage pumping stations. The diurnal course of TSS is separated, according to Figure 3, into erosion events (blue) and settling events (red). The diurnal course of the raw sewage inflow is presented by Q_{inflow} and the resulting pump flow by Q_{pipe} (right axes).

The TSS values during the night are relatively low. They reach a minimum of ≈200 mg/L at about 03:30 a.m. (in pump sequence) before starting to increase from 06:00 a.m. to 12:00 a.m. up to 400 mg/L. Peaks up to 600 mg/L may be the result of the sensor wiper, cleaning the sensor from heavy dirt (e.g., paper shreds). The TSS course follows the inflow course of Q_{inflow}. Hence, there is a relationship between TSS and Q_{inflow}. Low inflow results in low TSS values and vice versa. It results from the water usage and the hydraulic conditions in the upstream sewers. An increased solids amount reaching the PS by increased water consumption (stool, cooking, etc.). Furthermore, high water consumption raises the hydraulic performance in the upstream sewers and erodes deposits.

The TSS course is characterized by two long settling periods. A specific settling characteristic becomes clear within these two periods. The TSS first decreases rapidly but then slows down. This is also addressed in [2], where the same characteristic was found. Accordingly, an exponential function (see Equation (3)) describes this process most appropriate. Consequently, the TSS never decreases to full extend in pump pauses.

Figure 6b shows an exemplary settling course in detail, while Figure 6d shows the resulting settling rate. Although the pump pause only takes ≈17 min, the TSS inside the fluid reduces from ≈200 mg/L at about 27.5% to ≈145 mg/L. The effect of the exponential decrease shows the risk of forming a consolidated sediment layer inside the pressure pipe, even in short pump pauses. If sediments erode incompletely within the subsequent pump sequences, permanent deposits are likely to develop.

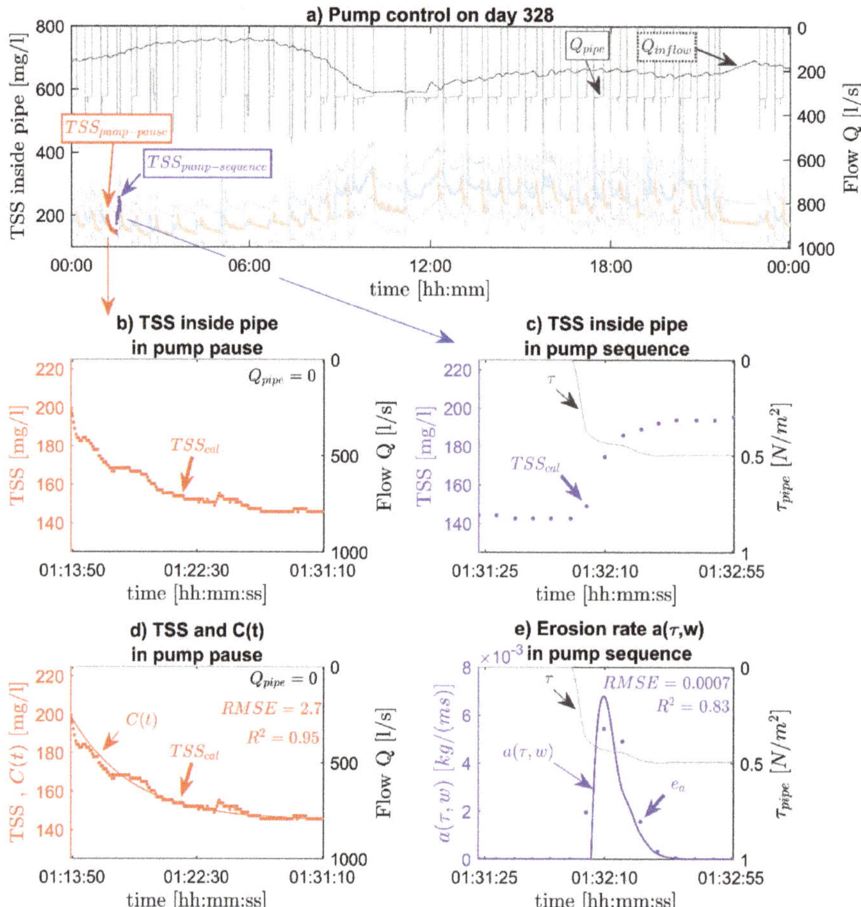

Figure 6. Monitored data in PS Rostock-Schmarl for day 328 (**a**) including exemplary erosion and settling determination scheme (**b**–**e**). (**a**) Q_{inflow} and Q_{pipe} (right axis) and TSS sensor data including 95% confidence levels (left axis). (**b**) Settling event: TSS values (TSS_{cal}) after pumps stop in the night. (**c**) Erosion event: TSS values (TSS_{cal}) after pumps start at night. (**d**) Settling event determination: TSS values (TSS_{cal}) and approximated settling rate $C(t)$ including fit results. (**e**) Erosion event determination: erosion rate e_a and approximated erosion rate $a(\tau,w)$ including fit results.

Figure 6c,e shows the subsequent erosion sequence and the determination of the erosion rate. The erosion process always forms an s-shaped curve (Figure 6c). This shape is characterized by a restrained start and is followed by an increased erosion. The inflection point of the s-curve marks the beginning of the decrease phase with a weakening erosion. The formerly settled solids are completely eroded from ≈145 mg/L up to ≈200 mg/L within 30 s. The resulting erosion rate (Figure 6e) shows an abrupt increase at the beginning, which is due to its moderate fitting results ($R^2 = 0.83$). Similar to the smoother decline at the end of the erosion event, a more gradual increase is assumed for the beginning. The maximum erosion appears at ≈0.43 N/m² and so, before the maximum shear stress level of ≈0.5 N/m² is reached. A further increase of shear stress (feasible by parallel pumping of P1 and P2) would not result in further solids erosion, as the maximum erosion level is already reached and the decline remarks the emptying of the sediment layer.

3.4. Comparison to Laboratory (Ex-Situ) Results

The results of the ex-situ laboratory experiments in [1] are similar to the in-situ measured erosion processes in this study. Both methods show the typical s-curve while eroding solids. Hence, the resulting erosion rates are quite similar. Especially the calculated duration for a complete resuspension in [1], with regard to similar hydraulic conditions (≈ 0.5 N/m^2 bed shear stress), is nearly equal to real world processes (duration ≈ 30 s). Thus, both methods (in-situ and ex-situ) are applicable to determine the erosion characteristics of raw sewage.

However, the in-situ characterization of the erosion events is much harder compared to the ex-situ method in [1], as the solids inside the pipe are moving in two main directions (upward and forward, micro-effects ignored). Therefore, next to the formerly settled solids directly under the TSS sensor, particles settled at the upstream section of the pressure pipe affecting the measurement. A closed reaction chamber, within the ex-situ experiments in [1], simplifies the measurement extremely, as the erosion process of a controlled suspension is detected. This may be another reason for the moderate fitting results (see previous chapter).

A direct comparison to the ex-situ settling experiments in [2] is not possible. The in-situ method measures the TSS reduction in the fluid phase (by a calibrated turbidity sensor) and the ex-situ method measures the total mass increase at the bottom of a cylinder (by weight loss). Furthermore, the mathematical description differs from [2]. However, by converting the cumulative growth from [2] into the fluids particle loss, an equal course to the settling rate is found (exponential shape). Furthermore, both methods counted approximately the same solids amount after similar settling durations. 25.4% of solids mass settle in a laboratory test within 17 min while approximately 27.5% of the solids settle within the same duration inside the pressure pipe.

A continuous measurement of solids decrease inside a fluid phase by a sensor is by far much easier and more worthwhile, than a manually ex-situ determination of solids mass growth. The laboratory experiment lags behind the sensor determination, because of the high effort in designing and construction, sampling, the experimental conduction and mass detection. The continuous and automated in-situ measurement scores by a unique installation, simple maintenance, and calibration, high-resolution measurement (5 s interval) and automated data processing.

3.5. Effect of Storm Water Inflow to Settling and Erosion Characteristics

Due to the connected road runoff, a changed erosion and sedimentation behavior is assumed by storm water inflow. Several storm events were measured during the study period. One example is shown in Figure 7. Figure 7a shows a rain event at 07:00 p.m. with 4.9 mm/h precipitation (right axis). The inflow curve shows the storm runoff slowly reaching the pump sump (see Q_{inflow}, left axis). The TSS course (left axis) is separated into erosion (blue) and sedimentation sequences (red). The TSS concentration increases significantly up to ≈ 600 mg/L after the peak runoff reaches the pump sump. Figure 7b compares a dry weather inflow erosion rate (1b) with a storm water erosion rate (2b). The maximum erosion increases with the storm inflow almost by a factor of five. One reason is the storm runoff composition. Solids (sand, tire abrasion, etc.) are washed off from roads and entering the sewer. Furthermore, the increased discharge erodes pre-settled and consolidated deposits and spills a mixture of runoff solids and sewer solids to the PS.

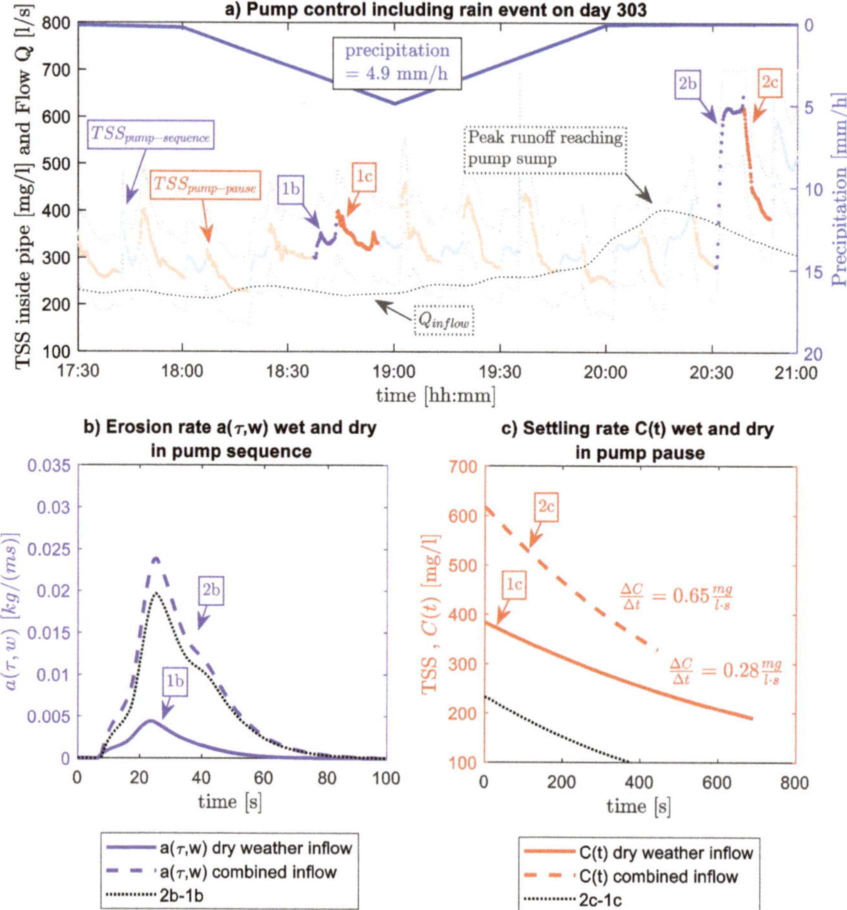

Figure 7. Effect of storm water inflow to erosion and sedimentation. (**a**) Q_{inflow} and TSS data (both on left axis) during rainfall event from 06:00 p.m. to 08:00 p.m. with a peak of 4.9 mm/h (right axis). (**b**) Erosion rate $a(\tau,w)$ during storm water inflow (2b) compared to dry weather inflow (1b). (**c**) Settling rate $C(t)$ during storm water inflow (2c) compared to dry weather inflow (1c) including mean decrease $\Delta C/\Delta t$. The German Weather Service (DWD) provides the precipitation data.

The following sedimentation process (2c) is significantly different compared to dry weather conditions (1c), accordingly. The increased TSS inflow raises the start value of sedimentation (C_0) from $\approx$400 mg/L up to >600 mg/L. A better comparison of the settling processes is provided by the mean decrease, which is defined as the difference quotient in the interval $[t_1;t_{end}]$, see Equation (9).

$$\frac{\Delta C}{\Delta t} = \frac{\left|C(t_{end}) - C(t_1)\right|}{t_{end} - t_1} \tag{9}$$

In this example, the TSS concentration decreases under dry weather inflow (1c) with 0.28 mg/(L s), while under storm water inflow (2c) twice as fast with 0.65 mg/(L s). Hence, next to the solids concentration, the solids composition changes as well. This indicated an increased inflow of heavier particles (runoff- and sewer solids). The total mean decrease under dry weather inflow, calculated over 4520 settling events, is 0.23 mg/(L s), while under storm water inflow 0.9 mg/(L s) (calculated over 9 rain events). Therefore, a more than 3.5-times faster settling can be assumed under storm water inflow. This increases the risk of blockages significantly. The risk is highest when the solids enter the pressure pipe. The storm runoff reaches the pump sump 1 h after the rain event and enters the pressure pipe usually 1.5 h after the rain event. The accumulation of particles on roads and the deposition of solids inside the sewers increases with longer dry periods. This is due to missing wash-off and limited discharge. Hence, the risk of blockages in the pressure pipe increases for short but intensive rain events after long dry weather periods. However, this effect is not investigated within this study.

3.6. Comparison to Laboratory (Ex-Situ) Results

A changed sedimentation behavior is already recognized in [2]. Samples, collected under storm water inflow settling significantly faster. Comparable laboratory tests showing a mean decrease for storm water samples by 1.1 mg/(L s), while 0.18 mg/(L s) for dry weather samples.

3.7. Diurnal Variation Settling and Erosion

Settling and erosion characteristics are changing not only with storm water inflow. As already mentioned previously, TSS follows the inflow course. As settling and erosion depends on TSS, their behavior follows TSS and consequently inflow. Hence, changes are also assumed due to the diurnal variation of inflow. This relation is shown in Figures 8 and 9. Both showing the average diurnal variation of inflow (black line), including boxplots for hourly mean decrease (Figure 8) and hourly maximum erosion per event (Figure 9).

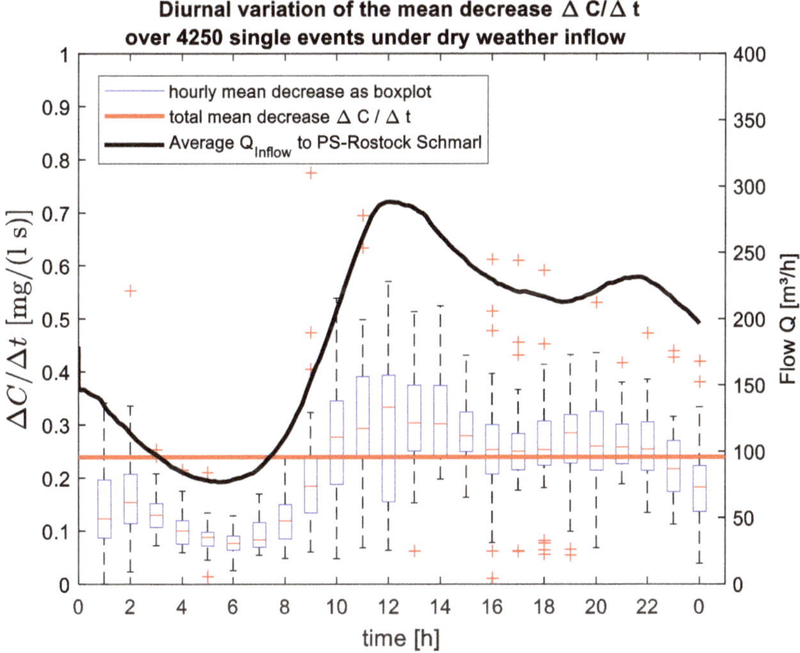

Figure 8. Diurnal variation of the mean decrease $\Delta C/\Delta t$ (boxplots) including total mean decrease over 4250 single settling events (red line) and the average Q_{inflow} for dry weather conditions (black line).

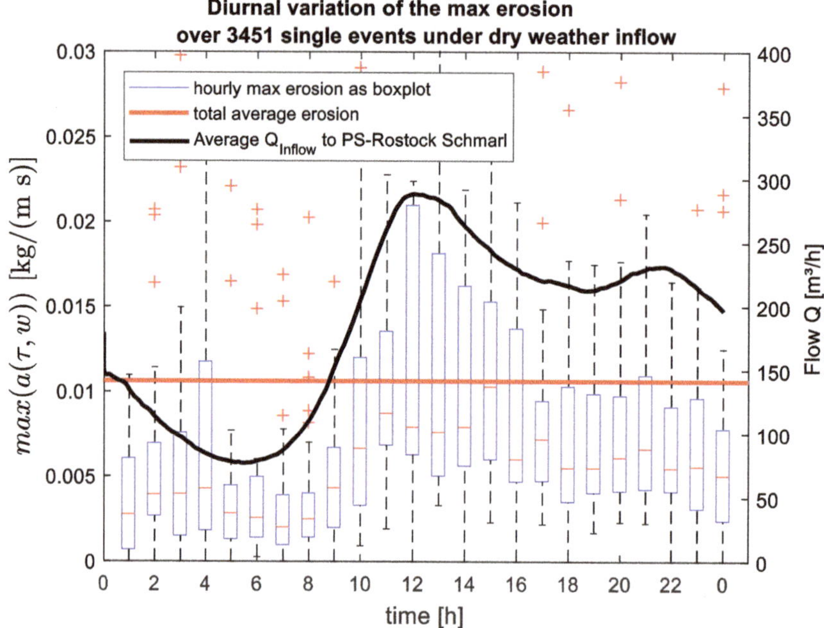

Figure 9. Diurnal variation of the max erosion (boxplots) including total average erosion over 3451 single erosion events (red line) and the average Q_{inflow} for dry weather conditions (black line).

Both indicators (mean decrease and maximum erosion) follow the up- and downward course of inflow. Especially in the morning, the swage is characterized by slow settling processes (<0.1 mg/(L s)) and low erosion rates (<0.0025 kg/(m s)). This can be explained by the reduced water usage in the night (reduced solids input, reduced hydraulic performance in upstream sewers). Vice versa, due to a high water usage in morning hours, peak inflows reaching the PS at lunch and changing the sewage to a faster settling mixture (up to 0.5 mg/(L s)). Accordingly, the erosion rate increases up to 0.02 kg/(m s). The increased erosion rate is a result of a faster settling process. The more solids settle within the pump pauses, the more solids can be eroded in the pump phases. Hence, the erosion rate depends on the pump pause duration. Longer pump pauses generally occur in the night or in the morning with low inflow rates (see [2]). Nevertheless, due to a slow settling sewage, the erodible amount of solids is low. The erodible amount is at its peak, when the settling process is fast (usually at midday), irrespective to pump pause duration. Concluding, the sewage settling characteristics (slow or fast) determine the deposits formation more than the duration of the pump pauses.

4. Conclusions

The paper presents a continuous in-situ TSS measurement system for raw sewage inside a pressure pipe and the determination and characterization of settling and erosion behavior based on high-resolution sensor data. Ultimately, the following findings are concluded:

- The installed sensors are suitable for supervision of TSS fluxes inside sewage pressure pipes;
- Periodically calibration and maintenance of TSS sensors result in reliable data;
- TSS sensor data allow for a characterization of solids sedimentation and erosion behavior;
- Measured in-situ erosion and settling results are similar to ex-situ (laboratory) results;
- Settling accelerates with high inflow rates (storm water inflow, diurnal inflow peaks) and decelerates with low inflow (reduced TSS inflow in night phases);

- Erosion rate increases and decreases based on the available amount of solids, hence, with changing settling behavior;
- Solids are eroded before maximum shear stress level reached

Within continuous sensor measurements, a huge amount of data is generated. Especially with regard to urban water simulations, this provides the opportunity for precise calibration up to specified scenarios. Hence, changes of solids erosion and sedimentation caused by storm water inflows of various intensity or by the diurnal inflow can be dynamically implemented into hydraulic models by providing a wide spectrum of appropriate calibration parameters. The presented results are primarily used for a sediment transport simulation inside the pressure pipe of PS Rostock-Schmarl, presented in Part II of this publication: "Sediment Transport in Sewage Pressure Pipes, Part II: 1D Numerical Simulation".

Author Contributions: Conceptualization, M.R., J.T. and T.K.; Data curation, M.R.; Formal analysis, M.R.; Funding acquisition, J.T.; Investigation, M.R.; Methodology, M.R.; Project administration, J.T.; Software, M.R.; Supervision, J.T.; Validation, M.R.; Visualization, M.R.; Writing—original draft, M.R.; Writing—review and editing, J.T. and T.K.

Funding: The presented studies are supported by Deutsche Bundesstiftung Umwelt (DBU), Germany (AZ 32253/01).

Acknowledgments: The precipitation data used are available from Germany's national meteorological service (Deutscher Wetterdienst (DWD)), http://www.dwd.de.

Conflicts of Interest: The authors declare no conflict of interest. The funders had no role in the design of the study; in the collection, analyses, or interpretation of data; in the writing of the manuscript, or in the decision to publish the results.

References

1. Rinas, M.; Tränckner, J.; Koegst, T. Erosion characteristics of raw sewage: Investigations for a pumping station in northern Germany under energy efficient pump control. *Water Sci. Technol.* **2018**, *78*, 1997–2007. [CrossRef] [PubMed]
2. Rinas, M.; Tränckner, J.; Koegst, T. Sedimentation of Raw Sewage: Investigations for a Pumping Station in Northern Germany under Energy-Efficient Pump Control. *Water* **2019**, *11*, 40. [CrossRef]
3. Seco, I.; Gómez Valentín, M.; Schellart, A.; Tait, S. Erosion resistance and behaviour of highly organic in-sewer sediment. *Water Sci. Technol.* **2014**, *69*, 672–679. [CrossRef] [PubMed]
4. Gromaire, M.C.; Kafi-Benyahia, M.; Gasperi, J.; Saad, M.; Moilleron, R.; Chebbo, G. Settling velocity of particulate pollutants from combined sewer wet weather discharges. *Water Sci. Technol.* **2008**, *58*, 2453–2465. [CrossRef] [PubMed]
5. Chebbo, G.; Gromaire, M.-C.; Lucas, E. Protocole VICAS: Mesure de la vitesse de chute des MES dans les effluents urbains. *TSM Tech. Sci. Methodes Génie Urbain Génie Rural* **2003**, *A98*, 39–49.
6. Regueiro-Picallo, M.; Naves, J.; Anta, J.; Suárez, J.; Puertas, J. Monitoring accumulation sediment characteristics in full scale sewer physical model with urban wastewater. *Water Sci. Technol.* **2017**, *76*, 115–123. [CrossRef] [PubMed]
7. Xu, Z.; Wu, J.; Li, H.; Liu, Z.; Chen, K.; Chen, H.; Xiong, L. Different erosion characteristics of sediment deposits in combined and storm sewers. *Water Sci. Technol.* **2017**, *75*, 1922–1931. [CrossRef] [PubMed]
8. Bersinger, T.; Le Hécho, I.; Bareille, G.; Pigot, T.; Lecomte, A. Continuous Monitoring of Turbidity and Conductivity in Wastewater Networks. *Rev. Des Sci. De L'eau* **2015**, *28*, 9. [CrossRef]
9. Bersinger, T.; Le Hécho, I.; Bareille, G.; Pigot, T. Assessment of erosion and sedimentation dynamic in a combined sewer network using online turbidity monitoring. *Water Sci. Technol.* **2015**, *72*, 1375–1382. [CrossRef] [PubMed]
10. Lacour, C.; Joannis, C.; Chebbo, G. Assessment of annual pollutant loads in combined sewers from continuous turbidity measurements: Sensitivity to calibration data. *Water Res.* **2009**, *43*, 2179–2190. [CrossRef] [PubMed]
11. Métadier, M.; Bertrand-Krajewski, J.-L. From mess to mass: A methodology for calculating storm event pollutant loads with their uncertainties, from continuous raw data time series. *Water Sci. Technol.* **2011**, *63*, 369–376. [CrossRef] [PubMed]

12. Métadier, M.; Bertrand-Krajewski, J.-L. The use of long-term on-line turbidity measurements for the calculation of urban stormwater pollutant concentrations, loads, pollutographs and intra-event fluxes. *Water Res.* **2012**, *46*, 6836–6856. [CrossRef] [PubMed]

13. Lacour, C.; Joannis, C.; Gromaire, M.-C.; Chebbo, G. Potential of turbidity monitoring for real time control of pollutant discharge in sewers during rainfall events. *Water Sci. Technol.* **2009**, *59*, 1471–1478. [CrossRef] [PubMed]

14. Sun, S.; Barraud, S.; Castebrunet, H.; Aubin, J.-B.; Marmonier, P. Long-term stormwater quantity and quality analysis using continuous measurements in a French urban catchment. *Water Res.* **2015**, *85*, 432–442. [CrossRef] [PubMed]

15. Bertrand-Krajewski, J.-L. TSS concentration in sewers estimated from turbidity measurements by means of linear regression accounting for uncertainties in both variables. *Water Sci. Technol.* **2004**, *50*, 81–88. [CrossRef] [PubMed]

16. Bertrand-Krajewski, J.-L.; Bardin, J.-P. Evaluation of uncertainties in urban hydrology: Application to volumes and pollutant loads in a storage and settling tank. *Water Sci. Technol.* **2002**, *45*, 437–444. [CrossRef] [PubMed]

17. Knubbe, A.; Fricke, A.; Ecktädt, H.; Neymeyr, K.; Schwarz, M.; Tränckner, J. Energieeffizienter Betrieb von Abwasserfördersystemen Energy efficient strategies for wastewater pumping systems. *Gwf. Wasser|Abwasser* **2014**, *155*, 640–646.

18. HACH-LANGE GmbH. *SOLITAX sc User Manual: Edition 4A*; HACH LANGE GmbH: Düsseldorf, Germany, 2009; Available online: https://de.hach.com/asset-get.download.jsa?id=25593604876 (accessed on 12 October 2019).

19. Deutsches Institut für Normung e.V. (DIN). *Deutsche Einheitsverfahren zur Wasser-, Abwasser- und Schlammuntersuchung; Summarische Wirkungs- und Stoffkenngrößen (Gruppe H); Bestimmung des Gesamttrockenrückstandes, des Filtrattrockenrückstandes und des Glührückstandes (H 1) (German Standard Methods for the Examination of Water, Waste Water and Sludge; General Measures of Effects and Substances (Group H); Determination of the Total Solids Residue, the Filtrate Solids Residue and the Residue on Ignition (H 1))*; DIN ISO 38414-S; Beuth Verlag GmbH: Berlin, Germany, 1987.

20. International Organization for Standardization (ISO). *Uncertainty of Measurement—Part 3: Guide to the Expression of Uncertainty in Measurement (GUM:1995)*; ISO/IEC Guide 98-3:2008(E); ISO: Geneva, Switzerland, 2008.

Article

Sediment Transport in Sewage Pressure Pipes, Part II: 1 D Numerical Simulation

Martin Rinas [1,*], **Alexander Fricke** [2], **Jens Tränckner** [1], **Kurt Frischmuth** [2] **and Thilo Koegst** [1]

[1] Department of Water Management, University of Rostock, Satower Straße 48, 18059 Rostock, Germany; jens.traenckner@uni-rostock.de (J.T.); t.koegst@stalums.mv-regierung.de (T.K.)

[2] Institute of Mathematics, University of Rostock, Ulmenstraße 69, 18055 Rostock, Germany; alexander.fricke@uni-rostock.de (A.F.); kurt.frischmuth@uni-rostock.de (K.F.)

* Correspondence: martin.rinas@uni-rostock.de

Received: 17 December 2019; Accepted: 15 January 2020; Published: 18 January 2020

Abstract: Urban drainage modelling is a state-of-the-art tool to understand urban water cycles. Nevertheless, there are gaps in knowledge of urban water modelling. In particular pressure drainage systems are hardly considered in the scientific investigation of urban drainage systems, although they represent an important link in its network structure. This work is the conclusion of a series of investigations that have dealt intensively with pressure drainage systems. In particular, this involves the transport of sediments in pressure pipes. In a real-world case study, sediment transport inside a pressure pipe in an urban region in northern Germany was monitored by online total suspended solids measurements. This in situ data is used in this study for the development and calibration of a sediment transport model. The model is applied to investigate sediments transport under low flow velocities (due to energy saving intentions). The resulting simulation over 30 days pumping operation shows that a transport of sediments even at very low flow velocities of 0.27 m/s and under various inflow conditions (dry weather and storm water inflow) is feasible. Hence, with the help of the presented sediment transport model, energy-efficient pump controls can be developed without increasing the risk of deposition formation.

Keywords: sediment transport model; numerical simulation; advection-dispersion equation; total suspended solids; erosion; sedimentation; erosion; pressure pipe

1. Introduction

Numerical simulations are state of the art in challenging problems in urban water management. Whether in wastewater treatment, trying to optimize single or multiple treatment processes, or in sewer systems, for hydraulic optimization or planning and design. Modelling is today's tool to solve complex problems efficiently.

Dealing with hydraulic problems, modelling focus lies on non-pressure systems (open-channel flow), mainly driven by heavy rain events, combined sewer overflows (CSO), pollutant loads, etc. Hence, the main effort of scientific research concentrates on gravity sewers. Certainly, hydraulic simulation of pressurized systems is mostly a part of urban drainage modelling software, but this becomes almost insignificant in engineering science. Currently, a reason might be due to more relevant problems occurring in gravity sewers such as overload, flooding, CSO's, or fat deposition. All these issues provide opportunities for research. Whereas pumping systems are considered to be safe and trouble-free systems, which may be due to their controllability (flow control by speed regulation, pumps in series or parallel, pumps switching, etc.). The possibility to control results in an almost absolute steady operation, thus, equal flow processes and subsequently the assumption of a uniform and unchanging environment inside the pressure pipe.

However, there are several disadvantages of pumping systems. One main drawback is the use of energy. For the transportation of fluids, pump power has to overcome, beside the geodetic height difference, the sum of friction losses. This sum increases with the square of the flow velocity (according to the calculation of friction losses by Darcy–Weisbach). Conversely, reducing the speed by half, quarters the dynamic friction losses and subsequently reduces energy consumption. Therefore, a large energy saving potential in urban drainage is related to the operation of sewage pumps. So, reducing flow velocity is the key to energy optimization. The reduction has several benefits (next to energy saving): It increases the pump duration, reduces the off/on switching frequency, and homogenizes the flow to the downstream sewer system. But it also comes with a significant disadvantage. It might increase the risks of sedimentation and subsequently blockages, as solids are settling when velocity and resulting bed shear stress is below a critical level. Another disadvantage may occur due to the increased retention time inside the pressure pipe. The decomposition of sewage may be increased, leading to the formation of toxic and corrosive gases (e.g., hydrogen sulfide). This problem can be engaged by a chemical precipitation.

Especially with regard to sedimentation and erosion, a numerical simulation is now of interest. Within a case study in an urban region in northern Germany (city of Rostock), solids transport inside a pressure pipe was investigated within several ex situ (laboratory) experiments [1,2] and continuously monitored by in situ turbidity measurements for one year under energy efficient pump control [3]. However, the energy efficient control was only permitted in certain range of operation conditions. Especially low flow velocities could not be realized due to the risk of blockages. Monitoring of solids transport under low flow velocities, or after longer pump pauses, could not be investigated. Hence, a sediment transport model was developed and calibrated, to extrapolate the observed data into the restricted range of operation.

In this work, the sediment transport model is introduced, calibrated, and used to simulate the above mentioned conditions. This publication implies the following main objectives:

- Derive a physical, but still simple, numerical model for solids transport inside sewage pressure pipes;
- Calibrate the model based on ex and in situ determined sedimentation and erosion characteristics;
- Determine the accuracy of the transport simulation;
- Investigate and evaluate solids transport under various flow regimes.

Literature Review

The simulation of non-cohesive sediment transport in sewers can be split up into morphological and mathematical models. Morphological (or detailed sediment transport) models uses, as the name suggests, the (mostly abstracted) morphology of the particles to be transported. Common morphological models are the "Bagnold Model" [4], the "Engelund–Hansen Model" [5], the "Ackers–White Model" [6], the "Engelund & Frodsøe Model" [7], and the "van Rijn Sediment Transport Model" [8]. Several morphological models found their way into hydrological and hydraulic modelling, whether in river modeling (e.g., HEC-RAS) or urban water modelling (e.g., DHI Mike Urban). All these models are driven by the main physical force reacting to the particles, which is shear stress. Hydraulic data is received by a hydrodynamic simulation. The sediment transport is then uncoupled (water flow and sediment transport not interacting), semi-coupled (water flow and sediment transport interacting by iteration of uncoupled model, e.g., [9–11]), or fully coupled (simultaneous computation of flow and sediment transport, e.g., [12,13]) to the hydraulic computation.

Mathematical models bases on the one-dimensional advection-dispersion equation (ADE), describing the mass conservation of substances transported in direction of the mean flow velocity. The use of the ADE is widely spread, e.g., for modelling the transport of dissolved substances in natural flow processes (e.g., inside ground water bodies or river flows) or modelling substances in urban drainage systems or water distribution.

Hydrodynamic urban drainage models are regarding 1 D channel flow by solving the Saint Venant equation. The ADE is then commonly used by urban water modelling software (DHI Mike Urban [14]) for modelling the transport of dissolved substances/pollutants (e.g., organic pollutions (biochemical oxygen demand)). Nevertheless, the ADE is not exclusively limited to the transport of dissolved substances. [14] mentioned the ADE "can also be used for the simulation of suspended (fine) fraction of particulate pollutants and sediments" ([14] p. 99).

However, irrespective of the substance to be transported or the transport formulation approach (detailed or mathematically), the sediment transport is usually described in literature/computed by software for non-pressure systems (open-channel flow, pipes). The transport of substances in urban drainage pressure pipes is commonly not computed by any transport simulation. Pumping systems are usually modelled without pressure mains, but rather as a direct connection between the sump and the end-node with a fixed pump capacity (see also [14]). As a result, the "dissolved matter is routed through such a system with no time lag between the pump and the end of the conduit" ([14] p. 108).

The application of ADE for modelling transport of dissolved/particulate substances inside pressure pipes is common for water distribution systems. Modelling water quality within the simulation software EPANET 2 provides possibilities to transport dissolved substances through a pressurized system [15] (also implemented in DHI MIKE URBAN WD tool). The basic idea of the substance transport within this study is similar to the water quality modelling idea in EPANET 2.

Next to numerical approximations, soft computing methods (e.g., artificial neural networks, fuzzy logic, and evolutionary computation) also tried to approximate real-life problems and so used to estimate the sediment transport in urban drainage systems ([16–21]).

The presented 1 D sediment transport simulation in this publication is computed by an uncoupled mathematical model that is based on the ADE.

2. Materials and Methods

2.1. Study Area, Pump Control, and Monitoring Total Suspended Solids

The supervision of solids transport under an energy efficient pump control was implemented in pumping station (PS) Rostock-Schmarl in the city of Rostock (northern Germany). PS Rostock-Schmarl conveys the raw sewage of ≈40,000 inhabitants directly, via two cast iron pipelines of 600 mm diameter and 4500 m length, to the central wastewater treatment plant (wwtp) in Rostock, by four pumps of 220 kW total pump power. The incoming sewage is filtered by a 20 mm rake at the inflow side of the PS. Under dry weather inflow the total suspended solids (TSS) concentration usually ranges from 150 mg/L up to 350 mg/L. The upstream, usually separating sewer sums up to 80 km. Under rainfall, the main roads storm runoff is connected to the upstream sewer (total suspended solids concentration then increases to >500 mg/L). A parallel storm sewer receives the residual surface and roof runoff. Additional information about the study area are provided by [1–3] which includes a detailed schematic view of the sewer system and PS Rostock-Schmarl.

The usual operation mode of PS Rostock-Schmarl is a conventional two-point operation, where pumps switch on and off at pre-defined water levels inside the pump sump (sloped, squared geometry with a volume of 178 m^3). When pumps switch on, the variable frequency drive guarantees a soft start of the pumps. After a one minute soft start, the pumps operate at a defined duty point. In full power mode, the pumps duty point is then at ≈166 L/s (head loss ≈ 18.3 m) by ≈0.6 m/s flow velocity, respectively. The usual operation mode is the reduced two-point control where the flow decreases down to ≈100 L/s (head loss ≈ 17 m) by ≈0.35 m/s flow velocity, respectively. For a studied period of one year, an energy saving operation was implemented, to control two pumps and enhance energy efficiency. At low inflow, the duty point decreases to ≈76.5 L/s (head loss ≈ 16.7 m) at 0.27 m/s flow velocity. Energy savings amount to 11% compared to conventional operation. The following constraints mainly hindered additional energy savings: (i) minimum flow rate of ≈53 L/s (flow velocity ≈ 0.2 m/s) and (ii) pumps forced were to start up to maximum flow in each pump sequence, before regulating down to energy efficient flow. These rules are defined by the operator to ensure solids transport and to avoid blockages. For detailed information about the control modes, read [1,22,23].

In parallel, measuring TSS by two turbidity sensors directly inside the pressure pipe (one at pumps pressure side and one at the outflow side in wwtp Rostock) ensured a continuous monitoring of solids transport. Furthermore, the measurements provided data for continuous determination of solids settling and erosion characteristics and subsequently data for model calibration. The monitoring system is explained in depth in [3]. Figure 1 shows a simple schematic view of the study side.

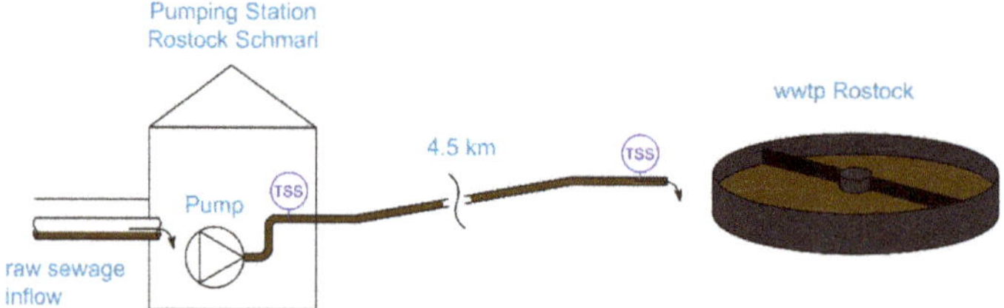

Figure 1. Schematic view of the study side. Total suspended solids (TSS) are measured inside the pressure pipe in pumping station (PS) Rostock-Schmarl and at the outflow side in the wastewater treatment plant (wwtp) Rostock after 4.5 km.

2.2. Sediment Transport Basics

The transportation of solids inside a fluid is mainly influenced by two main physical effects: (i) sedimentation and (ii) erosion. Equally to those two effects, two operation modes can be distinguished in pressurized flow: (i) pump pauses (shut-off mode), where only sedimentation occurs, and (ii) pump sequences (shut-on mode), where solids eroded and subsequently transported under adequately hydraulic conditions.

In pump pauses, solids are settling according to their settling velocity at different speeds, mainly determined by their density and size. The sediment layer at pipes invert is then a superposition of different solid fractions. The settling behavior of various particle fractions can be described as a settling velocity distribution, as conducted in [1]. The computation of several particle fractions is useful for dealing with pollutant transport when specific fractions contain more pollutants than other components (see [24]). However, modelling various velocity classes also requires larger coding and computing effort. In contrast, pure mass growth of solids, when only TSS fluxes are of interest, can be modeled and calculated easily as an appropriate solution of a usual differential equation, e.g., as an exponential decay of solids concentration inside the fluid, as introduced in [3].

In pump sequences, solids eroding due to the force exerted by the turbulent fluid flow inside the pressure pipe (Reynolds number of 240,000 at 0.4 m/s flow velocity). The force is expressed by the shear stress τ (N/m^2) or indicated more exactly by the bed shear stress. When pumps speed up, flow velocity increases until τ reaches the critical shear stress level of the particles τ_{crit} (N/m^2). Because of the different densities of particles, solids are eroding successively after τ_{crit} level of the lightest particle fraction is reached, until all particles are eroded. Solids are then transported either as bed load (sliding, rolling, and leaping of single particles) or suspended load, where particles are following the swirled streamlines of the turbulent flow. Suspended solids are then also moving transverse to the flow direction. The physical processes within the erosion are by far more complex as pure sedimentation of particles. So, the implementation into a simulation results in high computation effort. If the flow is always turbulent (here Reynolds number of 120,000 by a minimum flow velocity of 0.2 m/s) and only a mass balance is required for a simulation, swirls and micro effects can be neglected.

2.3. Mathematical Approximation

The solid transport is simulated for the above described pressure pipe (length $l = 4500$ m, diameter $d = 0.6$ m), conveying (mechanical pre-treated) raw sewage and sequentially combined sewage to the wwtp Rostock. The mathematical model is based on the advection-dispersion equation, Equation (1).

$$\frac{\delta u}{\delta t} = -v\cdot\frac{\delta u}{\delta x} + \frac{\delta}{\delta x}D_{xx}\frac{\delta u}{\delta x} + r \tag{1}$$

In Equation (1), the advective transport is represented by the first term $-v\cdot\frac{\delta u}{\delta x}$ (kg/(m^3 s)), the dispersive transport is represented by the second term $\frac{\delta}{\delta x}D_{xx}\frac{\delta u}{\delta x}$ (kg/(m^3 s)), and the reaction of a substance is represented by the third term r (kg/(m^3 s)). With u being the concentration of a substance to be transported (kg/m^3), t the time coordinate (s), v the flow velocity in flow direction (m/s), x the space coordinate (m), and D_{xx} the dispersion coefficient (m^2/s). The reaction of substances r is defined by Equation (2). With P being the production of a substance to be transported (kg/(m^3 s)) and S the sinking or degradation of a substance to be transported (kg/(m^3 s)).

$$r = P - S \tag{2}$$

The complex physical processes during sedimentation and erosion are unnecessary for pure modelling of the sediment flux. Hence, the following simplifications are assumed for the simulation. The sedimentation of different solid fractions, as described in [1], is not considered. The settling process is described as an exponential decay of solids, similar to [3]. This minimizes the particle fractions to simulate and approximates the settling of solids adequately. The erosion of solids is described by a single particle fraction as well. Once eroded, the particles are distributed homogeneously over the pipes bottom, as the mean flow velocity is uniformly distributed over the pipes cross section in the model. As already mentioned in [15] (p. 193) "longitudinal dispersion is usually not an important transport mechanism under most operating conditions" (with regard to dissolved substances). As a result, dispersion effects of contiguous grid sections is neglected completely (similar to [15] p. 193). Further simplifications are: No attention is paid to biogenic processes inside the fluid or deposits phase, the pipes geometrical shape is ignored as the 1 D sediment transport is computed in longitudinal direction, the pipes negative or positive slope, and curvature has been ignored too. As a result of the simplification, the advection-dispersion equation simplifies to the one-dimensional advection equation with only the production and sinking left, Equation (3) (similar to [15] p. 193).

$$\frac{\delta u}{\delta t} + v\cdot\frac{\delta u}{\delta x} = P - S \tag{3}$$

The transportation of solids is simulated as a mass balance (conservation of mass) for the suspended load and bed load along the pressure pipe. Equation (4) describes the conservation of mass for suspended load transport. Equation (5) describes the conservation of mass for bed load:

$$\frac{\delta u}{\delta t} + v\cdot\frac{\delta u}{\delta x} = \frac{a(w,v)}{A} - s(u) \tag{4}$$

$$\frac{\delta w}{\delta t} = A\cdot s(u) - a(w,v) \tag{5}$$

The suspended load transport is defined as feeding minus loss. Therefore, the production P is replaced by the erosion of solids $a(w,v)$ (kg/(m s)) and the degradation S is replaced by the particle loss inside the fluid $s(u)$ (kg/(m^3 s)). Thus, eroded mass per pipe length and time $a(w,v)$ divided by pipes cross section A (m^2) minus the particle loss inside the fluid $s(u)$ represents the suspended load.

The bed load (Equation (5)) is as well defined as feeding minus loss. Here, the feeding is described by the particle loss $s(u)$ multiplied by A minus the eroded mass $a(w,v)$. Erosion $a(w,v)$ is described by

Equation (6), with $\tau_{pipe}(v)$ (N/m^2) the current bed shear stress inside the pressure pipe, $\tau_{crit}(w)$ (N/m^2) the critical shear stress level of the raw sewage dependent on the particle mass on pipes invert w (kg/m) and the erosion parameter d (s), which describes the strength of the erosion (see [2]).

$$a(w,v) = max\left(0, d\left(\tau_{pipe}(v) - \tau_{crit}(w)\right)\cdot w\right) \tag{6}$$

The sedimentation is described as a first order decay process modeled by Equation (7).

$$\frac{du}{dt} = -\alpha u \tag{7}$$

The differential equation is approximated by the particle loss $s(u)$ by Equation (8), with the settling parameter α (1/s), which determines the exponential decay and the particle concentration inside the fluid section, u (kg/m^3) (see also [3,25]):

$$s(u) = u\cdot\alpha \tag{8}$$

2.4. Numerical Method

The partial differential equations (PDE) are solved by a finite difference method (FDM) (partial derivatives are approximated as finite differences). FDM's within water-quality modelling were i.a. investigated by [26]. The authors conclude, that the FDM method is, next to others, "capable of adequately representing observed water-quality behavior" ([26], p. 146).

The FDM method applied here is an explicit/implicit finite difference scheme centered in time (discretization in time) and backward in space (discretization in space). The centered in time scheme taking the average value between time steps n and $n + 1$ (also known as the Crank–Nicolson scheme). The first value n is calculated based on the previously computed value $n - 1$ (explicit), while the second value $n + 1$ is calculated based on the formerly computed value n within the same time step (implicit). The advective transport with the mean flow velocity results in the backward in space (or upwind) scheme, where only transport in the flow direction (from backward grid point) is allowed.

For the 1 D sediment transport model, the pipe is separated into 900 segments, with a length of Δl = 5 m. For a stable solution, the Courant–Friedrichs–Lewy condition is defined as $\left|\frac{v\cdot\Delta t}{\Delta l}\right| \leq 1$. The time increment Δt for a simulation step is then defined as $\Delta t \leq \frac{\Delta l}{|v|}$. Hence, the advective transport cannot be faster than one grid point per time step. Assuming a flow velocity of $v = 0.5$ m/s, the time increment of a simulation step is then calculated to $\Delta t = 10$ s.

The numerical simulation is realized in several Matlab functions. The pump control strategies of PS Rostock-Schmarl (several regular and energy efficient control modes) are implemented into the transport simulation to investigate the particle transport under various control modes. Hence, the sediment transport simulation is based on the mean flow velocity, computed by a previous pumping simulation.

2.5. Calibration Parameters

As described by Equations (6) and (8), the parameters α, d, and τ_{crit} are mainly responsible for the sedimentation and erosion behavior and subsequently important for model calibration. The calibration of the transport simulation is based on two parameter sets: (i) ex situ parameters, resulting from laboratory experiments and (ii) in situ parameters, resulting from continuous turbidity measurement.

Ex situ parameters are provided by laboratory experiments, dealing with sedimentation [1] and erosion [2] of raw sewage samples from PS-Rostock Schmarl. Both experiments are explained in a few words: (i) the sedimentation tests are conducted in a vertical cylinder, the deposited mass is determined after various settling durations, and the main outcome are growth curves for settled solids mass. (ii) the erosion is tested in a vertical cylinder, the raw sewage is stirred until particles eroding from the ground level, the resulting erosion rates are detected by a continuous turbidity measurement, and the calibration parameters are derived from the growth curves and erosion rates. Hence, the parameters are based on a great simplification of real-world conditions. For the settling parameter α, values between 0.0036 s^{-1} and 0.032 s^{-1} were determined, for settling periods of 24 h. The erosion parameter d was determined between 0.057 s and 0.56 s after settling for 24 h.

Another calibration parameter is τ_{crit}. The critical bed shear stress defines the erosion limit and depends on the previous settling duration (long settling periods resulting in high τ_{crit} values). The determination of τ_{crit} results in values of 0.08 N/m^2 for settling periods up to one hour and <0.2 N/m^2 for settling periods of up to three days (see [2]). The bed shear stress inside the studied pressure pipe is calculated by Equation (9), based on the fluid density ρ (kg/m^3), the flow velocity v (m/s), and the friction factor λ (calculated by the Colebrook–White equation).

$$\tau_{pipe} = \rho \cdot \frac{v^2}{2} \cdot \frac{\lambda}{4} \tag{9}$$

τ_{pipe} is calculated for the minimum flow velocity of $\approx$0.2 m/s to $\approx$0.1 N/m^2. Hence, the minimum flow velocity reaches the τ_{crit} value for one hour prior settling (0.08 N/m^2). At PS Rostock-Schmarl, pump pauses above one hour are prevented by a control regulation: At least one pump starts per hour to avoid blockages. But due to pumps soft start, the motor speed is regulated from zero flow up to full flow within 60 s, which results temporarily in flow velocities <0.2 m/s. To always have the correct τ_{crit} value before pumps start, τ_{crit} is implemented into the simulation as a function of the prior settling duration. The present value of τ_{crit} is received during the simulation by an interpolation between the laboratory τ_{crit} results of [2].

The second parameter set are the in situ parameters, provided by a continuous turbidity measurement, as described in [3]. Mass growths and erosion rates are determined inside the pressure pipe, as the pump pauses and pump sequences representing the settling and erosion experiments, respectively. By this, a wide range of calibration parameters are derived under real world conditions. After analyzing 6733 single settling events (or pump pauses), the settling parameter α is calculated to an average value of 0.0026 s^{-1}. The erosion parameter d is calculated to an average value of 0.018 s, after an analysis of 6653 single erosion events (or pump sequences).

However, due to the large number of events, a much more precise classification than pure average values can be achieved. In [3], a diurnal variation of the settling and erosion behavior along with the variation of the inflow rate (and accordingly the TSS inflow) was detected. The changing settling and erosion behavior are reflected by the settling parameter α and the erosion parameter d. Hence, if the transport parameters dynamically change with the TSS inflow during simulation, the model always computes the appropriate settling or erosion process. After combining the settling parameter α of each single settling event (6733 in total) to the present TSS value, a simple linear function derives, Equation (10), with u_{inflow} (mg/L), the TSS inflow concentration.

$$\alpha\left(u_{inflow}\right) = 1.06 \cdot 10^{-5} \cdot u_{inflow} \tag{10}$$

The erosion parameter d changes proportionally with the settling parameter α. The ratio is determined to 1/0.1902. d is then calculated by Equation (11).

$$d(\alpha) = \frac{\alpha}{0.1902} \tag{11}$$

To illustrate the continuous adaptation of the model parameters, Figure 2 shows, exemplarily, the status bar of the sediment transport simulation. At the current simulation time 01:36 p.m. the pumps generate a flow velocity of 0.3 m/s. Figure 2a–d show the simulated values for suspended load (Figure 2a), bed load (Figure 2b), the settling parameter α (Figure 2c), and the erosion parameter d (Figure 2d) over pipe length (4500 m).

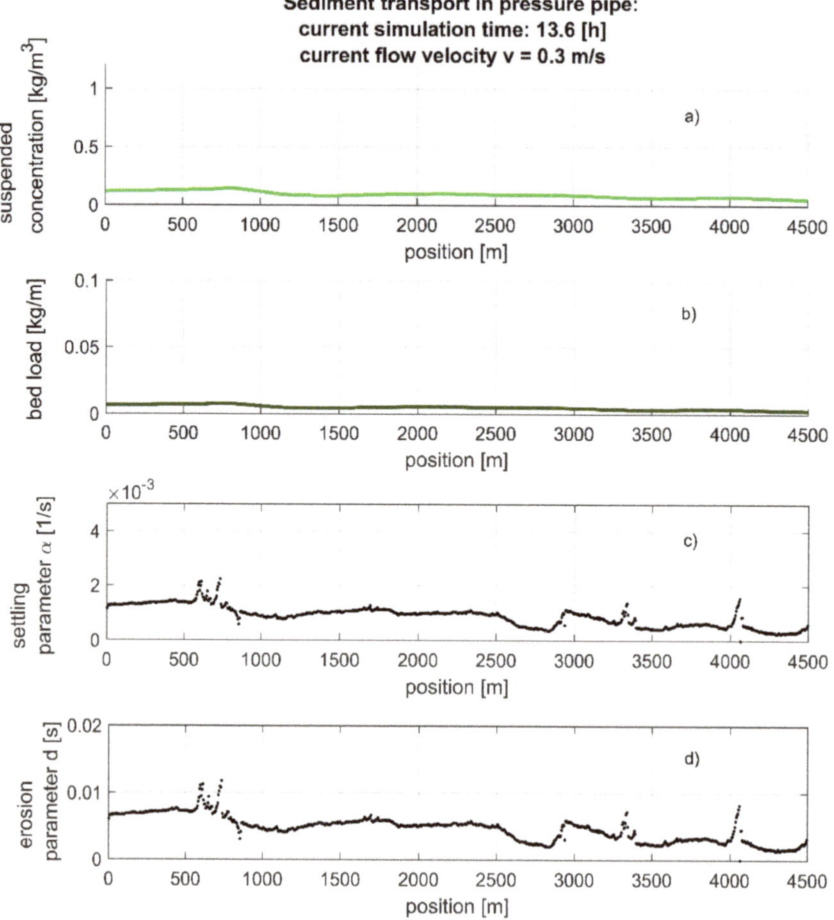

Figure 2. Status panel of the sediment transport simulation. (**a**) present suspended load inside the pressure pipe at each grid position. (**b**) present bed load inside the pressure pipe at each grid position. (**c**) present settling parameter α inside the pressure pipe at each grid position. (**d**) present erosion parameter d inside the pressure pipe at each grid position.

3. Results and Discussion

3.1. Evaluation of Model Accuracy

The evaluation of the numerical solution is based on a simple comparison. The reference values were compared to the simulation results. The best solution (minimal deviation to reference values) provides the best parameter set for later simulation of scenarios. The reference values were measured at the pipes end by the calibrated monitoring, see [3]. The simulation was conducted with (i) static ex situ parameters, (ii) static in situ parameters, and (iii) dynamic in situ parameters.

Figure 3 depicts the results of the sediment transport simulation for a typical dry weather inflow at PS-Rostock Schmarl. With the original particle concentration measurement at pipes end representing the reference values (black line). The simulation provides representative results after 13 h, as the pipe is filled with clear water at the beginning of the computation. All three simulations generally follow the measured TSS course. The variable parameter set (green line, $\alpha = 1.06 \cdot 10^{-5} \cdot u_{inflow}$, $d = \alpha/0.1902$)

only differs marginally from the reference values. The simulation with in situ parameters (blue line, α = 0.0017 s^{-1}, d = 0.0189 s) and laboratory parameters (red line, α = 0.00033 s^{-1}, d = 0.0362 s) differ to larger extent. In particular, particles are transported too fast and reach the end of the pressure pipe too early. This is related to an underestimation of the settling process. The simulated TSS by laboratory parameters never decreases below 0.25 kg/m^3, while the original measurement decreases partially below 0.15 kg/m^3. So the particles, mostly settle more completely in real life than expected by the laboratory results from [1]. One reason might be in the method (ex situ) itself, as several uncertainties overlay (sampling, experimental setup, and laboratory analysis) and increase the measurement error. The computation with in situ parameters resulted, by far, in better settling prediction, but contains a time shift as well. The simulated curve illustrates the advantage of the real-life measurement with less uncertainties, results in a more precise computation. With the variable parameter set (derived from all in situ parameter sets), a realistic particle concentration profile is simulated. This parameter set will further be used to compute several scenarios.

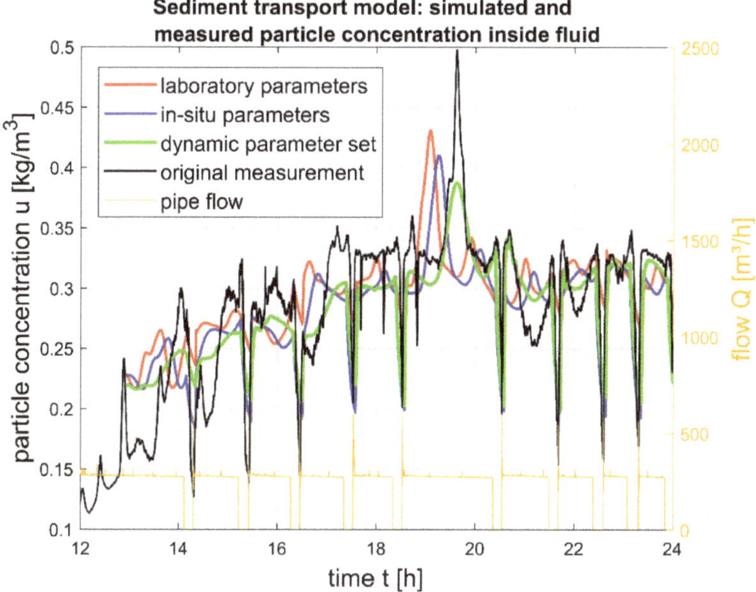

Figure 3. Results of the sediment transport simulation. Particle concentration measured and simulated with laboratory, in situ, and dynamic parameters with slightest deviation.

The present deviation to the reference values are shown in Figure 4a and calculated as $(u_{model} - u_{reference})/u_{reference}$. A positive deviation within the settling sequences ($Q = 0$) remarks an underestimation of the settling process, while reversely a positive deviation within the erosion sequence remarks an overestimation of the real erosion process. Both, under and overestimation occurs within the settling and erosion sequences, with maximum values of $\approx$+109% (for laboratory parameters) (underestimation of settling sequence). The variable parameter set simulation deviated at a maximum of +75% (underestimation of settling sequence) at the beginning and reduced to $\approx$+50% during the later simulation.

The largest deviations occur during settling sequences. The negative deviation at the beginning of the settling sequence (see Figure 4a) remarks an overestimation, the particles are settling too fast. Later they are settling too slow, remarked by the positive deviation. Figure 4b shows the simulated particle concentration for an exemplary settling sequence. As one can see, the real settling process (black line) seemed to be delayed. The settling process started significantly after pumps stop ($Q = 0$). This behavior can be explained by the turbidity measurement. The reference values were measured only in the sensor section (in pipes central region). If the pump process stops, particles settled immediately over the complete pipe section. The particle concentration in the complete fluid section decreases, while it stagnated in the sensor section. Particles settling out of the sensor section were compensated by particles settling into this section. As a result, the reference particle concentration decreases later than expected by the computation.

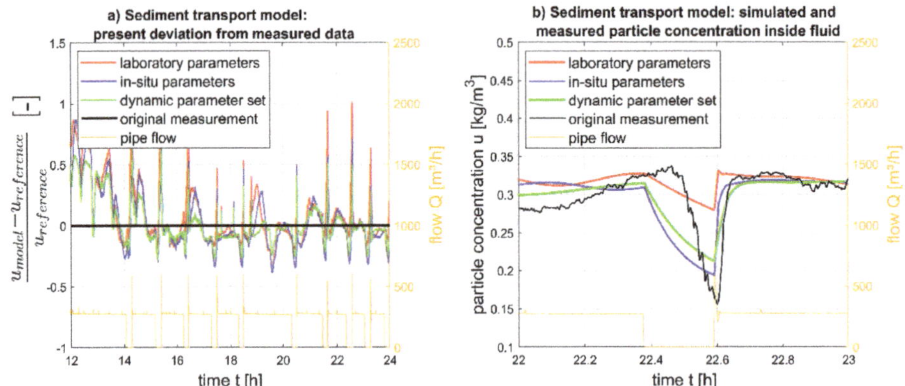

Figure 4. Results of the sediment transport simulation. (**a**) present deviation from all three simulated parameter sets. (**b**) settling sequence detail.

However, the simulated settling illustrates real life conditions, as the computed mass balance relates to the complete fluid section. This results, especially during the pump pauses, in a varying deviation. A trend cannot be detected visually. Therefore, Figure 5 shows the cumulated particle mass, calculated from the measurement and all three simulations. The particle mass, transported through the pipe, amounts to 1164 kg (reference values). All three simulations were within a 10% deviation (±164 kg). The in situ and variable parameter sets were within 5% deviation (±82 kg). Hence, the sediment transport model can be used with the variable parameter set to simulate the particle fluxes inside a pressure pipe appropriately.

3.2. Sediment Transport under Various Regimes

The calibrated sediment transport model offers various possible applications in the field of urban drainage. We concentrate on the investigation of sediment transport inside pressure pipe under different flow regimes. Hence, the simulation of different pump control modes is the basis. Step one: We computed the mean flow velocity over a defined period, by a pump simulation. Step two: We computed the particle transport based on the previously simulated mean flow velocity, by a sediment transport simulation. The simulated pump control modes are summarized in Table 1.

The main control rules from Table 1 represent real life pump control modes of PS Rostock-Schmarl appropriately. The control modes differ mainly in the reduction of the duty. Furthermore, some pump restrictions were added or left out, see Table 1.

The measured inflow and TSS hydrographs at PS Rostock-Schmarl represent the incoming sewage and TSS flow for the pumping simulation. To compare the different pumping strategies, all five control modes were simulated over the same period of 30 days. The results were evaluated for bed load, suspended load, and resulting energy consumption.

The resulting bed load and suspended load transport is shown in Figure 6. Figure 6a shows the cumulative bed load while Figure 6b,c shows the cumulative suspended load over 30 days simulation. Table 2 summarizes the simulation results for all five pump control modes including the calculated energy demand.

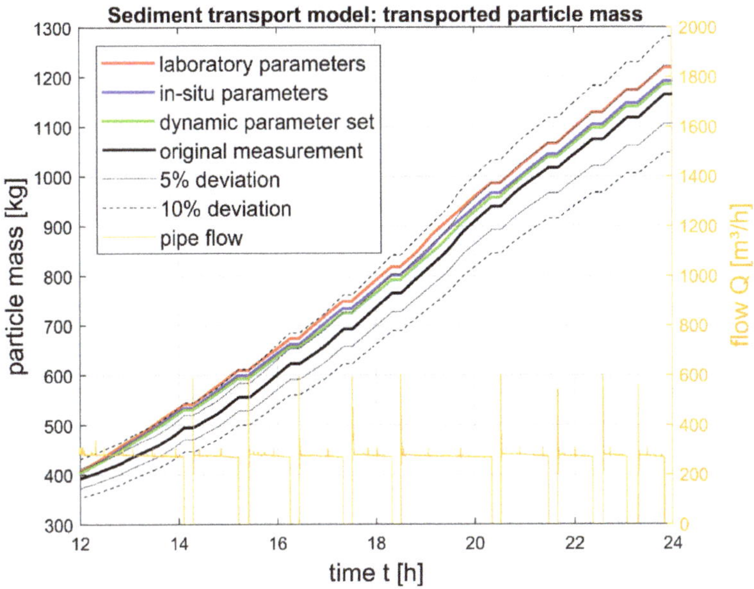

Figure 5. Results of the sediment transport simulation. Cumulative particle mass transport measured and simulated with laboratory, in situ, and dynamic parameters.

Table 1. Simulated pump control modes with significant settings.

Pump Data	Simulated Pump Control Modes				
	Full Power 2-Point Control	**Reduced 2-Point Control**	**Energy Optimal 2-Point Control**	**Energy Efficient Control**	**Reduced Energy Efficient Control**
Flow in duty point (L/s)	166.5	100	76.5	76.5	76.5
Flow velocity in duty point (m/s)	0.6	0.35	0.27	0.27	0.27
Head loss in duty point (m)	18.3	17	16.7	16.7	16.7
Bed shear stress in duty point (N/m²)	0.95	0.32	0.2	0.2	0.2
Power Input in Duty Point (kW)	55	33.5	22.5	22.5	22.5
Power start-up *	yes	yes	yes	yes	no
Adjusted pump flow to inflow **	no	no	no	yes	no
Parallel pumping from critical sump level	yes	yes	yes	yes	yes
Forced operation at least 1/hour	yes	yes	yes	yes	yes

* Pumps start-up to maximum power over one minute before regulating down to duty point in each pump sequences,
** If the inflow exceeds the present duty point, pump regulates up to the inflow rate.

The total transported sediment mass is similar in all five control modes and only differs by ≈2%. In contrast, the ratio between bed load and suspended load varies between control modes. As to be expected, the least amount of bed load was transported within the strongest operation mode: Full power two-point control with ≈360 kg in sum. The suspended load amounts to ≈27,800 kg in full power mode. Hence, the bed load takes only ≈1.3% of the total transported mass (= 28,160 kg) while 98.7% are transported within the suspended load at a flow velocity of ≈0.6 m/s.

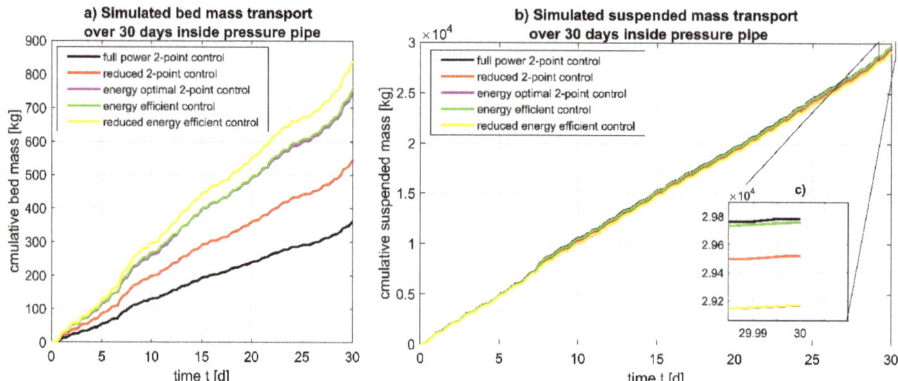

Figure 6. Sediment transport simulation for five different pumping modes over 30 days. Cumulative bed mass (**a**) and suspended mass (**b**) transport.

Table 2. Resulting bed and suspended load for five different pumping modes.

	Simulated Pump Control Modes				
	Full Power 2-Point Control	**Reduced 2-Point Control**	**Energy Optimal 2-Point Control**	**Energy Efficient Control**	**Reduced Energy Efficient Control**
Total sediment transport (kg)	30,140 (100%)	30,066 (100%)	29,918 (100%)	30,517 (100%)	30,011 (100%)
Bed load transport (kg)	360 (1.2%)	546 (1.8%)	750 (2.5%)	757 (2.5%)	841 (2.8%)
Suspended load transport (kg)	29,780 (98.8%)	29,520 (98.2%)	29,170 (97.5%)	29,760 (97.5%)	29,170 (97.2%)
Power consumption (kWh)	11,982	10,960 (−9%)	9868 (−18%)	9874 (−18%)	7974 (−33%)

This ratio changes due to the weaker control modes on behalf of the bed load. The bed load transport increases with a decreasing duty point (compare to Table 1). However, even at very low flow velocities of ≈0.27 m/s within the energy efficient control modes, the proportion of the bed load transport only amounts up to ≈2.8% of the total transported sediments. This means, a reduction of >50% flow velocity and a loss of ≈80% bed shear stress (from full power two-point control down to energy efficient control, see Table 1) does not results in significant changes of the bed or suspended load transport.

A critical transport limit is reached when the present bed shear stress falls below the critical bed shear stress for eroding sediments. The erosion rate then becomes zero. The least critical limit was calculated based on the results from [2] to ≈ 0.1 m/s flow velocity (≈ 28 L/s flow rate) at a critical bed shear stress of 0.02 N/m^2 after 20 min settling. This is below the technical minimum flow possible at PS Rostock-Schmarl of ≈ 53 L/s, which at the same time represents the absolute mean inflow value to the PS. However, the simulation at a duty point of 0.2 m/s flow velocity showed that only 77% of the total sediment mass is transported (compared to 100% sediment transport in full power mode). Applying the calibrated model at a duty point below the technical limitation showed that having specific pure sewage flow velocities down to 0.1 m/s are feasible, but gradually lead to a decrease in transported sediment mass. Apart from this, such a reduction is not feasible for reasons of capacity loss. It is therefore recommended to set the duty point at least above the absolute mean inflow value, or even better to the energy-optimal value. In addition, a flow adjustment should be installed to compensate for inflow peaks, especially when storm runoff is connected. This guarantees both sediment transport and energy savings.

In order to put the sediment transport into context with the energy savings, these were recorded during pump simulation. The simulation of the energy consumption shows a good correlation of the pump simulation with the real system. The power consumption simulated for 30 days under reduced two-point control amounts to ≈10,960 kWh. In the same period, 11,179 kWh are measured in

PS Rostock-Schmarl under reduced two-point control. The deviation is only −2%. Concluding, the computed energy consumption can be used to make reliable statements for all five control modes.

Already ≈−9% energy savings were achieved in PS Rostock-Schmarl's usual operation mode (reduced two-point control) compared to full power two-point control. Further energy savings were obtained by the energy optimal two-point and energy efficient control with ≈−18%. The maximum energy savings of ≈−33% could be achieved with the reduced energy efficient control.

It should be noted that these energy savings were computed for a PS with a very flat system curve. The share of dynamic losses in total pressure losses was only 11% in full power duty point (≈16 m static head to 18.3 m total head loss in duty point = 2.3 m dynamic loss). This proportion decreased with the energy saving control modes to 4%. So, the energy saving potential was low and resulted in moderate energy savings compared to the usual operation mode. The ratio changes in favor of energy saving when greater friction losses occur in the usual duty point of a simple two-point control (e.g., with smaller pipe diameter). However, the reduced flow regime in energy efficient control contribute to the reduction of CO_2 emissions and operation costs without dangerously increasing the sedimentation of sewage solids.

3.3. Sediment Transport under Storm Water Inflow

Storm water inflow alter the transport characteristics of the raw sewage, see [1,3]. Sedimentation and erosion are intensified. Storm water inflows are considered by the dynamical adjustment of calibration parameters α and d to the TSS inflow. Hence, the settled mass in the specific grid cell increases faster within pump pauses when storm water inflow enters the cell boundary. Subsequently erosion is intensified when the pumps start up. The critical bed shear stress τ_{crit} increases with the duration of the pump pause, but is not affected by the storm inflow. As the mechanical pre-treatment cleans the inflow from coarse material, it is assumed that the critical bed shear stress value after the pre-treatment is not modified by storm inflow. This assumption was supported by the results from [1]. The laboratory experiments with raw sewage samples from PS Rostock-Schmarl have shown that "the particle spectrum did not change from dry- to wet-weather inflow (. . .), but rather the proportion of particles, especially in the medium speed fraction" ([1], p. 10). The proportion of the fastest particle class (> 40 mm/s) only increased by 1.19% from dry to wet weather sewage samples. An aggravated erosion was considered by the increased sediment amount on pipes bottom. The erosion process then simply takes longer. An example calculation is given in [2], where the erosion process is prolonged with increasing settling duration.

The effect of storm water inflow to sediment transport inside pressure pipe were evaluated for the same 30 days period as described in the previous section. Several rain events are recorded during this period in the catchment area of PS Rostock-Schmarl. Figure 7 shows the present amount of sediments inside the pressure pipe, simulated over 30 days for all five control modes. The present particle mass was calculated as the total sediments quantity in all grid cells per time step (including both suspended and bedload). Figure 6 shows the present particle mass inside the pressure pipe for full power two-point control in detail (grey line) and for better visualization the 24 h simple moving average of the present particle mass of all five control modes (black to yellow lines).

The recorded rain events are shown on the right axis (blue line). Due to the size of the catchment area and the limitation of the connected areas on main roads (see also [3]), not all storm runoffs finally reach the PS. A good example was recorded at day 6. The storm runoff quickly enters the upstream sewer and subsequently the pressure pipe. The present particle mass increases up to ≈270 kg.

Under dry weather inflow, the peaks show the maximum sediment load inside the pipe during the day (at lunch) at ≈160 kg, while the sediment mass decreases in the night down to ≈115 kg.

Due to the 24 h average values, peaks were dampened but it provided the opportunity of supervision. It shows the results of long-lasting trends more clearly. If the moving average value increases constantly without decreasing, permanent deposits forming. However, such a trend cannot be detected. Even the increase in sediment volume due to storm runoff does not lead to a long-term formation of deposits. The sediments mass leveled out at ≈145 kg.

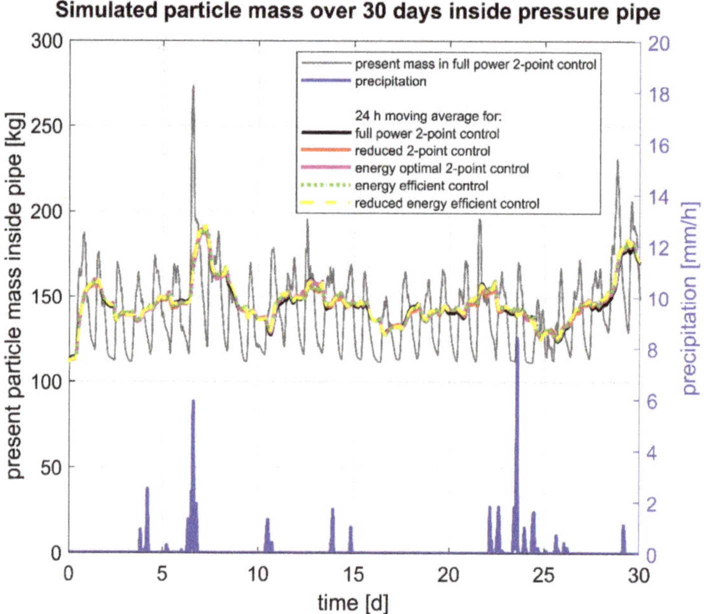

Figure 7. Sediment transport simulation for five different pumping modes over 30 days. Present particle mass inside pressure pipe on left axis and precipitation data on right axis (rain events are measured by a German Weather Service (DWD) meteorological station located inside the catchment area).

This is, off course, a result of the model calibration with settling parameters α, d, and τ_{crit}. These parameters, calculated from in situ measurements in [3], representing real life conditions inside the pressure pipe. The simulation shows, that even under very low flow velocities, the sediments were transported safely. A permanent deposits formation is prevented by the continuous pump operation regardless of the control mode or the duty point.

4. Conclusions

The paper presents a transport model for 1 D numerical simulation of the sediment transport inside a pressure pipe. The model aims at an exact description of the sediment transport by a limited but case-specific set of transport equations. It is quite simple from a physical point of view. However, an easy but descriptive assessment tool for sediment transport in pressure pipes is available, based on calibration parameter determination by in situ measurements.

The following fields of application are conceivable: Energy-efficient pump control, optimization of sewage disposal and treatment supply to wwtp's, pollution transport to urban drainage facilities, planning and design of urban drainage systems, and optimization of pipe flushing.

The investigation of hydraulic boundary conditions in pressure pipes under dry weather and storm water inflow have shown in case specific that very low flow velocities and shear stress values are tolerable. Accordingly, the developed energy-efficient pump control with low flow velocities can guarantee the safe disposal of wastewater streams.

Author Contributions: Conceptualization, M.R., J.T. and K.F.; Data curation, M.R.; Formal analysis, M.R. A.F. and K.F.; Funding acquisition, J.T.; Investigation, M.R.; Methodology, M.R., K.F. and J.T.; Project administration, J.T.; Software, A.F., K.F. and M.R.; Supervision, J.T.; Validation, M.R. and A.F.; Visualization, M.R.; Writing—original draft, M.R.; Writing—review & editing, J.T., T.K., A.F. and K.F. All authors have read and agreed to the published version of the manuscript.

Funding: The presented studies are supported by Deutsche Bundesstiftung Umwelt (DBU), Germany (AZ 32253/01).

Acknowledgments: The precipitation data used are available from Germany's national meteorological service (Deutscher Wetterdienst (DWD)), http://www.dwd.de.

Conflicts of Interest: The authors declare no conflict of interest. The funders had no role in the design of the study; in the collection, analyses, or interpretation of data; in the writing of the manuscript, and in the decision to publish the results.

References

1. Rinas, M.; Tränckner, J.; Koegst, T. Sedimentation of Raw Sewage: Investigations for a Pumping Station in Northern Germany under Energy-Efficient Pump Control. *Water* **2019**, *11*, 40. [CrossRef]
2. Rinas, M.; Tränckner, J.; Koegst, T. Erosion characteristics of raw sewage: Investigations for a pumping station in northern Germany under energy efficient pump control. *Water Sci. Technol.* **2018**. [CrossRef] [PubMed]
3. Rinas, M.; Tränckner, J.; Koegst, T. Sediment Transport in Sewage Pressure Pipes, Part I: Continuous Determination of Settling and Erosion Characteristics by In-Situ TSS Monitoring inside a Pressure Pipe in Northern Germany. *Water* **2019**, *11*, 2125. [CrossRef]
4. Bagnold, R.A. *An Approach to the Sediment Transport Problem from General Physics*; U.S. Government Publishing Office: Washington, DC, USA, 1966.
5. Engelund, F.; Hansen, E. *A Monograph on Sediment Transport in Alluvial Streams*; Teknisk Forlag: Copenhagen, Denmark, 1967.
6. Ackers, P.; White, R. Sediment Transport: New Approach and Analysis. *J. Hydraul. Div.* **1973**, *99*, 2041–2060.
7. Engelund, F.; Fredsøe, J. A Sediment Transport Model for Straight Alluvial Channels. *Hydrol. Res.* **1976**, *7*, 293–306. [CrossRef]
8. van Rijn, L.C. Sediment Transport, Part I: Bed Load Transport. *J. Hydraul. Eng.* **1984**, *110*, 1431–1456. [CrossRef]
9. Campisano, A.; Creaco, E.; Modica, C. Numerical modelling of sediment bed aggradation in open rectangular drainage channels. *Urban Water J.* **2013**, *10*, 365–376. [CrossRef]
10. Shirazi, R.H.S.M.; Campisano, A.; Modica, C.; Willems, P. Modelling the erosive effects of sewer flushing using different sediment transport formulae. *Water Sci. Technol.* **2014**, *69*, 1198–1204. [CrossRef]
11. Creaco, E.; Bertrand-Krajewski, J.-L. Numerical simulation of flushing effect on sewer sediments and comparison of four sediment transport formulas. *J. Hydraul. Res.* **2009**, *47*, 195–202. [CrossRef]
12. Correia, L.P.; Krishnappan, B.G.; Graf, W.H. Fully Coupled Unsteady Mobile Boundary Flow Model. *J. Hydraul. Eng.* **1992**, *118*, 476–494. [CrossRef]
13. Bonakdari, H.; Ebtehaj, I.; Azimi, H. Numerical Analysis of Sediment Transport in Sewer Pipe. *Int. J. Eng.* **2015**, *28*, 1564–1570. [CrossRef]
14. Danish Hydraulic Institute (DHI). *MOUSE—Pollution Transport Reference Manual*; Danish Hydraulic Institute: Hørsholm, Denmark, 2017.
15. Rossman, L.A. *EPANET 2 Users Manual*; U.S. Environmental Protection Agency: Cincinnati, OH, USA, 2000.
16. Ebtehaj, I.; Bonakdari, H. Evaluation of Sediment Transport in Sewer using Artificial Neural Network. *Eng. Appl. Comput. Fluid Mech.* **2013**, *7*, 382–392. [CrossRef]
17. Azamathulla, H.M.; Ab Ghani, A.; Fei, S.Y. ANFIS-based approach for predicting sediment transport in clean sewer. *Appl. Soft Comput.* **2012**, *12*, 1227–1230. [CrossRef] [PubMed]
18. Bizimana, H.; Altunkaynak, A. A novel approach for the prediction of the incipient motion of sediments under smooth, transitional and rough flow conditions using Geno-Fuzzy Inference System model. *J. Hydrol.* **2019**, *577*, 123952. [CrossRef]
19. Safari, M.J.S.; Danandeh Mehr, A. Multigene genetic programming for sediment transport modeling in sewers for conditions of non-deposition with a bed deposit. *Int. J. Sediment Res.* **2018**, *33*, 262–270. [CrossRef]
20. Roushangar, K.; Ghasempour, R. Estimation of bedload discharge in sewer pipes with different boundary conditions using an evolutionary algorithm. *Int. J. Sediment Res.* **2017**, *32*, 564–574. [CrossRef]

21. Ebtehaj, I.; Bonakdari, H.; Khoshbin, F. Evolutionary design of a generalized polynomial neural network for modelling sediment transport in clean pipes. *Eng. Optim.* **2016**, *48*, 1793–1807. [CrossRef]

22. Fricke, A. Optimale Steuerung und Ihre Anwendungen in der Abwassertechnik (Optimal Control and Its Application in Wastewater Technology). Ph.D. Thesis, University of Rostock, Rostock, Germany, 2015.

23. Knubbe, A.; Fricke, A.; Ecktädt, H.; Neymeyr, K.; Schwarz, M.; Tränckner, J. Energieeffizienter Betrieb von Abwasserfördersystemen [Energy Efficient Strategies for Wastewater Pumping Systems]. *GWF Wasser Abwasser* **2014**, *155*, 640–646.

24. Tränckner, J.; Bönisch, G.; Gebhard, V.R.; Dirckx, G.; Krebs, P. Model-based assessment of sediment sources in sewers. *Urban Water J.* **2008**, *5*, 277–286. [CrossRef]

25. Rossman, L.A.; Boulos, P.F.; Altman, T. Discrete Volume-Element Method for Network Water-Quality Models. *J. Water Resour. Plan. Manag.* **1993**, *119*, 505–517. [CrossRef]

26. Rossman, L.A.; Boulos, P.F. Numerical Methods for Modeling Water Quality in Distribution Systems: A Comparison. *J. Water Resour. Plan. Manag.* **1996**, *122*, 137–146. [CrossRef]

 *water*

Article

Simulating Flow of An Urban River Course with Complex Cross Sections Based on the MIKE21 FM Model

Qianxun Wang [1,2], Wenqi Peng [1,*], Fei Dong [1], Xiaobo Liu [1] and Nan Ou [2]

[1] State Key Laboratory of Simulation and Regulation of River Basin Water Cycle, China Institute of Water Resources and Hydropower Research, Beijing 100038, China; wqxbjut@163.com (Q.W.); dongfei@iwhr.com (F.D.); xbliu@iwhr.com (X.L.)

[2] Capital Urban Planning and Design Consulting Development Company Limited, Beijing 100038, China; ounan_214@126.com

* Correspondence: pwq@iwhr.com; Tel.: +86-10-6878-1885

Received: 24 January 2020; Accepted: 7 March 2020; Published: 10 March 2020

Abstract: In this case study, the method of unsteady flow was used to study the flow characteristics of a planned urban river course, River A, with complex cross sections, based on the MIKE21 FM hydrodynamic module, which is an important tool for analyzing and solving hydrodynamic problems. First, the rationality and feasibility of the planning scheme were verified by building a two-dimensional numerical model, which can provide a scientific basis for the river course planning. Then, the flow characteristic of the river course was analyzed and summarized, to give several suggestions and improvement measures for follow-up river planning. Finally, on the basis of the case study, the general rules of hydraulic factors in river courses with complex cross sections were summarized, which can facilitate the understanding of the genesis and evolution of river courses.

Keywords: trapezoidal stretch; transition stretch; culvert; open channel; hydraulic factors

---

1. Introduction

Generally, river courses can be divided into natural river courses and artificial channelized river courses by genesis [1–3]. Artificial channelized river courses can be further divided into farmland river courses and urban river courses. Urban river courses usually have regular planned cross sections, such as a trapezoidal section, rectangular section, compound section and other single-form sections [4]. Hence, in terms of flow characteristics, flow in an urban river course can be considered as constant uniform flow in an open channel by approximating [5]. However, in particular cases, due to compact layout of urban land, a river course may have to be rerouted away from the construction land, resulting in various kinds of cross sections (trapezoidal section, rectangular section, transition section and curved section) occurring alternatively in a single river course. The River A (see Section 2.1) studied in this paper is a typical example of urban channelized river with complex cross sections.

Many studies have been conducted on the flow characteristics of a natural river course [6–9]. Xia et al., studied the relationship between cross-sectional velocity and the maximum velocity under natural conditions by collecting the cross section velocity data of the Mississippi River [10]; Lane et al., explored the adjustment between flow processes, sediment transport and river channel morphology by using computational fluid dynamics and evaluated the extent to which three-dimensional (3D) models can improve the predictive ability and prediction utility in comparison with two-dimensional (2D) applications [11]; Yu et al., established a planar 2D flow and sediment mathematical model for natural branching rivers, based on triangular grids with the finite element method, and the model was applied to a branching river of Tianxingzhou in a successful way [12]; Jia et al., developed a three-dimensional (3D) dynamic numerical method of meander migration to simulate the vertical migration in the Jingjiang reach of the middle Yangtze River and established a method based on the mechanism of bank failure to simulate the erosion of composite banks [13]. The evolution trend of natural course is usually influenced by factors such as the degree of deposition, the depth of erosion and sediment transport. As for the urban river course this article refers to the narrow concept, the influence of sediment transport does not generally need to be considered under the premise of ensuring a certain velocity as the section of the river is protected by the revetment. However, the backwater, the deposition that will lead to the increase of the roughness coefficient in the cross sections of the urban river course and water depth difference between the two sides of the urban river course should also be considered, and the effect of erosion should be paid more attention to if the flow velocity of the river is over the limitation.

Up to now, more existing studies have been focused on the landscape restoration issue of urban rivers [14–18], while fewer studies have been focused on the flow characteristics of the urban channelized rivers with complex cross sections. For river courses with complex cross sections, the flow characteristics should be considered as unsteady flow. Garcia-Navarro et al., described the use of the McCormack explicit finite difference scheme and the treatment of the boundary problem in the development of a one-dimensional simulation model that solves the St. Venant equations of the unsteady open channel flow [19]. Hersberger et al., analyzed the wall-roughness effects on flow and scouring in curved channels with gravel beds and discussed the optimal macroroughness configuration in terms of scour reduction [20]. An et al., studied the flow characteristics of a compound channel with different cross sections and different roughness under the different flow rate based on a numerical simulation model [21]. The researchers studied and analyzed the changes of water level on both sides of the curved stretch and the distribution of water level and velocity in the straight stretch under the premise of single-form sections of the urban river course. In the practical urban river course planning and designing, the urban river courses with complex cross sections sometimes are designed in particular cases, and thus we need to pay more attention to the change rules of hydraulic factors of these rivers.

The research object in this study, River A, contains complex cross sections, including a trapezoidal section, rectangular section and transition section, and in the rerouted area with a total river course length of 1300 m, there are four right-angle turns and 3-channel culverts and open channels alternatively arranged, which makes the hydraulic conditions of the river more complex. It is obvious that the method of constant uniform flow on river hydraulic factors is no longer applicable in this case; in order to predict and evaluate the influence of the complex cross sections on the hydraulic factors of the river in a more accurate way, the hydraulic characteristics in the river should be analyzed with the method of unsteady flow with a two-dimensional numerical model [22,23]. Therefore, the MIKE21 FM model, which was developed by Hydraulic Research Institute of Denmark, was used to simulate the hydraulic conditions of the channelized river with complex cross sections, and further to provide a scientific basis for river course planning and decision-making, and offer some rational guides and suggestions for follow-up project construction and river operation and management [24,25].

The MIKE21 FM model is applied widely for analysis on flow characteristics. Chen et al., evaluated floods in urban development scenarios with a secondary development of GIS and the MIKE21 FM model, which provided a scientific basis for the development of corresponding flooding measures [26]. Uddin et al., studied the flow field structure and the flow direction in the northern Bay of Bengal coastal waters with the MIKE21 FM model [27]. Kaergaard et al., simulated the evolution of large curvature coastline by using the MIKE21 FM model [28]. Abily et al., studied the flood runoff depth on industrial sites by using three evaluation models (the MIKE21 FM, MIKE21 and 3D FVM) to indicate the stability, difference and limitation of different evaluation models in runoff simulation and calculation [29]. Kimiaghalam et al., analyzed wave current erosion of cohesive banks in north Manitoba, Canada, by using the MIKE21 FM model [30].

Based on the simulation and calculation conducted using the MIKE21 FM model, from the perspective of the river course safety, to analyze whether the hydraulic factors of the river course with complex cross sections can meet the planning requirements, the following contents should be mainly analyzed:

(1) Under the designed conditions for once-in-20/50-years floods, judge whether the flow state in the river is consistent with that in the plan and analyze the change rules of hydraulic factors in each stretch of the river. Furthermore, evaluate whether the river's water depth and flow velocity can meet the planning requirements, verify the rationality and feasibility of the plan, and finally, propose rational suggestions on slope protection and dredging in local river stretches.

(2) On this basis, analyze the relationship between the flow velocity (along with the river course) and the water depth, to reveal the causes and change rules of the backwater, and quantitate the level and length of the backwater in the river; study the rules of hydraulic factors in a single river course, especially at turns of the river course, to reveal the causes and change rules of the water level difference between the two sides of the river; summarize the influence of the change of roughness coefficient and the downstream clogging of the planned river, to reveal the importation of the dredging of the planned river.

2. Materials and Methods

2.1. River Planning Scheme

River A currently contains a trapezoidal section with an ecological embankment. Currently, the top section width is around 15 m, the slope coefficient is about 2.0, the river depth is about 2 m, and the average bottom longitudinal slope is about 0.0005.

College B will be constructed near River A. There was an arrangement contradiction between the construction location of College B and the current route of River A. In order to ensure the construction of College B and the flood control in the river, after discussion, the related departments finally decided to reroute the river as follows (see Figure 1):

(1) Excavate concrete-lined 3-channel culverts (each channel of the culverts has a 3500×2500 mm^2 rectangular cross section, with a bottom longitudinal slope of 0.001), along the south side of Road X, the west side of Road Y, and the north side of Road Z; reroute River A to the east side of College B, and merge the new river course into the downstream of current River A on the south side of College B.

(2) In order to facilitate dredging, every 100 m along the culverts, arrange a concrete-lined open channel that has a rectangular section, with a length of 30 m, a width of 12 m, a depth of 2.5 m and a bottom longitudinal slope of 0.001.

(3) Arrange a transition stretch (horn mouth stretch 1) at the junction of River A and Road X, with a length of 110 m, a top section width changing from 30 m to 12 m, a bottom section width changing from 18 m to 12 m, a depth changing from 3.0 m to 2.5 m, and a longitudinal slope of 0.001; arrange another transition stretch (horn mouth stretch 2) at the junction of River A and Road Z, with a length of 110 m, a top section width changing from 12 m to 30 m, a bottom section width changing from 12 m to 18 m, a depth changing from 2.5 m to 3.0 m, and a bottom longitudinal slope of 0.001. The two transition stretches shall both be lined with ecological revetment, and the cross section schematic diagram is shown in Figure S1.

(4) The planned cross sections upstream of the junction of River A and Road X and downstream of the junction of River A and Road Z are both trapezoidal with ecological protection, and the two river stretches are both planned to have a top width of 30 m, a bottom width of 18 m, a depth of 3.0 m, a slope coefficient of 2 and a bottom longitudinal slope of 0.001.

(5) The total length of the rerouted stretch of River A is about 1300 m and the planned design standard is for once-in-20-years floods. The planned river flow rate for once-in-20-years floods is 29.4 m^3/s, and the planned river flow rate for once-in-50-years floods is 34.5 m^3/s.

It is generally believed that the river flow is steady uniform flow for a single cross section of the urban river course, which greatly simplifies the hydraulic calculation work. In the practical urban river course planning and designing, the rationality and feasibility of the planning scheme with the model method (unsteady flow) are rarely verified because of the shortage of model funding and the limitation of the project cycle. However, the planned cross sections in River A are complex and diverse, and there are four right-angle turns and several alternately-arranged culverts and open channels in River A's rerouted area, which lead to backwater, erosion and water depth difference between the two sides of the river. Hence, the change rules of the hydraulic factors of the river course cannot be predicted accurately using a conventional urban river course planning method (steady uniform flow). Therefore, in order to analyze the flow characteristics of the river more clearly and accurately in this study, River A was divided into four stretches and each stretch was discussed and analyzed separately with the MIKE21 FM model. Based on the analysis and summary of flow characteristics of the river course with complex cross sections, it provides technical support and practical experience for the successful completion of the subsequent similar planning and designing work.

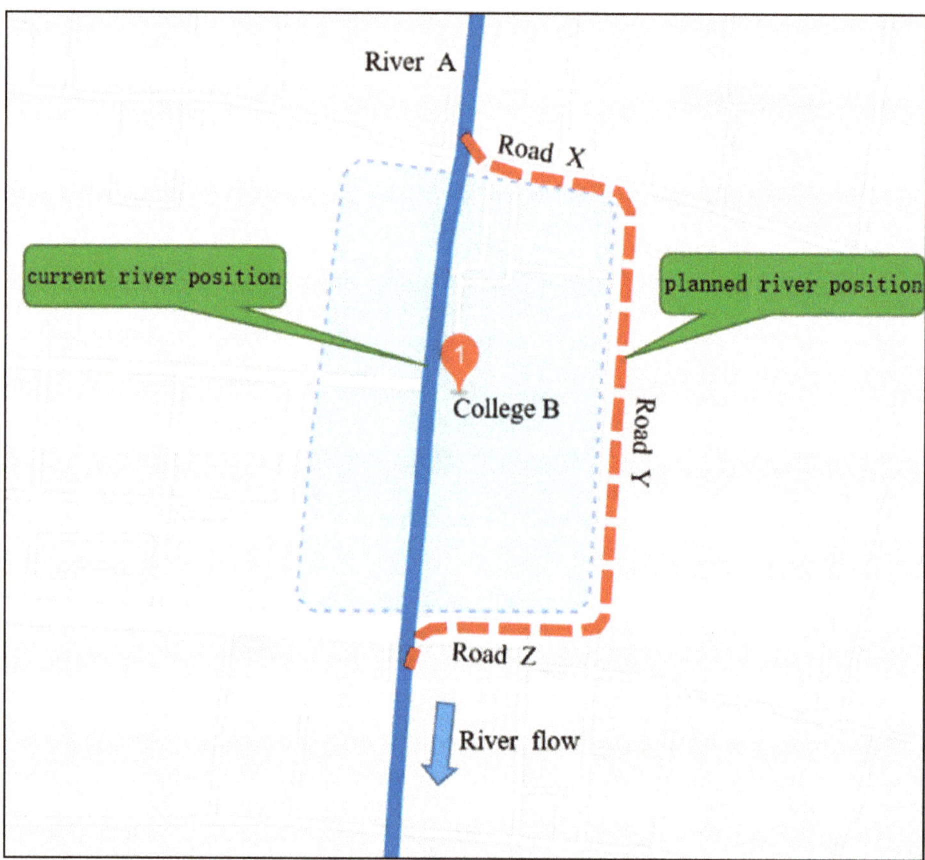

Figure 1. River position schematic diagram.

2.2. Model Description

The MIKE21 FM model is a 2D shallow water equation based on numerical solutions and an incompressible Reynolds-averaged Navier–Stokes Equation integrated along the water depth (see Formulas 1–4) [31]. Therefore, the model integrates continuity, momentum, temperature, salinity and density equations, which can use the Cartesian coordinates or spherical coordinates to simulate changes of water level and flow due to various acting forces and any 2D free surface flow without considering stratification, by using unstructured grids on a plane [31].

$$\frac{\delta h}{\delta t} + \frac{\delta h\overline{u}}{\delta x} + \frac{\delta h\overline{v}}{\delta y} = hs \tag{1}$$

$$\frac{\delta h\overline{u}}{\delta t} + \frac{\delta h\overline{u}^2}{\delta x} + \frac{\delta h\overline{u}\overline{v}}{\delta y} = f\overline{v}h - gh\frac{\delta\eta}{\delta x} - \frac{h\delta p_a}{\rho_0\delta x} - \frac{gh^2\delta\rho}{2\rho_0\delta x} + \frac{\tau_{sx}}{\rho_0} - \frac{\tau_{bx}}{\rho_0} - \frac{1}{\rho_0}\left(\frac{\delta s_{xx}}{\delta x} + \frac{\delta s_{xy}}{\delta y}\right) + \frac{\delta}{\delta x}\left(hT_{xx}\right) + \frac{\delta}{\delta y}\left(hT_{xy}\right) + hu_s s \tag{2}$$

$$\frac{\delta h\overline{v}}{\delta t} + \frac{\delta h\overline{u}\overline{v}}{\delta x} + \frac{\delta h\overline{v}^2}{\delta y} = f\overline{u}h - gh\frac{\delta\eta}{\delta y} - \frac{h\delta p_a}{\rho_0\delta y} - \frac{gh^2\delta\rho}{2\rho_0\delta y} + \frac{\tau_{sy}}{\rho_0} - \frac{\tau_{by}}{\rho_0} - \frac{1}{\rho_0}\left(\frac{\delta s_{yx}}{\delta x} + \frac{\delta s_{yy}}{\delta y}\right) + \frac{\delta}{\delta x}\left(hT_{xy}\right) + \frac{\delta}{\delta y}\left(hT_{yy}\right) + hu_s s \tag{3}$$

$$h\overline{u} = \int_{-d}^{\eta} u\,dz, \; h\overline{v} = \int_{-d}^{\eta} v\,dz \tag{4}$$

where x, y and z are the Cartesian co-ordinates in m; $\overline{u}$ and $\overline{v}$ are the velocity components in the x and y direction in m/s; t is the time in s; η is the bottom elevation in m; d is depth of water in m; $h = \eta + d$ is the total water depth in m; u, v are the velocity components in the x and y direction in m/s; g is the gravitational acceleration in m/s^2; $f = 2\Omega\sin\varphi$ is the Coriolis parameter (Ω is the angular rate of revolution and φ is the geographic latitude) in s^{-1}; ρ is the density of water in kg/m^3; ρ_0 is the reference density of water in kg/m^3; s_{xx}, s_{xy}, s_{yx} and s_{yy} are components of the radiation stress tensor in kg/s^2; τ_{sx} and τ_{sy} are the surface stress in kg/s^2·m; τ_{bx} and τ_{by} are the bottom stress in kg/s^2·m; p_a is atmospheric pressure in kg/s^2·m; s is the magnitude of the discharge due to point sources in s^{-1}; u_s and v_s are the velocity by which the water is discharged into the ambient water in m/s; T_{xx}, T_{xy} and T_{yy} are the lateral stresses in m^2/s^2, which are estimated using an eddy viscosity formulation based on the depth average velocity gradients:

$$T_{xx} = 2A\frac{\delta\overline{u}}{\delta x}, \; T_{xy} = A\left(\frac{\delta\overline{u}}{\delta x} + \frac{\delta\overline{v}}{\delta y}\right), \; T_{yy} = 2A\frac{\delta\overline{v}}{\delta y} \tag{5}$$

$$A = c_s^2 l^2 \sqrt{2S_{ij}S_{ij}}, \; S_{ij} = \frac{1}{2}\left(\frac{\delta u_i}{\delta x_j} + \frac{\delta u_j}{\delta x_i}\right)(i, j = 1, 2) \tag{6}$$

where A is the horizontal eddy viscosity in m^2/s; c_s is a constant, which should be chosen within the range from 0.25 to 1.0; l is a characteristic length in m.

The numerical calculation method used in the MIKE21 FM model is the finite volume method in cells [32,33]. The finite volume method is to divide a continuous body into triangle and quadrilateral non-overlapping cells. The normal vector can be obtained by establishing a cell hydraulic model in the outer normal direction and solving the one-dimensional Riemann problem. This method has a good property of integral conservation and can be used to handle supercritical flows and discontinuous solutions accurately.

The numerical calculation accuracy of the MIKE21 FM model is becoming more and more recognized. Gayer et al., simulated the influence of eddy current and roughness coefficient in the tsunami model with the MIKE21 FM model, and the calculated results were nicely consistent with the measured data [34]; Sokolov et al., carried out a simulation analysis with the measured data of the Baltic Sea collected by the Polish Academy of Sciences and found out that the correlation between the simulation results and the measured results was up to 0.86 [35]; Pedersen et al., carried out a simulation analysis with the measured tsunami data along the northern coast of Sumatra and found out that the simulation results were basically consistent with the measured data after several times of debugging [36]. Xie et al., determined the model parameters based on flooding data of the Luanhe River measured in 2012 to make the simulated flood level agree well with the measured flood level hydrograph [37]. Feng et al., elaborated the modeling process of the ocean flow field in simulation with the MIKE21 FM module and found that the measured results and the simulation results matched within an allowable error range for model verification [38]. Guo et al., simulated the flood evolution of the Pajiang River flood storage area, and demonstrated that the MIKE21 FM model had high simulation accuracy and credible simulation results, which could be used for numerical simulation of floods in more complex flood storage areas [39].

2.3. Calculation Conditions

2.3.1. Model Parameters

River A is proposed to be rerouted, and there are currently no measured data and similar planning data for reference. In order to do it in a scientific and accurate way, the selection of model parameters in this study was conducted based on relevant references published by domestic and foreign researchers [34–40]. On the other hand, relevant model parameters (see Table 1) were determined by considering the actual conditions of the project and the engineering safety. At the same time, considering the complex cross sections and various structures along the river course, in order to avoid the setting of model parameters this study described the relevant complex cross sections and structures in the actual terrain as much as possible, which can improve the accuracy of numerical calculation; for example, the local terrain correction methods [38] were selected to describe the 3-channel culverts and the transition stretches in the digital elevation model, to avoid setting related parameters such as the water head loss coefficient and the diffusion coefficient.

Table 1. Model parameters table [31].

Model Parameters	Recommended Range	Module Default Value	Value in This Study
Time step	0.01–30 s	0.01–30 s	0.01–0.03 s
Wet and dry boundaries	-	$h_{dry} = 0.005$ $h_{flood} = 0.05$ $h_{wet} = 0.1$	$h_{dry} = 0.01$ $h_{flood} = 0.1$ $h_{wet} = 0.3$
Manning coefficient (n)	0.01–0.05 $m^{1/3}/s$	0.03 $m^{1/3}/s$	Roughness coefficient of ecological slope section (n): 0.025; Roughness coefficient of 3-channel culverts (n): 0.017
Vortex viscosity coefficient	0.25–1.0	0.28	0.28
Structures	-	-	Described in the terrain DEM

For roughness coefficients, relevant hydraulic studies were referred to [41] (see Table 2). The roughness coefficient of the trapezoidal stretches with ecological revetment was taken as 0.025 and the roughness coefficient of the concrete-lined 3-channel culverts was taken as 0.017.

Table 2. Roughness coefficients table [41].

No.	Boundary TYPE and Conditions	n
1	Thoroughly planed wood boards and freshly cleaned pig iron pipes and cast-iron pipes with smooth lining and joints	0.01
2	Dirty water supply and drainage pipes; ordinary concrete surfaces; ordinary brickworks	0.014
3	Old brickworks; very rough concrete surfaces; carefully excavated smooth rock faces	0.017
4	Canals in solid clay; loess with continuous silt layers; well-maintained large canals in earth	0.0225
5	Ordinary large earth canals; well-maintained small canals in earth; natural rivers under excellent conditions	0.025
6	Earth drains under particularly bad conditions; natural rivers under poor conditions (with much wild grasses and stones, irregular and curved riverbeds and many collapses and deep pools, etc.)	0.04

There are still no mature methods for selecting roughness coefficients. If the roughness coefficient value is too small, the water flow resistance may be underestimated and the drainage capacity of the proposed river course may be overestimated, which may result in overflow and other disasters [41].

2.3.2. Terrain Processing and Meshing

Terrain processing and meshing are important and challenging technical issues in the simulation analysis. Terrain data are the basic data for the model. The quality of the terrain data directly determines the reliability of the model's results. There are complex cross sections in the planned river course, which need to be described in the terrain. Therefore, there are high requirements for the fineness of the terrain. Based on the previous modeling experiences and related references [42], particular attention should be paid to the effects of terrain interpolation in areas with large terrain fluctuations. According to the planned river, the terrains should be specially treated at the junctions of the side slope to the river bottom, culverts, open channels and transition stretches.

At the same time, the areas with significantly changing terrain elevation should be meshed separately to avoid divergence in the model calculation. It should be noted that the solution to this terrain was completely based on the section size and elevation of the river section (see Figure S2), so as to predict actual flow characteristics of the planned River A in a more scientific and accurate way.

2.3.3. Boundary Conditions

Two boundary conditions needed to be set in this project: the water level boundary condition and the flow rate boundary condition. The water level boundary was set at about 840 m in the downstream away from the junction of River A to Road Z, and the flow rate boundary was set at about 540 m in the upstream away from the junction of River A to Road X. The specific values should be based on different simulation scenarios (see Section 2.4).

2.4. Scenario Settings

In order to simulate and analyze the changes of hydraulic factors in River A in a better way, this study set four simulation scenarios (see Table 3). The simulation scenarios 1 and 2 are planned design working conditions that are used to analyze and summarize the changes of hydraulic factors, propose pointed opinions on planning and verify the rationality of the planned river course further, to provide a scientific basis for river planning. The simulation scenario 3 is check and comparison working conditions that are used to simulate and analyze the changes of the hydraulic factors in the culvert and those in other river stretches due to the changes of the roughness coefficient in this stretch, when the culvert's clogging leads to an increase in the roughness coefficient. Furthermore, based on the results obtained with simulation scenario 3, targeted suggestions for follow-up management of the river can be proposed, to ensure successful implementation of the long-term river pollution control system of China [43,44]. The simulation scenario 4 is design check working conditions that are used to simulate and analyze the changes of the hydraulic factors in the river when the whole stretch of the river is concrete-lined, which can provide a scientific basis for the comparison and selection of river planning schemes.

Table 3. Simulation scenarios.

Scenario	Roughness Coefficient	Water Level Boundary	Flow Rate Boundary
1	The roughness coefficient of the ecological revetment section was taken as 0.025 and the roughness coefficient of the culvert was taken as 0.017	Water depth for once-in-20-years floods (1.46 m)	Flow rate for once-in-20-years floods (29.4 m^3/s)
2		Water depth for once-in-50-years floods (1.58 m)	Flow rate for once-in-50-years floods (34.5 m^3/s)
3	0.025	Water depth for once-in-50-years floods	Flow rate for once-in-50-years floods
4	0.017	Water depth for once-in-50-years floods	Flow rate for once-in-50-years floods

3. Results and Discussion

Based on the simulation results, and taking into account the influence of backwater in the river, from the downstream to the upstream of the river, the downstream trapezoidal stretch, the transition stretch (horn mouth stretch 2), the culvert and the open channel stretch, the upstream stretch (horn mouth stretch 1 and the upstream trapezoidal stretch).

3.1. Downstream Trapezoidal Stretch

(1) According to the simulation results of Scenario 1, the water depth and flow velocity of the downstream trapezoidal stretch are clarified as Figure 2. As shown in Figure 2a, the upstream water depth of the bottom is lower than that in the downstream and the water depth at the bottom of the river is higher than the water depth on both sides in a single cross section (see Figure S3); the water depth on the right side averages 0.61 m and changes within the range from 0~1.26 m; the water depth on the left side averages 0.83 m and changes within the range from 0~1.57 m; the water depth at the bottom of the river averages 1.45 m and changes within the range from 0.90~1.57 m.

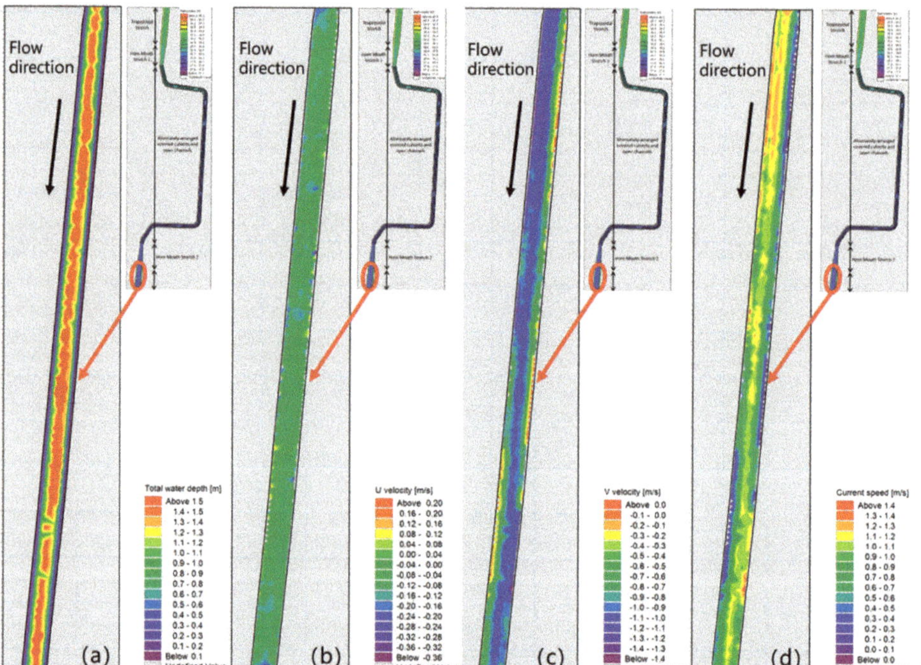

Figure 2. Hydraulic factors map of the downstream trapezoidal stretch (simulation scenario 1). (**a**) Water depth map; (**b**) Horizontal flow velocity; (**c**) Vertical flow velocity; (**d**) Current flow velocity (From bottom to top and left to right, the flow velocity was set to be positive).

As shown in Figure 2b, the flow velocity on the right side averages −0.08 m/s and changes within the range from −0.26~0.19 m/s; the flow velocity on the left side averages −0.07 m/s and changes within the range from −0.40~0.19 m/s; the flow velocity at the bottom of the river averages −0.09 m/s and changes within the range from −0.37~0.14 m/s.

As shown in Figure 2c, the flow velocity on the right side is larger than that on the left side in a single cross section in the upstream, while the flow velocity on the left side is larger than that on the right side in a single cross section in the downstream. The flow velocity on the right side averages

−0.72 m/s and changes within the range from −1.52~0 m/s; the flow velocity on the left side averages −0.73 m/s and changes within the range from −1.48~0 m/s; the flow velocity at the bottom of the river averages −1.0 m/s and changes within the range from −1.31~0 m/s.

As shown in Figure 2d, the current flow velocity on the right side (the current flow velocity is the combination of the horizontal flow velocity and the vertical flow velocity) is greater than that on the left side in a single cross section in the upstream, while the flow velocity on the left side is greater than that on the right side in a single cross section in the downstream; the flow velocity on the right side averages 0.73 m/s and changes within the range from 0~1.54 m/s; the flow velocity on the left side averages 0.73 m/s and changes within the range from 0~1.47 m/s; the flow velocity at the bottom of the river averages 1.0 m/s and changes within the range from 0.07~1.33 m/s.

From the above results, it can be seen that the water depth in the downstream trapezoidal stretch gradually increases from upstream to downstream, and the maximum water depth at the bottom of the river is 1.57 m. The planned water depth in this stretch is 3 m. Therefore, the water depth meets the planning requirement of the design standard for once-in-20-years floods; the flow velocity of the river does not vary dramatically, and the average flow velocity at the bottom of the river is 1.0 m/s, which meets the flow velocity requirement. The Froude number is about 0.3, which indicates that the flow is slow and meets the planning requirements; the average water depth on the right side of the river is lower than that on the left side, which leads to horizontal circulation in the river, thus causing a water depth difference of about 0.22 m between the two sides of the river. The water depth difference is below the designed safe super elevation and meets the planning requirements; there are some flow velocity differences between the upstream and downstream, and between the left and right sides of the river. In the next design stage, local bank reinforcement on the right side in the upstream and on the left side in the downstream should be considered, to prevent local erosion and deposition in the river.

(2) Simulation scenario 2 was analyzed in the same way. The variations of the hydraulic factors in the downstream trapezoidal stretch were basically the same as those in scenario 1 (see Figure S4).

Based on the statistical analysis, the water depth in the downstream trapezoidal stretch gradually increases from upstream to downstream, and the maximum water depth at the bottom of the river is 1.71 m. The planned river depth in this stretch is 3 m. Therefore, the water depth meets the planning requirement that no overflow occurs in once-in-50-years rains. The average water depth on the right side is about 0.18 m lower than that on the left side. The water depth difference in this scenario is lower than that in scenario 1. This is mainly because the dominant effect of vertical flow velocity in scenario 2 is stronger than that in scenario 1, thus weakening the vertical circulation effect of the river.

(3) The simulation conditions of the downstream trapezoidal stretch in scenario 3 are the same as those in scenario 2, but different from those on the upstream culverts. According to the above analysis, the water flow in the entire stretch is slow, so the hydraulic conditions in the downstream trapezoidal stretch are not affected by the changes in the hydraulic conditions of the upstream culvert, which will not be analyzed in this study.

(4) The comparison of the simulation results between scenario 4 and scenario 2 is shown in Figure 3. The water depth difference in the downstream trapezoidal stretch changes within the range from −0.14~0 m. The current flow velocity difference changes within the range from 0~0.75 m/s. The local flow velocity difference is less than 0 m/s.

By analysis, it can be seen that the roughness coefficient in this stretch decreases from 0.025 to 0.017, which means that the ecological revetment in the river's planned section is replaced by concrete revetment. The current flow velocity in the river increases but the water depth decreases, as shown in Figure 3. The water depth difference, as shown in Figure 3a, increases gradually from upstream to downstream, until it increases to 0 m/s at the downstream water level boundary and the maximum water depth decreases by 0.14 m. In Figure 3b, the current flow velocity difference between the two locations ② and ③ is less than 0 m/s, but the current flow velocity difference between the two locations ① and ④ is relatively larger, which indicates that the dredging of the river stretch at ② and ③ should be improved and that the river revetment of the river stretch at ① and ④ should be strengthened.

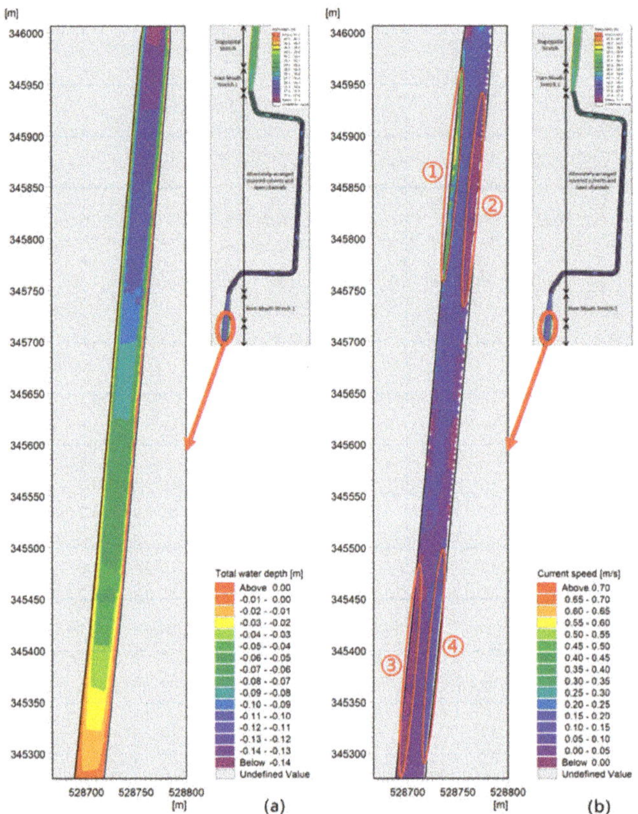

Figure 3. Comparison of hydraulic factors between scenario 4 and scenario 2. (**a**) Water depth difference; (**b**) Current flow velocity difference.

3.2. Transition Stretch (Horn Mouth Stretch 2).

(1) Based on the simulation results of scenario 1, the water depth and flow velocity results in the transition stretch are shown in Figure S5.

As shown in Figure S5a, the water depth at the bottom of the river from upstream to downstream first decreases and then increases. The water depth at the bottom of the river is higher than that at the both sides in a single cross section (see Figure S6); the water depth on the right side averages 0.53 m and changes within the range from 0~1.33 m; the water depth on the left side averages 0.47 m and changes within the range from 0~1.11 m; the water depth at the bottom of the river averages 1.40 m and changes within the range from 1.30~1.50 m.

From the results shown in Figure S5b, it can be seen that the flow velocity at the bottom of the river is greater than that on both sides in a single cross section; the flow velocity on the right side averages −0.14 m/s and changes within the range from −0.27~0.45 m/s; the flow velocity on the left side averages −0.19 m/s and changes within the range from −1.26~0.30 m/s; the flow velocity at the bottom of the river averages −0.17 m/s and changes within the range from −0.85~0 m/s.

From the results shown in Figure S5c, it can be seen that the flow velocity at the bottom of the river is greater than that on both sides in a single cross section; the flow velocity on the right side averages −0.75 m/s and changes within the range from −2.25~0 m/s; the flow velocity on the left side averages −0.55 m/s and changes within the range from −1.48~0 m/s; the flow velocity at the bottom of the river averages −1.77 m/s and changes within the range from −2.82~−0.83 m/s.

From the results shown in Figure S5, it can be seen that the current flow velocity distribution in the transition stretch is as follows: the flow velocity at the bottom of the river gradually decreases from upstream to downstream, and the flow velocity at the bottom of the river is higher than that on the both sides (see Figure S7); the flow velocity on the right side averages 0.77 m/s and changes within the range from 0~2.31 m/s; the flow velocity on the left side averages 0.60 m/s and changes within the range from 0~1.85 m/s; the flow velocity at the bottom of the river averages 1.78 m/s and changes within the range from 0.84~2.83 m/s.

According to the results above, the water depth in the transition stretch from upstream to downstream first decreases and then increases. It is because the gradually growing cross sections of the transition stretch cause a decrease in flow velocity (see Figure S7a). In the beginning, the water depth in the stretch decreases; when the water depth of the transition stretch reaches 1.38 m, backwater in the transition stretch occurs due to the backwater effect of the downstream trapezoidal stretch and the water depth increases. The level and length of the backwater are about 0.04 m and 60 m, respectively (see Figure S6a); the maximum water depth at the bottom of the river is 1.50 m. The planned water depth in this stretch is 3 m. Therefore, the water depth meets the planning requirement of design standard for once-in-20-years floods. The average water depth on the right side of the river is 0.06 m higher than that on the left side, and the difference is below the design safe super elevation and meets the planning requirement.

(2) Simulation scenario 2 was analyzed in the same way, and the changes of hydraulic factors in the downstream trapezoidal stretch were basically the same as those in scenario 1 (see Figure S8).

According to the results above, from upstream to downstream, the water depth at the bottom of the river in the transition stretch first decreases and then increases and the maximum water depth at the bottom of the river is 1.79 m. The planned water depth in this stretch is 3 m. Therefore, the water depth meets the planning requirements that no overflow occurs in once-in-50-years rains; the average water depth on the right side of the river is about 0.05 m higher than that on the left side and the water depth difference under this scenario is still lower than that in scenario 1; the maximum flow velocity in this stretch is 3.26 m/s. Concrete revetment rather than ecological grass revetment should be adopted for the section lining.

(3) Compared to scenario 2, the simulation conditions in the transition stretch in scenario 3 are similar, so the basic hydraulic conditions in the transition stretch are not affected by the changes of hydraulic conditions in the upstream culverts, which will not be analyzed in this study.

(4) The comparison of the simulation results between scenarios 4 and 2 is shown in Figure 4. The water depth difference in the transition stretch changes within the range from −0.23~−0.11 m, the current flow velocity difference changes within the range from 0~0.43 m/s and the local current flow velocity difference changes within the range from −1.8~0 m/s.

By analysis, it can be seen that due to the roughness coefficient in this stretch decreasing from 0.025 to 0.017, the current flow velocity increases, the water depth decreases and the water depth and flow velocity at a single position change in opposite directions (shown in Figure S9). The water depth decreases by 0.23 m at maximum in Figure S9a, and the current flow velocity difference between ① and ② in Figure 4b is less than 0 m/s. The two locations should be managed and dredged in the future; especially in location ①, deposition will easily occur when the flow velocity changes.

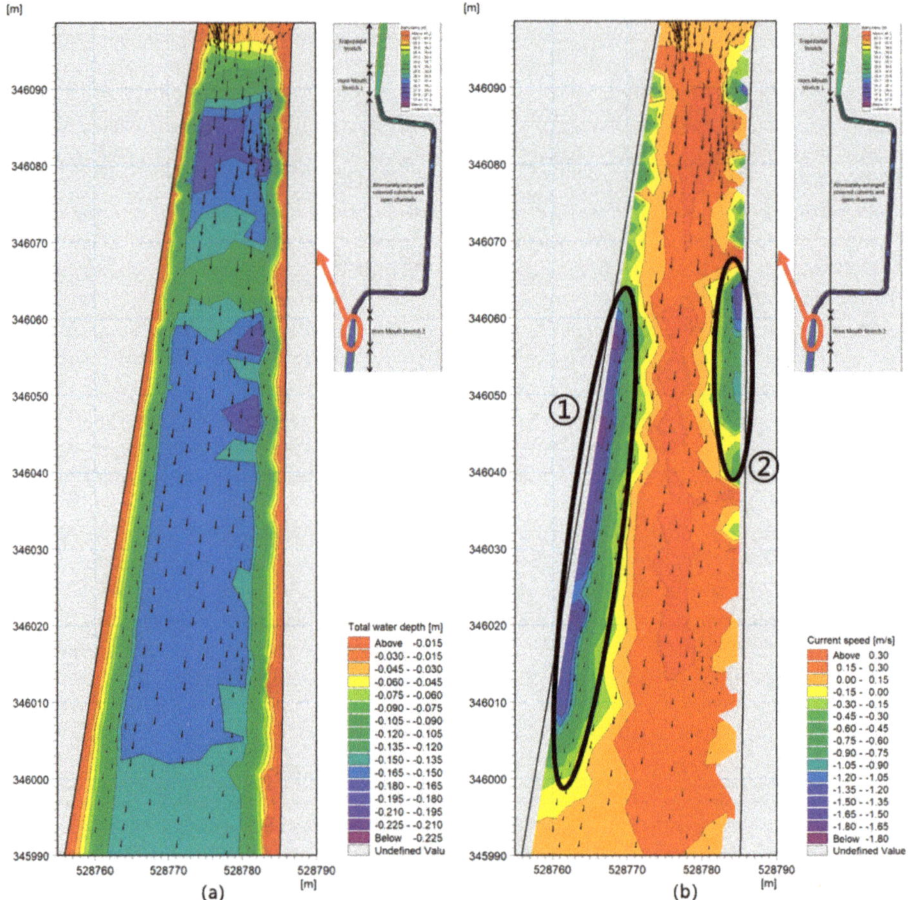

Figure 4. Water depth differences and current flow velocity differences in scenario 4 and scenario 2. (**a**) Water depth difference; (**b**) Current flow velocity difference.

3.3. Culverts and Open Channels

(1) Based on the simulation results of scenario 1, the water depth and flow velocity are shown in Figure 5. The four locations marked with red circles in Figure 5a are culverts and the six locations marked with black circles in Figure 5d are open channels. Because the hydraulic conditions in the sections are relatively complicated, in order to analyze in a convenient way, three channels of the culverts have been defined as the left side, middle side and right side, along the flow direction, and the water level, flow velocity and flow rate are analyzed in their vertical directions, respectively (see Figures S10 and S11).

In Figures S10 and S11, the water depth in the 3-channel culverts from upstream to downstream increases with slight fluctuation; the water depth changes within the range from 1.30~1.74 m; the height of the culverts in this stretch is 2.5 m. Therefore, the water depth meets the planning requirement of water depth of the culverts. The flow velocity in the 3-channel culverts decreases from upstream to downstream, and the flow velocity changes within the range from 1.33~2.79 m/s, which meets the planning requirements of flow velocity control. The red number represents the changes of the hydraulic factors in the turning position of the culverts, and the black number represents the changes of the hydraulic factors in the open channels, as below:

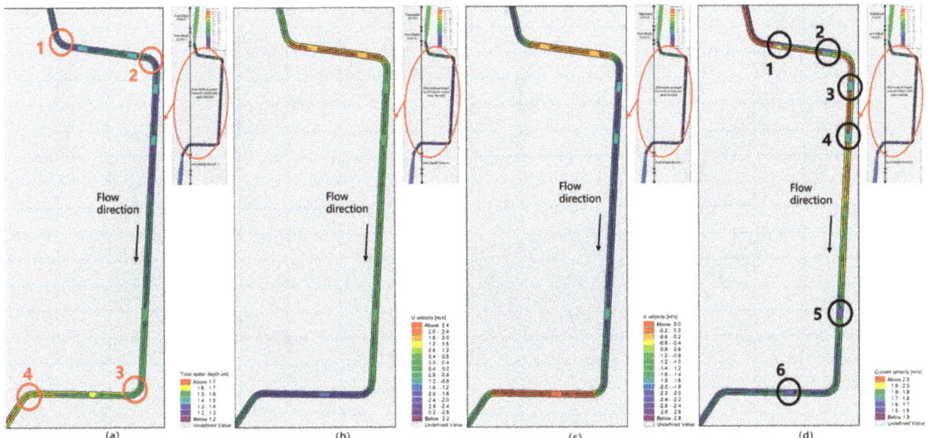

Figure 5. Hydraulic factors map in the culverts and open channels (simulation scenario 1). (**a**) Water depth map; (**b**) Horizontal flow velocity; (**c**) Vertical flow velocity; (**d**) Current flow velocity.

In turn 1 of the culverts, the current flow velocities in the 3 channels are in the sequence $V_{left} < V_{right} < V_{middle}$, the water depths are in the sequence $H_{left} < H_{right} < H_{middle}$, and the flow rates are in the sequence $Q_{left} < Q_{right} < Q_{middle}$. The middle culvert channel in this stretch has relatively larger flow velocity and flow rate, which should be protected from scouring; the left culvert channel has lower flow velocity and flow rate, which should be dredged in time; the horizontal flow velocity (relative flow velocity between the two sides) in the culvert suddenly increases and the vertical flow velocity suddenly decreases, which indicates that the horizontal circulation effects in all culverts are enhanced and the height difference between the two sides at the turns of culvert increases.

In the open channels 1 and 2, the current flow velocity suddenly decreases. The flow velocities are in the sequence $V_{left} < V_{middle} < V_{right}$ and the flow rates are in the sequence $Q_{left} < Q_{middle} < Q_{right}$. Therefore, dredging in the two open channels need to be enhanced, especially in the left channel of the culvert between two open channels; the current flow velocity in the open channel suddenly increases, and the average backwater level is about 0.04 m, which meets the planning requirements of safe super elevation for height.

In turn 2 of the culverts, the current flow velocities in the 3 channels are in the sequence $V_{left} < V_{right} \approx V_{middle}$, and the flow rates are in the sequence $Q_{left} < Q_{middle} < Q_{right}$; therefore, in-time dredging is still necessary in the left culvert channel. The horizontal flow velocity in the section suddenly decreases and the vertical flow velocity (relative flow velocity between the both sides) suddenly increases, which indicates that the horizontal circulation effects in all culverts are enhanced and that the height difference between the two sides at the turns of culverts increases.

In the open channels 3, 4, 5 and 6, the current flow velocity suddenly decreases, the flow rate basically remain the same, and the water depth suddenly increases; the average backwater level is about 0.05m, which meets the planning requirements of safe super elevation.

In turns 3 and 4 of the culverts, the flow rates in the 3 channels are almost the same. The current flow velocities are in the sequence $V_{right} < V_{left} \approx V_{middle}$, and the water depths are in the sequence $H_{middle} \approx H_{left} < H_{right}$. Therefore, the right culvert channel needs to be dredged in time. The horizontal flow velocity (relative flow velocity between the both sides) suddenly increases, and the vertical flow velocity suddenly decreases, which indicates that the horizontal circulation effects in all culverts are enhanced and that the height difference between the two sides at the turns of culvert increases.

In addition, in turns 1, 2, 3, and 4 of the culverts, the water depths on the concave side are higher than those on the convex side, as shown in Figure 6.

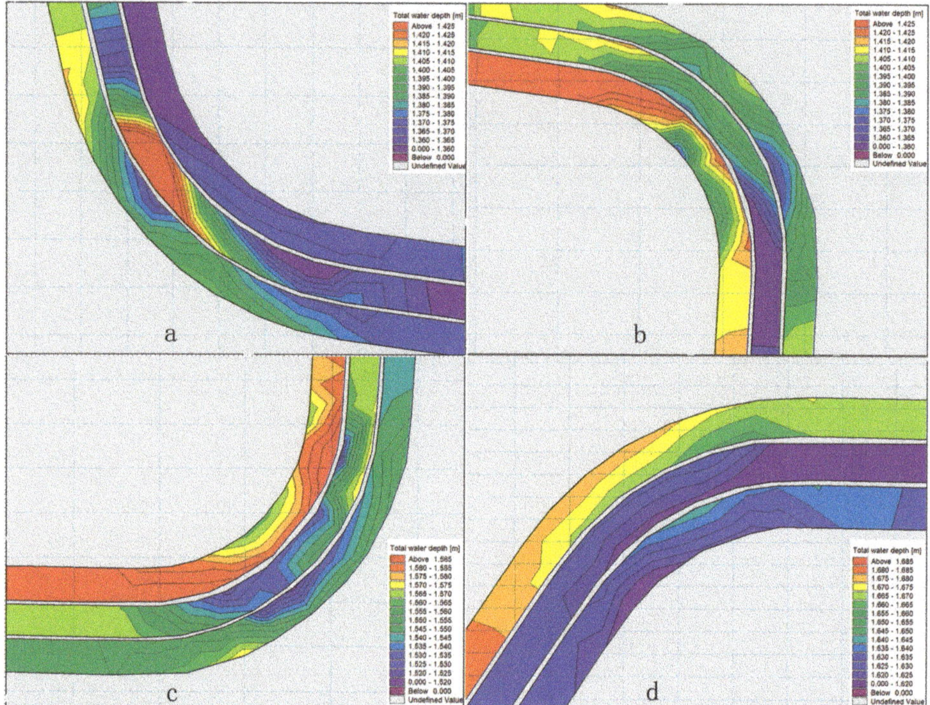

Figure 6. Enlarged water depth map of the four turns of 3-channel culverts (simulation scenario 1). (**a**) Enlarged water depth map of the locations marked with red circles 1 in Figure 5a; (**b**) Enlarged water depth map of the locations marked with red circles 2 in Figure 5a; (**c**) Enlarged water depth map of the locations marked with red circles 3 in Figure 5a; (**d**) Enlarged water depth map of the locations marked with red circles 4 in Figure 5a.

(2) Simulation results of scenario 2 were analyzed in the same way. This study conducted analysis on the water level, flow velocity and flow rate in the vertical direction of the 3-channel culverts (see Figures S12 and S13) and the water depth map at the four turns (see Figure S14).

As shown in the figures, the water depths in the 3-channel culverts increase from upstream to downstream, changing within the range from 1.48~1.90 m. The height of the culverts in this stretch is 2.5 m. Therefore, the water depth meets the planning requirement of water depth of the culverts; the flow velocity decreases from upstream to downstream, changing within the range from 1.45~2.85 m/s, which meets the planning requirements of flow velocity control. The water depth on the concave side is higher than that on the convex side in the 3-channel culverts, and the changes of the hydraulic factors in the culverts and the open channels are also basically consistent with the results in scenario 1.

(3) The culverts should be dredged regularly in follow-up maintenance. The long-term clogging of the culverts might result in inconsistency between the actual hydraulic conditions and the planning hydraulic conditions. The local poor discharge capability in the stretch may cause overflow. Therefore, this planning compared the simulation scenario 3 with the simulation scenario 2, by changing the roughness coefficient of the culverts from 0.017 to 0.025, to predict changes of the water level, current flow velocity, and flow rate (see Figure S15).

As shown in Figure S15, the changes of the water depth, the flow velocity and the flow rate in the culverts gradually decrease from upstream to downstream; the flow velocity change in the middle culvert is the largest, with the maximum change of 0.64 m/s; the flow rate change in the left culvert is the largest, with the maximum change of 1 m/s^3; and the water depth change in the 3-channel culverts basically remains the same, with the maximum change of 0.34 m and the maximum water depth of 2.24 m, Therefore, the water depth does not meet the planning requirement of the safe super elevation (at least 0.3 m). Hence, regular dredging is rather necessary. It is also shown that the setting of the six open channels is not only convenient for dredging, but also can be used to adjust the flow rate in the culverts.

(4) The simulation conditions in scenario 4 are the same as those in scenario 2, but the downstream hydraulic conditions are inconsistent. In order to study the influence of the downstream hydraulic factors on the hydraulic factors in the culverts and the open channels, this study compared the simulation results of scenarios 4 and 2 and the analysis results are shown in Figure S16.

As shown in Figure S16, the water depth, flow velocity and flow rate in the culverts do not change dramatically, the water depth difference changes within the range from −0.026~ −0.002 m, the flow velocity difference changes within the range from 0.002~0.028 m/s and the flow rate difference changes within the range from −0.013~0.006 m/s^3. Therefore, when the roughness coefficient of the downstream stretch changes from 0.025 to 0.017, the water level in the culverts decreases slightly, but the flow rate, water depth and flow rate in the culverts and open channels do not change significantly. From the drainage function perspective, the concrete lining in the downstream stretch is beneficial to rapid discharge; however, from the overall appearance and return of investment perspective, this planning study still recommends the ecological slope lining for the downstream stretch.

3.4. Upstream Stretch (Horn Mouth Stretch 1 and the Upstream Trapezoidal Stretch)

(1) According to the simulation results of scenario 1, the water depth and flow velocity in the upstream stretch are shown in Figure S17. According to the results in Figure S17a, the water depth distribution at the bottom of the river first increases and then decreases from upstream to downstream (see Figure S18); the water depth on the right side averages 0.71 m and changes within the range from 0~1.72 m; the water depth on the left side averages 0.81 m and changes within the range from 0~1.31 m; the water depth at the bottom of the river averages 1.58 m and changes within the range from 0.56~1.75 m.

It can be seen from the results in Figure S17b that the horizontal flow velocity on the right side of the upstream stretch averages −0.07 m/s and changes within the range from −0.26~0.18 m/s; the horizontal flow velocity on the left side averages −0.09 m/s and changes within the range from −0.62~0.19 m/s; the horizontal flow velocity at the bottom of the river averages −0.1 m/s and changes within the range from −0.38~0 m/s.

It can be seen from the results in Figure S17c, the vertical velocity on the right side of the upstream section averages −0.73 m/s and changes within the range from −2.44~0 m/s; the vertical velocity on the left side averages −0.90 m/s and changes within the range from −2.29~0 m/s; the vertical flow velocity at the bottom of the river averages −1.1 m/s and changes within the range from −2.33~−0.47 m/s.

It can be seen from the results in Figure S17d that the current flow velocity at the bottom of the river increases from upstream to downstream (see Figure S16). The overall flow velocity on the left side is greater than that on the right side; the flow velocity on the right side averages 0.74 m/s and changes within the range from 0~2.46 m/s; the flow velocity on the left side averages 0.91 m/s and changes within the range from 0~2.30 m/s; the flow velocity at the bottom of the river averages 1.1 m/s and changes within the range from 0.47~2.35 m/s.

It can be seen from the above results that the water depth at the bottom of the river in this stretch first increases and then decreases from upstream to downstream. It is because the section area at the horn mouth stretch decreases, which results in a sudden increase in the current flow velocity and decrease in the water depth; because the water depth is affected by horn mouth stretch 1 and the downstream stretch, backwater with a level of about 0.27 m and a length of about 420 m is generated in the upstream trapezoidal stretch (see Figure S18). The water depth at the bottom of the river is 1.75 m at maximum and the planned water depth in this stretch is 3 m, which meets the planning requirements to endure once-in-20-years floods. The average horizontal flow velocity (relative flow velocity between the both sides) is less than 0, which indicates that the horizontal flow flows from left to right. The horizontal circulation in the cross section results in a water depth difference of about 0.1 m between the two sides, which meets the planning requirements of safe super elevation; at the same time, the flow rate on the left side is larger than on the right side. Therefore, deposition can easily form on the right side, which needs in-time dredging.

(2) Simulation scenario 2 was analyzed in the same way. The changes of hydraulic factors in the upstream stretch are similar to those in scenario 1 (see Figure S19).

Based on the analysis, in the upstream stretch, the water depth increases first and then decreases from the upstream to the downstream. The maximum water depth at the bottom of the river is 1.91 m and the planned water depth is 3 m. Therefore, water depth meets the planning requirement of design standard for once-in-50-year floods. The average water depth on the right side is about 0.02 m higher than that on left side, and the water depth difference in the scenario is still lower than that in scenario 1.

(3) The simulation conditions in this stretch for scenarios 3 and 2 are the same, but the downstream hydraulic conditions are inconsistent. In order to study the influence of the downstream hydraulic factors on the hydraulic factors in this stretch, this study compared simulation results in scenario 3 with those in scenario 2. The analysis results are shown in Figure S20.

As shown in the figure above, the water depth difference in the upstream section changes within the range from 0~0.24 m, the maximum water depth at the bottom of the river is 2.15 m, the horizontal flow velocity difference changes within the range from 0~0.04 m/s, the average vertical velocity difference mainly changes within the range from 0~0.30 m/s, and the current flow velocity difference mainly changes within the range from −0.35~0 m/s. The clogging of the downstream culverts might increase the risk of deposition in the upstream stretch. Therefore, the river planning should be studied as a whole, and the river management should not only strengthen the management of this stretch but also pay attention to the dredging of the downstream trapezoidal stretch.

(4) The comparison of simulation results between scenarios 4 and 2 is shown in Figure S21. The water depth difference in the upstream section changes within the range from −0.1~0 m, the maximum water depth at the bottom of the river is 1.81 m, the horizontal flow velocity difference mainly changes within the range from −0.002~0 m/s, the average vertical flow velocity difference mainly changes within the range from −0.18~0 m/s, and the current flow velocity difference mainly changes within the range from 0.01~0.20 m/s.

By analysis, due to the roughness coefficient in the section decreasing from 0.025 to 0.017, the current flow velocity in the section increases and the water depth decreases. The water depth and flow velocity in a single location change reversely, which may facilitate the discharge of the river. Therefore, the concrete revetment and ecological revetment for the stretch both meet the planning requirements. From the perspective of landscape ecology and return of investment, this planning still recommends the ecological revetment.

4. Conclusions

In this study, the planned river course was divided into four stretches for separate analysis and discussion. Each stretch has its own unique characteristics, and the change rules of hydraulic factors also vary from stretch to stretch. Based on a summary of the rules and feasibility verification of the planning scheme, this study proposed some suggestions and improvement measures for planning (see Sections 3.1–3.4).

For River A, the main suggestions and guidance for follow-up river planning can be summarized as follows:

(1) The simulation scenarios 1 and 2 are used to analyze and summarize the changes of hydraulic factors, propose pointed opinions on planning, and further verify the rationality of the planned river course, which provides a scientific basis for river planning. Based on the analysis of the rationality and feasibility of the verified planning scheme, the planners should take pertinent measures to protect the stretches that are prone to erosion and deposition and water level differences at turns. At the same

time, the river sections with excessive velocity call for revetment. On this basis, similar planning projects in the future can be quickly responded to and the planners can take measures to prevent these issues. Meanwhile for rivers with other cross section forms, it can provide useful know-how and guide the planners in the future.

(2) The simulation scenarios 3 and 4 are used to simulate and analyze the changes of the hydraulic factors in the culvert and those in other river stretches due to the changes of the roughness coefficient in this stretch. For the downstream trapezoidal stretch and upstream trapezoidal stretch and transition stretch, the roughness coefficient in this stretch decreases from 0.025 to 0.017, which means that the ecological revetment in the river's planned section is replaced by concrete revetment. The current flow velocity in the river increases, but the water depth decreases, and the local velocity of the river decreases. For culverts and open channels, the roughness coefficient in this stretch increases from 0.017 to 0.025. The current flow velocity in the river decreases, but the water depth increases, and the water depth cannot meet the planning requirement of the safe super elevation (at least 0.3 m). Hence, regular dredging is rather necessary to ensure constancy of the roughness coefficient, and culverts should not be designed where open channels can be designed. For the upstream trapezoidal stretch, the clogging of the downstream culverts might increase the risk of deposition in the upstream trapezoidal stretch. Therefore, the river planning should be studied as a whole, and the river managers should not only strengthen the management of this stretch but also pay attention to the dredging of the downstream stretch.

Therefore, in the follow-up river planning, we should consider the overall situation, and not only pay attention to the study and analysis of the hydraulic factors of the planned river, but also pay attention to the dredging and management of the downstream river, and we should consider the changes of the hydraulic factors in the river stretches due to the changes of the roughness coefficient in this stretch.

At the same time, on the basis of this river planning research, the general rules of hydraulic factors in a river course with complex cross sections can be summarized as follows:

(1) In a single vertical section of the river course, water depth and current flow velocity change in the opposite directions, that is to say, when the current flow velocity changes, the water level will change reversely, causing a water depth difference in the vertical direction, thus resulting in backwater.

(2) In the trapezoidal and transition sections of a single river course, the flow velocity in the middle of the river course is greater than that on both sides, and the water depth in the middle is also greater than that on the two sides; in the rectangular section of a river course, the water depth and the flow velocity are basically the same as that on both sides; the water depth difference between the two sides is obvious at turns and the water depth on the concave side is greater than that on the convex side; at the junction between the culvert and the open channel, when the water flows from culvert to open channel its flow velocity decreases suddenly, while the water depth increases suddenly, which results in an increase in water head loss, thus resulting in backwater.

(3) The horizontal flow velocity and vertical velocity abruptly change at turns of the river course, enhancing the horizontal circulation in the river course, thus resulting in an increase in the water depth difference between the two sides of the river course.

(4) For river courses with complex cross sections, the flow characteristics should be considered as unsteady flow. When the flow is slow, the changes of hydraulic factors in the downstream stretch will affect the hydraulic factors in the upper stretch; therefore, the management of the river should not only strengthen the management of local stretch but also pay attention to the dredging of downstream stretch.

On this basis, when we analyze the hydraulic factors of a natural river course in the future, we can compare and analyze the hydraulic factors of the urban river course and natural river course, and then find more general rules of change. Further, the urban river course and natural river course have different evolution trends in the same river section. As the form of revetment is different, the erosion degree of the river on both sides is different, even if the same form of revetment can eventually lead to different evolution trends of the river due to the differences in flow, river upstream conditions and human activities. Based on the detailed research and analysis on the hydraulic factors of River A, the general rules of the urban river course are summarized, which establishes a solid foundation for understanding the genesis and evolution of the river course in the future.

Supplementary Materials: The following are available online at http://www.mdpi.com/2073-4441/12/3/761/s1: Figure S1: Transition stretch schematic diagram, Figure S2: Schematic diagram of terrain elevation and meshing, Figure S3: Water depth map in vertical and cross sections of the downstream trapezoidal stretch (simulation scenario 1), Figure S4: Hydraulic factors map of the downstream trapezoidal stretch (Simulation scenario 2), Figure S5: Hydraulic factors map in the transition stretch (simulation scenario 1), Figure S6: Water depth map in the vertical and cross sections in the transition stretch (simulation scenario 1), Figure S7: Current flow velocity in the vertical and cross sections in the transition stretch (simulation scenario 1), Figure S8: Hydraulic factors map in the transition stretch (simulation scenario 2), Figure S9: Reverse changes of water depth and flow velocity in a single position, Figure S10: Comparison of water depth and flow velocity in the left, middle and right culverts (simulation scenario 1); The cyan-blue curve represents the changes of the hydraulic factors in the left culvert, the red curve represents the changes of the hydraulic factors in the middle culvert, and the blue curve represents the changes of the hydraulic factors in the right culvert, Figure S11: Comparison of flow velocities in the left, middle and right culverts (simulation scenario 1), Figure S12: Comparison of the water level and flow velocity in the left, middle and right culvert channels (simulation scenario 2), Figure S13: Comparison of the flow rates in the left, middle and right culvert channels (simulation scenario 2), Figure S14: Enlarged water depth map at the four turns of the 3-channel culverts (simulation scenario 2), Figure S15: Comparison of hydraulic factors in scenarios 3 and 2, Figure S16: Comparison of hydraulic factors in scenarios 4 and 2, Figure S17: Hydraulic factors map in the upstream stretch (simulation scenario 1), Figure S18: Comparison of the water depth and current flow velocity in the upstream stretch (simulation scenario 1), Figure S19: Hydraulic factors map in the upstream stretch (simulation scenario 2), Figure S20: Comparison of hydraulic factors in scenarios 3 and 2, Figure S21: Comparison of hydraulic factors in scenarios 4 and 2.

Author Contributions: Q.W. performed model simulations, data analysis and wrote the manuscript; W.P. conceived and supervised the study; F.D., X.L. and N.O. helped the establishment of model and data analysis. All authors have read and agreed to the published version of the manuscript.

Funding: This study was jointly supported by the Major Science and Technology Program for Water Pollution Control and Treatment of China (2017ZX07101004-001), National Natural Science Foundation of China (51809288, 51861135314), National Key R&D Program of China (2018YFC0407702), Basic Research Program of the China Institute of Water Resources and Hydropower Research (WE0145B532017).

Conflicts of Interest: The authors declare no conflict of interest.

References

1. Brookes, A.; Gregory, K.J.; Dawson, F.H. An assessment of river channelization in England and Wales. *Sci. Total Environ.* **1983**, *27*, 97–111. [CrossRef]

2. Hickin, E.J. The development of meanders in natural river-channels. *Am. J. Sci.* **1974**, *274*, 414–442. [CrossRef]

3. Wohl, E.; Merritts, D.J. What is a natural river? *Geogr. Compass* **2007**, *1*, 871–900. [CrossRef]

4. Findlay, S.J.; Taylor, M.P. Why rehabilitate urban river systems? *Area* **2006**, *38*, 312–325. [CrossRef]

5. Cardoso, A.; Graf, W.H.; Gust, G. Uniform flow in a smooth open channel. *J. Hydraul. Res.* **1989**, *27*, 603–616. [CrossRef]

6. Bridge, J.S. A revised model for water flow, sediment transport, bed topography and grain size sorting in natural river bends. *Water Resour. Res.* **1992**, *28*, 999–1013. [CrossRef]

7. Darby, S.E.; Alabyan, A.M.; Van de Wiel, M. Numerical simulation of bank erosion and channel migration in meandering rivers. *Water Resour. Res.* **2002**, *38*, 1163–1186. [CrossRef]

8. Zhang, M. Study on the Hydrodynamic and Water Quality Model in River. Ph.D. Thesis, Dalian University of Technology, Dalian, China, 2007.

9. Liu, X. Gravel Bed-Load Transport and its Modelling. Ph.D. Thesis, Sichuan University, Chengdu, China, 2004.

10. Xia, R. Relation between mean and maximum velocities in a natural river. *J. Hydraul. Eng.* **1997**, *123*, 720–723. [CrossRef]

11. Lane, S.; Bradbrook, K.; Richards, K.; Biron, P.; Roy, A. The application of computational fluid dynamics to natural river channels: Three-dimensional versus two-dimensional approaches. *Geomorphology* **1999**, *29*, 1–20. [CrossRef]

12. Yu, X.; Tan, G.; Zhao, L.; Wang, J. Planar 2-D flow and sediment numerical modeling of branching river. *J. Sichuan Univ. (Eng. Sci. Ed.)* **2007**, *39*, 33–37.

13. Jia, D. Three-Dimensional Numerical Simulation of Lateral Migration of Alluvial Channels with Composite Banks. Ph.D. Thesis, Tsinghua University, Beijing, China, 2010.

14. Chen, X.; Wang, D.; Tian, F.; Sivapalan, M. From channelization to restoration: Sociohydrologic modeling with changing community preferences in the Kissimmee River Basin, Florida. *Water Resour. Res.* **2016**, *52*, 1227–1244. [CrossRef]

15. Wohl, E.; Lane, S.N.; Wilcox, A.C. The science and practice of river restoration. *Water Resour. Res.* **2015**, *51*, 5974–5997. [CrossRef]

16. Palmer, M.A.; Bernhardt, E.; Allan, J.; Lake, P.S.; Alexander, G.; Brooks, S.; Carr, J.; Clayton, S.; Dahm, C.; Follstad Shah, J. Standards for ecologically successful river restoration. *J. Appl. Ecol.* **2005**, *42*, 208–217. [CrossRef]

17. Wohl, E.; Angermeier, P.L.; Bledsoe, B.; Kondolf, G.M.; MacDonnell, L.; Merritt, D.M.; Palmer, M.A.; Poff, N.L.; Tarboton, D. River restoration. *Water Resour. Res.* **2005**, *41*, W10301. [CrossRef]

18. Beechie, T.; Imaki, H. Predicting natural channel patterns based on landscape and geomorphic controls in the Columbia River basin, USA. *Water Resour. Res.* **2014**, *50*, 39–57. [CrossRef]

19. Garcia-Navarro, P.; Saviron, J. McCormack's method for the numerical simulation of one-dimensional discontinuous unsteady open channel flow. *J. Hydraul. Res.* **1992**, *30*, 95–105. [CrossRef]

20. Hersberger, D.S.; Franca, M.J.; Schleiss, A.J. Wall-roughness effects on flow and scouring in curved channels with gravel beds. *J. Hydraul. Res.* **2016**, *142*, 04015032. [CrossRef]

21. An, M. The Numerical Simulation of the River Flow on Urban Landscape River. Master's Thesis, North China University of Water Resources and Electric Power, Zhengzhou, China, 2016.

22. Costabile, P.; Costanzo, C.; De Bartolo, S.; Gangi, F.; Macchione, F.; Tomasicchio, G.R. Hydraulic characterization of river networks based on flow patterns simulated by 2-D shallow water modeling: Scaling properties, multifractal interpretation and perspectives for channel heads detection. *Water Resour. Res.* **2019**, *55*, 7717–7752. [CrossRef]

23. Williams, R.D.; Brasington, J.; Hicks, M.; Measures, R.; Rennie, C.; Vericat, D. Hydraulic validation of two-dimensional simulations of braided river flow with spatially continuous aDcp data. *Water Resour. Res.* **2013**, *49*, 5183–5205. [CrossRef]

24. Chin, A.; Gregory, K.J. Managing urban river channel adjustments. *Geomorphology* **2005**, *69*, 28–45. [CrossRef]

25. Gregory, K.J. Urban channel adjustments in a management context: An Australian example. *Environ. Manag.* **2002**, *29*, 620–633. [CrossRef] [PubMed]

26. Chen, A.S.; Evans, B.; Djordjević, S.; Savić, D.A. Multi-layered coarse grid modelling in 2D urban flood simulations. *J. Hydrol.* **2012**, *470*, 1–11. [CrossRef]

27. Uddin, M.; Alam, J.B.; Khan, Z.H.; Hasan, G.J.; Rahman, T. Two dimensional hydrodynamic modelling of Northern Bay of Bengal coastal waters. *Comput. Water Energy Environ. Eng.* **2014**, *3*, 140–151. [CrossRef]

28. Kaergaard, K.; Fredsoe, J. Numerical modeling of shoreline undulations part 1: Constant wave climate. *Coast. Eng.* **2013**, *75*, 64–76. [CrossRef]

29. Abily, M.; Duluc, C.; Faes, J.; Gourbesville, P. Performance assessment of modelling tools for high resolution runoff simulation over an industrial site. *J. Hydroinform.* **2013**, *15*, 1296–1311. [CrossRef]

30. Kimiaghalam, N.; Goharrokhi, M.; Clark, S.P.; Ahmari, H. A comprehensive fluvial geomorphology study of riverbank erosion on the Red River in Winnipeg, Manitoba, Canada. *J. Hydrol.* **2015**, *529*, 1488–1498. [CrossRef]

31. Yi, X. *Application and Research of DHI Mike Flood Simulation Technology*, 1st ed.; China Water Power Press: Beijing, China, 2014; pp. 11–253.

32. Eymard, R.; Gallouët, T.; Herbin, R. Finite volume methods. *Handb. Numer. Anal.* **2000**, *7*, 713–1018.

33. Barth, T.; Herbin, R.; Ohlberger, M. Finite volume methods: Foundation and analysis. In *Encyclopedia of Computational Mechanics Second Edition*; John Wiley & Sons: New York, NY, USA, 2018; pp. 1–60.

34. Gayer, G.; Leschka, S.; Nöhren, I.; Larsen, O.; Günther, H.; Sciences, E.S. Tsunami inundation modelling based on detailed roughness maps of densely populated areas. *Nat. Hazards* **2010**, *10*, 1679–1687. [CrossRef]

35. Sokolov, A.; Chubarenko, B.J.A.o.H.-E. Wind influence on the formation of nearshore currents in the Southern Baltic: Numerical modelling results. *Nephron Clin. Pract.* **2012**, *59*, 37–48. [CrossRef]

36. Pedersen, N.H.; Rasch, P.S.; Sato, T. Modelling of the Asian tsunami off the coast of northern Sumatra. In Proceedings of the 3rd Asia-Pacific DHI Software Conference, Kuala Lumpur, Malaysia, 21–22 February 2015.

37. Xie, X. Application of MIKE21 FM in Flood Hazard Mapping-a Case Study on Luanhe Right-Bank Flood Protected Zone. Master's Thesis, Dalian University of Technology, Dalian, China, 2016.

38. Feng, J. MIKE21FMApplication of mIKE21FM Numerical Model in Environmental Impact Assessment of Ocean Engineering. Master's Thesis, Ocean University of China, Qingdao, China, 2011.

39. Guo, F.; Qu, H.; Zeng, H.; Cong, F.; Geng, X. Flood routing numerical simulation of flood storage area based on MIKE21 FM model. *Int. J. Hydroelectr. Energy* **2013**, *31*, 34–37.

40. Zhang, W. A 2-D numerical simulation study on longitudinal solute transport and longitudinal dispersion coefficient. *Water Resour. Res.* **2011**, *47*, 128–136. [CrossRef]

41. Ye, Z.; Wen, D. *Hydraulics and Hydrology of Bridge and Culvert*, 1st ed.; China Communications Press: Beijing, China, 1998; p. 87.

42. Wang, Q.; Zhao, S.; Zhou, Y.; Lu, Y.; Zhou, L.; Wang, Z.; Yang, W.; Gao, L. Probe into urban drainage and waterlogging prevention plan based on modeling-case study on ShaTauKok district of Shenzhen. *Water Supply Drain.* **2015**, *51*, 34–38.

43. Ren, M. The river-chief mechanism: A case study of china's inter-departmental coordination for watershed treatment. *J. Beijing Adm. Inst.* **2015**, *3*, 25–31.

44. Wang, S.; Cai, M. Critique of the system of river-leader based on the perspective of new institutional economics. *China Popul. Resour. Environ.* **2011**, *21*, 8–13.

MDPI
St. Alban-Anlage 66
4052 Basel
Switzerland
Tel. +41 61 683 77 34
Fax +41 61 302 89 18
www.mdpi.com

Water Editorial Office
E-mail: water@mdpi.com
www.mdpi.com/journal/water